Organic Chemistry
A Guided Inquiry
for Recitation, Volume 1

a process oriented guided inquiry learning course

Andrei Straumanis

Organic Chemistry: A Guided Inquiry, Volume 1
Andrei Straumanis

Publisher: Mary Finch
Acquisitions Editor: Chris Simpson
Editorial Assistant: Laura Bowen
Marketing Manager: Barb Bartoszek
Marketing Assistant: Julie Stefani
Project Manager, Editorial Production: Michelle Clark
Art Director: John Walker
Print Buyer: Linda Hsu
Permissions Editor: Dean Dauphinais
Manufacturing Manager: Marcia Locke
Production Service: PreMediaGlobal
Content Project Management: PreMediaGlobal

ISBN-13: 978-1-111-57399-7

ISBN-10: 1-111-57399-9

Brooks/Cole
20 Davis Drive
Belmont, CA 94002-3098
USA

Cengage Learning is a leading provider of customized learning solutions with office locations around the globe, including Singapore, the United Kingdom, Australia, Mexico, Brazil, and Japan. Locate your local office at **www.cengage.com/global.**

Cengage Learning products are represented in Canada by Nelson Education, Ltd.

To learn more about Brooks/Cole, visit **www.cengage.com/brookscole**

Purchase any of our products at your local college store or at our preferred online store **www.cengagebrain.com**

Printed in the United States of America
1 2 3 4 5 6 7 14 13 12 11 10

This book is dedicated to my friend and colleague, Rick Moog, and to his tireless, selfless work in support of student learning.

Acknowledgements

Huge thanks go to the organic chemistry students at the University of Washington and College of Charleston. In preparing these ChemActivities nothing could have substituted for watching real students work in a real classroom setting. Thanks also to my colleagues who have been generous with their knowledge of chemistry and chemistry students.

Particular thanks go to chairs in both my departments: Jim Deavor at the College of Charleston, and Paul Hopkins and Philip Reid at the University of Washington. Their support of this project is yet further proof of their dedication to chemistry instruction, and their willingness to take risks to keep their departments on the cutting edge. It is not a coincidence that under their leadership both departments sit atop their respective institutional categories in the key metric of number of chemistry majors.

Many of the over 100 faculty who used the full version of this workbook (Organic Chemistry: A Guided Inquiry, 2e) have contributed to improvements in this recitation edition. Thanks for all of your suggestions and corrections.

Thanks to the POGIL Project and to its growing number of participants. Special thanks to Rick Moog, whose contagious enthusiasm for guided inquiry inspired me and many others to embark on this path.

This work would not have been possible without the POGIL Project and its primary funder, the National Science Foundation. The United States Department of Education's Fund for Improvement of Post-secondary Education (FIPSE) supported my work with very large classes at the University of Washington under grant number P116B060026.

Words cannot express the depth of my love and appreciation for my wife, Allyson Chambers, MD, and for my three sons, Milo, Luca, and Nico, purveyors of the thunder of tennis shoes that rolls through my house each afternoon.

Comments from Faculty

'Organic Chemistry: A Guided Inquiry' was a true revelation to me. In adopting a POGIL format in a large classroom my day-to-day preparations were comparable in intensity and duration to the time I spent preparing for traditional lectures. In my years as a college educator, I have not seen anything as pedagogically powerful as a POGIL class using Straumanis' workbook. I believe the future of college 'teaching' lies in this type of 'learning.' I give Straumanis my highest rating.

Dr. Stefan Kraft, Kansas State University

This workbook has revolutionized the way I teach organic chemistry. The students process the material in logical steps, are active learners in the classroom, and the end result is a deeper understanding of organic compounds and reactions. I highly recommend this book!

Dr. Bruce J. Heyen, Tabor College

The guided inquiry helps me [the professor] think more like a student and it helps me cover material more efficiently.

Dr. Dan Esterline, Heidelberg College

Organic Chemistry: A Guided Inquiry is a great way to teach and learn organic chemistry. The students love the interactive nature of class time. I become better acquainted with each of my students, which enables me to tailor my teaching to maximize each student's learning. It has transformed the way I teach.

Dr. Timothy M. Dore, University of Georgia

It is fabulous to see the increase in understanding of reaction mechanisms. The students learn the mechanisms by working together, discussing, and sometimes arguing about them, but they don't memorize them. What I really like about the approach is that when class is ending, the students are still working. There is no clock watching!

Dr. William Wallace, Barton College

I am thankful that I decided to use the Guided Inquiry approach in my class. Students have responded in a very positive manner. Other faculty, too, see a change in the students' attitude towards learning Organic Chemistry.

Dr. Karen Glover, Clarke College

I have been using the Guided Inquiry Organic Chemistry materials since they were in manuscript form, and I am delighted about the second edition. The ChemActivities do an excellent job enabling students to build on their knowledge to develop new knowledge and to apply chemical concepts to new situations. Students enjoy organic chemistry and they learn it well because they engage directly with the material and with their peers.

Dr. Laura Parmentier, Department Chair, Beloit College

Watching my students engaged in discussing organic chemistry in their groups during organic class has convinced me that I made the right decision to change to POGIL after twenty years of brilliant lecturing. Straumanis' ChemActivities really do help the students learn organic much better than my lectures ever did.

Dr. Barbara Murray, University of Redlands

Comments from Students

I didn't get tired during class because I was constantly thinking and working instead of in a lecture class where I just listen and get easily tired.

The act of explaining the concept forced me to clarify the concept in my own head.

Class time was actually learning time, not just directly-from-ear-to-paper-and-bypass-brain-writing-down time. Learning the material over the whole term is far easier than not "really" learning it until studying for the tests.

I wasn't just blindly copying notes on the board but actually working through problems and learning.

Overall, I was far less stressed than many of my friends who took the lecture class. They basically struggled through everything on their own.

Group work has helped me find motivation for studying.

It was hugely beneficial to be able to discuss through ideas as we were learning them; this way it was easy to immediately identify problem areas and work them out before going on.

The method of having us work through the material for ourselves— as opposed to being told the information and trying to absorb it—makes it seem natural or intuitive. This makes it very nice for learning new material because then we can reason it out from what we already know.

I felt like I was actually learning the information as I received it, not just filing it for later use. The format helped me retain much more material than I have ever been able to in a lecture class, and the small, group atmosphere allowed me to feel much more comfortable asking questions of both other students and the professor.

Through working in groups it was nice to see where everyone else was in understanding the material (i.e. to know where other people were having trouble too).

We were able to discover how things happened and why for ourselves…instead of being told.

Advice from Students to Students

Don't let yourself take the course lightly just because class is fun and relaxed (and goes by fast!). Do the homework and reading.

Give yourself some time to settle into group learning. Lots of us did not think we would like it or that it would work. It does.

Don't fall behind. Playing catch up is not fun. Don't be afraid to ask questions and argue in your group. That is the way learning is done in this class.

Find a study group ASAP and meet regularly [outside of class] every week. I wish I had done this sooner.

Overview of the POGIL Method

Process Oriented Guided Inquiry Learning (POGIL) is used to teach tens of thousands of science students each semester, and to date, over 150 colleges and universities have used POGIL to teach organic chemistry.

These materials are flexible and can be used in a variety of settings, but during most POGIL sessions…

- Students work in teams of three or four to answer carefully designed **Construct Your Understanding Questions**, which guide student groups toward discovery of a chemical concept.
- The teaching assistant, peer leader, or instructor serves as the facilitator of learning, not as the primary source of information. Though effective facilitators spend much of class time observing student group work, they do not sit on the sidelines. During each class they ask and answer questions, lead whole-class discussions, and deliver just-in-time mini-lectures.

For more information, read the chapter in this book entitled Introduction to POGIL, visit www.pogil.org/straumanis, or contact the POGIL Project at www.POGIL.org.

Another good resource is the instructor-only Yahoo group moderated by the author of these materials, and dedicated to POGIL Organic Chemistry (find it by searching "**giorganic**" at groups.yahoo.com).

Forward on Process Oriented Guided Inquiry Learning (POGIL)

by Rick Moog, Jim Spencer and John Farrell, authors of *Chemistry: A Guided Inquiry* (for General Chemistry) and two volumes of *Physical Chemistry: A Guided Inquiry*.

These guided activities were written because much research has shown that more learning takes place when the student is actively engaged and when ideas and concepts are developed by the student, rather than being presented by an *authority*—a textbook or an instructor. The *ChemActivities* presented here are structured so that information is presented to the reader in some form (an equation, a table, figures, written prose, etc.) followed by a series of questions that lead the student to the development of a particular concept or idea. Learning follows the scientific process as much as possible throughout. Students are often asked to make predictions based on the model that has been developed up to that point, and then further data or information is provided that can be compared to the prediction. In this way, students simultaneously learn course content and key process skills that constitute scientific thought and exploration

How to Use This Book

To Instructors

This book was developed for faculty who want to take advantage of the benefits of Process Oriented Guided Inquiry Learning (POGIL) during a supplemental class. Consistent with POGIL's focus on skills and team building, **research indicates POGIL increases a student's success in a subsequent lecture**.[1] The proposed mechanism is that the learning skills, confidence, and study group behavior developed in a POGIL environment help students get more out of lecture, readings, homework, etc.

The activities in this book are designed to be a student's first introduction to a topic. This means your TAs will be *ahead* of you, and your lecture on a topic should come *after* students have encountered that topic in recitation. This may be the opposite of what you are used to, but I think you will find that giving students a chance to discover a topic via a POGIL ChemActivity is superior to preparing for lecture by reading (or *not* reading) a traditional textbook chapter. I have found that a POGIL introduction to a topic greatly increases students' appreciation of lectures. You may even find that you can spend less time working through basic examples, or move more quickly through topics that gave your students trouble in the past.

This workbook is designed to complement any textbook. Most courses will not use all the activities herein. Choose the activities you like, and use them in whatever order works best for your course.

The once-per-week, student-friendly activities are designed for supplemental classes, but can also be used in lab, for homework, or as the basis for a hybrid POGIL-lecture approach.

The first chapter (**Introduction to POGIL**, starting on page 1) serves as a brief training for teaching assistants, undergraduate peer leaders, or instructors. POGIL is a *learning* method, so both students and instructors need to be familiar with POGIL, and will benefit from reading this chapter.

To Students

This book is designed to make organic chemistry more enjoyable and less intimidating, but without sacrificing depth. Too many organic chemistry students memorize facts, only to forget them after the exam. This workbook guides you toward a deep understand so you learn more, retain it longer, and do better in subsequent courses and on standardized exams.

This book is designed to be used during class. For each activity, read the **Model** then work with your group to answer the **Construct Your Understanding Questions** that follow. Stay together! For each question, it is a good idea to compare answers within the group before moving onto the next question.

If you are unsure of an answer after checking with your group, some good strategies are to read the next question, ask a nearby group, or pose a question to the TA or instructor, but instead of asking "Is our answer right?" ask a question that helps the instructor understand the source of your confusion.

It is your collective responsibility to manage your time so that you finish these questions and even get to some of the **Extend Your Understanding Questions** before the end of class. Before the next class, finish the *entire* activity including the **Confirm Your Understanding Questions**.

Be sure to read the *Advice from Students to Students* section on page vi, and the Introduction to POGIL (starting on page 1) so you can make the most of your efforts in this course. Many students find it useful to read these sections at the start of the course, and again two weeks into the course.

---

[1] 1) Moog, R.S.; Spencer, J.N., eds., American Chemical Society Symposium Series, American Chemical Society, Washington, DC, 2008. 2) Straumanis, *unpublished results*.

Contents

Notes

Introduction to POGIL

Permission to use material in this chapter has been generously provided by the POGIL Project.

Key Issues

This first page is a list of key considerations regarding a use of the POGIL method. The second page is a chronological account of a typical POGIL session. Both are provided in recognition of busy schedules, though students and instructors will likely benefit from reading this entire chapter.

- Each activity in this book is designed to be the introduction to a topic. This means each topic covered using POGIL should appear in recitation *before* it is covered in lecture.
- Start each POGIL session with a one-question **quiz** covering material from the *previous* class.
- Basic format: instructor-assigned groups of 3 or 4 complete a ChemActivity by reading the **Model** and working together to answer the **Construct Your Understanding Questions**.
- On the first day of class, many student groups are shy about talking. It helps to designate one person as **manager** and have this person read each question aloud or, at the very least, ask if everyone is in agreement before moving onto the next question. Manager duties should rotate.
- Students who participate in their group during class are far more successful on quizzes and exams than students who work without interacting with their group.
- Many students have had bad experiences with group grading. It is therefore worth emphasizing that in this course, **group work does not mean group grading**! The purpose of group work is to help each student prepare for quizzes and exams, taken *individually*.
- The Instructor should not spend more than a few minutes explaining how POGIL works. Students figure it out quickly, and can read this chapter for more information.
- POGIL instruction is three parts listening and one part talking. Spend the time to figure out what is causing a group's confusion and you may only need a few sentences to clear it up.
- The instructor should not sit on the sidelines and expect the materials to work without help. When not answering a question, rotate through the class listening to each group, monitoring progress, especially on key questions (marked with a key), and intervening when necessary. Get to know the students so you can guide and inspire them to exceed their own expectations.
- Reading on a topic in the regular textbook should follow (not come before) a POGIL activity.
- Student groups should complete as many questions as possible during class, and may choose to meet after class, but each student must finish the *entire* activity before the next class.
- Students who feel rushed during a POGIL session should preview the activity by reading the models and questions *before* class.
- Students: Asking the instructor the question "Is our answer right?" is not as useful as asking questions like: "We are concerned our answer is wrong because…"
- Students are *not* provided an answer key to the in-class Construct Your Understanding Questions because doing so short-circuits *the* critical step in the learning process. Learning is largely the act of confirming your answers via discussion with your group, other groups, or the instructor; doing homework problems (for which there *is* a key); and reading in your textbook.
- Instructors: When faced with the question, "Is our answer right?" be supportive but firm. Find out if the group is in agreement, or refer them to the relevant part of the Model. Ask students to explain their reasoning or rate their confidence level. If students are frustrated, stay and help them talk it out or promise to return and check that they achieved closure.

Account of a Typical Day in a POGIL Recitation Section

The following is a description of a typical day in a POGIL class of about 25 students. Most items on the list below are described in greater detail on the pages that follow.

- Seating chart with names (and roles) is posted on-line and on the door.
- On the first day of class, bring copies of the activity for students who do not bring a book.
- Each group has a folder (numbered or color coded). The first person in a group to arrive picks up the group folder, which contains graded work from the previous class. As each group member arrives, she retrieves her graded quiz from the folder.
- The group folder also contains blank copies of the upcoming quiz (face down in the folder). When the instructor gives the signal to begin the quiz, the Manager distributes a copy of the quiz to each group member. (Quizzes are taken individually.)
- After three minutes, instructor calls time and the completed quizzes are turned in or placed in the group folder.
- Instructor briefly goes over the quiz. (Alternatively, a longer period of time can be devoted to more open-ended group discussion of the quiz.)
- Instructor briefly (< 3 minutes) reviews the key concepts from the previous class, and very briefly (< 1 minute) previews the upcoming activity.
- Students are instructed to begin work on the new activity. It is best to write on the board how many minutes groups have to work on a given set of questions. (For example: "You have 5 minutes to work on Questions 1-3." This tells students that Questions 1-3 are fairly easy or review, and helps them manage their time effectively.
- The instructor circulates to each group and briefly checks that work has begun.
- On a second pass, the instructor takes more time to listen to groups, assess progress, and examine answers to the Construct Your Understanding Questions, intervening if necessary. Many times, a group will find their own answer, so not all errors require intervention.
- The instructor responds to questions when the manager of a group raises her hand. (Before responding, make sure the group has already discussed this question *within* the group.)
- At the end of the time allotted for the first group of questions (e.g. Question 1-3) the instructor pauses group work, and asks one or more groups to report an answer to a key question. If necessary, the instructor can ask for student commentary on this answer.
- The directive on the board is replaced with a new directive (e.g. "You have 20 minutes to work on Questions 4-8.") These directives can be compiled into a PowerPoint or overhead presentation that might also include pre-prepared examples or auxiliary questions.
- One or more times during class, but always at the end of class, the instructor interrupts group work for a mini-lecture or whole-class discussion. This can consist of a group's Presenter reporting an answer followed by instructor-led discussion of that answer. This type of summary is especially necessary when students are having difficulty with a topic.
- If you have time, class can end with preparation of a recorder's report. This report may be focused on content (e.g. "Write down the three most important concepts you learned today, and any questions that remain unanswered.") or process ("What is one strength and one area for improvement for your group's performance today?").
- The class ends on time.
- Student groups often choose to stay after class to complete the activity. If the classroom is occupied during the following class period, the instructor may want to suggest or arrange for a space for continued work (e.g. a nearby classroom or in the hall outside class).

How People Learn

Research on learning tells us that students learn best when they are...

- actively engaged and thinking in class. [1]
- given an opportunity to construct their own understanding.[2,3]
- discussing ideas and asking questions as part of a dynamic and social team. [4,5]

The research that inspired POGIL is not new, and POGIL is not the first classroom method that has put this research into practice. If you have been working to make your classroom more active or participated in group learning, it is likely you will find some POGIL techniques familiar. Yet it is best to suspend any doubts, preconceptions, or fears about organic chemistry or group learning. They will only interfere with your success and enjoyment of this class. At the end of a POGIL class, less than 10% of students are negative about POGIL,[2] but many students take several weeks to come to this conclusion. To get the most out of this course, read this chapter once before the start of class, and then again a few weeks into the course. For further reading, you may wish to consult the following excellent resources on teaching and learning.[1-10]

What is POGIL?

A POGIL classroom differs dramatically from a traditional classroom in that there is little formal lecture. The instructor serves as the facilitator of learning rather than the primary source of information and students work in self-managed teams to analyze data and draw conclusions, modeling the way a team of scientists function in the research laboratory.

POGIL's central claim is that it helps students simultaneously develop content knowledge and key process skills. The hypothesized mechanism is two-fold, stemming from marriage of the PO (Process Oriented) and GI (Guided Inquiry) methodologies implied in the name. The "GI" part is achieved via use of carefully designed learning cycle[6,7] activities that guide students toward construction of their own understanding. Such discovery experiences have been shown to improve confidence while helping students to understand and remember more.[4,5]

The "PO" part comes from the frequent or exclusive use of small groups. There is a large body of evidence suggesting that cooperative learning fosters positive attitudes toward the subject matter, as well as growth in process skills such as critical thinking, teamwork, and metacognition.[1] Perhaps most importantly, the positive interdependence generated in a small group setting has been shown to attenuate the feelings of isolation, disorientation and competition that often correlate with underachievement or failure in a traditional classroom environment, especially for women and minorities.[8,9,10]

---

[1] Bransford, J.D.; Brown, A.L.; Cocking, R.R., *How People Learn*; National Academy Press: Washington DC, 1999.

[2] Abraham, M. R., Inquiry and the learning cycle approach. In Chemists' guide to effective teaching (Vol. 1), ed.; Pienta, N.J.; Cooper, M.M.; Greenbowe, T.J., eds. Prentice Hall: Upple Saddle River, NJ, 2005.

[3] Lawson, A. E., "What Should Students Learn About the Nature of Science and How Should We Teach It?" Journal of College Science Teaching 1999, 401-411.

[4] Bruffee, Kenneth A. *Collaborative Learning: Higher Education, Interdependence, and the Authority of Knowledge.* Baltimore: Johns Hopkins Press, 1993.

[5] Johnson, D. W.; Johnson, R. T; Smith, K. A. *Active Learning: Cooperation in the College Classroom*; Interaction: Edina, MN, 1991.

[6] Karplus and Thier. *A New Look at Elementary School Science.* Chicago, Rand McNally (1967).

[7] Piaget, J.; *J. Res. Sci. Teach.* 1964, 2, 176.

[8] Hewitt, N.A.; E. Seymour, *Factors Contributing to High Attrition Rates Among Science, Mathematics, and Engineering Undergraduate Majors: A Report to the Sloan Foundation.* 1991, Denver: U. of Colorado Press.

One Explanation for Why POGIL Works (the "Aha!" moment)

"Aha! I get it!" The power of the POGIL method is manifest in the frequency with which such "Aha!" moments take place during class. Consider a student who at first struggles with a concept and becomes frustrated, but–through a combination of answering guiding questions, analyzing data, discussion with group mates, and (as needed) instructor facilitation–the concept finally "clicks" for the student. The culmination of this dissonance-resolution process is often marked by an utterance: "Aha!" or "I get it!" Flushed with the excitement of discovery, this student becomes an advocate for the concept. As if she were the first person to ever discover it, the student works to help her group also appreciate it. In explaining the concept to them she deepens her own understanding, perhaps developing a question for the facilitator about its implications, limitations, or connections to other material.

Such cognitively rich experiences (discovery, discussion, learning by teaching, asking questions, etc.) are established pathways to conceptual understanding and long-term retention, and help explain POGIL's effectiveness.[1] However, the most lasting outcomes of such an experience may be related not to content, but to skills, particularly metacognitive skills (learning how to learn). The learning experiences in a POGIL classroom enhance a student's experience of lecture, homework, and other course offerings because it awakens students to the idea that learning is not memorization of facts and rules (boring), but a creative process requiring thought, participation, and discussion (fun). The next time this student encounters frustration in her studies, she is less likely to give up, isolate herself, or resort to memorization. Knowing what it feels like to figure out and eventually own a concept will motivate her to use the resources at her disposal, especially discussion with others. The pathway to understanding provided by a POGIL learning environment appears to rewire many young learners' natural fear of dissonance—and transform it into the hallmark of a lifelong learner: the confidence that *not* understanding is just an exciting (though sometimes frustrating) prelude to the satisfaction of understanding something new.

The Structure of a POGIL Activity: The Learning Cycle

The learning cycle, developed by Karplus and Piaget,[6,7] is similar to the scientific method.

We have found that activities that follow a learning cycle are very effective at generating Ah ha! moments, and encouraging development of targeted process skills such as critical thinking and self-assessment.

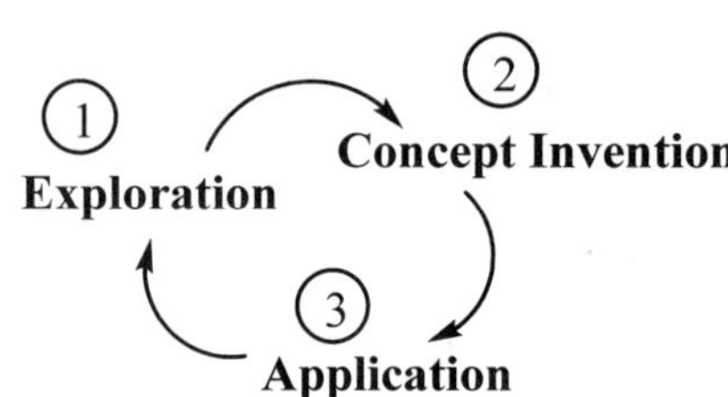

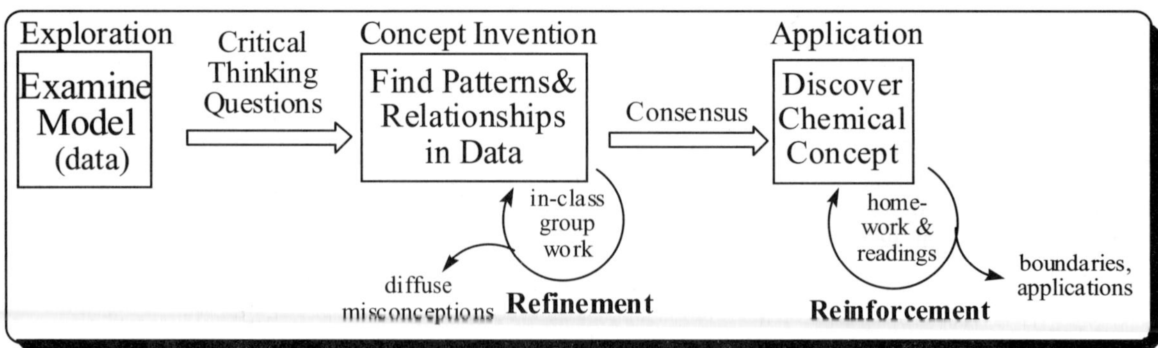

[9] Tobias, S., "Women in Science - Women and Science" *J. of Col. Science Teach*. 1992, 21, 276.
[10] Treisman, P.U. Innovations in Educating Minority Students in Math and Science; Dana Foundation: 1988.

Exploring the Model: Directed Questions

A POGIL learning cycle activity begins with a Model. This model often contains enough information such that a group of students could extract from it the target concept. However, to help students, Construct Your Understanding Questions are provided to guide students toward this concept. The questions usually begin with the very simple. Such questions help with the Exploration phase, and may simply direct students to look at the appropriate part of the model (for this reason such questions are called directed questions). For example, a directed question for the figure below might be: "What aspect of POGIL implementation helps diffuse student misconceptions?"

Coming to Consensus: Convergent Questions

After a few directed questions, the activity usually contains a question that requires students to process the data and find patterns. This is not a clean, linear process, but good questions will bring most student groups to a consensus that resembles the targeted concept. These types of questions are termed convergent questions since most student groups will converge on the same answer.

Though many student groups will converge on the same answer to a convergent question, they still may lack confidence in their answer. Confirmation by the instructor is often not the most valuable next step. As described below, it can be much more useful for the instructor to allow groups to confirm or correct their own answers via the application phase.

Confirming Your Understanding: Application Questions

The final stage of the learning cycle is Application. The purpose of application questions in a successful POGIL activity is to help students assess their current understanding, then assimilate this student-level understanding with expert explanations, including familiarizing themselves with expert terminology. This is best done with a combination of group discussion, teacher talk (e.g. a summary mini-lecture), homework, and reading from a complementary textbook. Textbooks provide a concise explanation of key concepts in expert terminology, but most students are not ready to read this explanation until after they have had a chance to discover the concepts and generate explanations in their own words. For this reason, most POGIL courses ask students to read the assigned sections of the text *after* they have completed the activity on that topic.

In general, forcing students to figure out if they have the concepts is far superior to simply publishing the answers to the in-class Construct Your Understanding Questions. Any time students work toward their own understanding they will learn more and retain it longer. When students are given an answer key, there is a great temptation to look at the expert answers without fully constructing their own answers, and this circumvents the critical part of the learning process.

Content and Process Skill Goals

Most instructors have clearly defined content goals for their course. Their syllabi might list topics like "Nucleophilc Substitution" or "Acid-Base Reactions." A teacher's content goals are often tied to subsequent courses, for which their course may be a prerequisite.

A course can also be designed with specific process skill goals in mind. Development of the key process skills listed at right will help students learn course content, and create new knowledge, applicable to other courses and contexts.

POGIL Process Skills[11]
Information Processing
Critical Thinking
Problem Solving
Communication
Teamwork
Management
Self-assessment

[11] Moog, R. S.; Creegan, F. J.; Hanson, D. M.; Spencer, J. N.; Straumanis, A., Process Oriented Guided Inquiry Learning. In *Chemists' guide to effective teaching (volume 2)*; Pienta, N.J.; Cooper, M.M.; Greenbowe, T.J., eds. Prentice Hall: Upper Saddle River, NJ, 2009.

Is my answer the *correct* answer?

Perhaps the most important but least self-explanatory skill on the list above is self-assessment: the ability to asses if you have understood a concept. A student does not need self-assessment skills to memorize facts and algorithms, to churn out answers to familiar exercises, or to check his answers on an answer key; so many students are not familiar with this skill.

Of course, on the exam *and in life*, there is no answer key. Many students are at first resistant to the frustration inherent in practicing self-assessment. This is the frustration of not knowing if your answer is the right answer. Experiencing and ultimately overcoming this frustration is a key part of becoming a mature learner.

A common student complaint in a POGIL classroom is the following: "How do we know if we are learning the right things if you don't tell us the right answers?" Advice on answering this question is given in the FAQ entry at the end of this chapter entitled **How do you answer the question: "How do I know if my answers are correct?"** One answer is that during an exam it is useful to know if you are right or wrong.

Experienced students sometimes describe self-assessment as the feeling you get when you finally figure something out…the "Ah ha! moment." Or, conversely, the ability to recognize the absence of that feeling: the confusion and frustration associated with not understanding *yet*. **This may be the central skill of all science.** Without this skill a student does not know when to keep asking questions and when his preparations for the exam are complete, and a researcher does not know when to do more experiments and when to publish her findings.

Assigning Group Membership

At the beginning of a semester start with alphabetical group assignments. Then, for the first few weeks, change weekly, mixing randomly so that students get to know a variety of other students and learning styles. Many instructors strive, by mid-course, to have students in a set of static and functional groups so they can begin building specific dynamics. Others continue changing groups throughout the course.

There are three basic strategies for assigning groups:

- Instructor assigns heterogeneous groups based on quiz or exam scores
- Students are allowed to self-select into groups (usually resulting in homogeneity)
- Instructor assigns groups organically based on a variety of criteria (e.g. functionality)

Instructor assignment of group membership is an opportunity to gain valuable insight into the personalities of your students. The more you know about your students, the better you will be able to place them in functional groups. For even a class of twenty this can be a time consuming but rewarding process.

When assigning group membership based on skills and personalities, a useful trait to think about is assertiveness. (This may be more important than gender or other variables.) In general, do not put one quiet, unassertive person in a group with three loud, pushy people. Some of the most successful groups result from putting four quiet, thoughtful students together, or four loud, talky students together.

Formal Roles

One of the great challenges of group work is achieving equal participation among group members. Equal participation is important both in terms of content goals (students who do not participate may not learn as much), and process goals (students who do not participate will not have an opportunity to develop key process skills).

Many POGIL classrooms employ the formal roles described below. Often roles are rotated every class meeting so that every student experiences each role and its responsibilities. Some instructors have had success rotating roles less frequently. It can be helpful to distribute color coded cards, pins, or lanyards, each with the name and description of a role. You can also simply designate that the person with the group folder (see section on Course Management) is the manager, the person to the right of the manager is the presenter, etc..

Even without formal roles, an instructor can informally encourage equal participation. For example, a question may be posed directly to a less active student. Similarly, the instructor can call on certain individuals to present information at the board, or to serve as spokesperson for the group. The following are commonly used roles.[12]

Manager Manages the group. Ensures that members are working together, no one is left behind, and that assigned tasks are being accomplished on time, including that all members of the group participate in activities and understand the concepts. The instructor responds only to questions from the manager who must raise his or her hand to be recognized. This encourages groups to do some internal processing of a question instead of immediately asking the instructor.

Presenter Presents oral reports to the class. These reports should be as concise as possible; the instructor will normally set a time limit.

Recorder Recorder keeps a record of the group's official answers in his or her workbook. This allows the instructor to keep tabs on all the groups and encourages groups to come to consensus about each answer.

Reflector Observes and comments on group dynamics and behavior with respect to the learning process. These observations should be made to the manager on a regular basis (no more than 15 minutes between reports) in an effort to constantly improve group performance. The reflector may be called upon to report to the group (or the entire class) about how well the group is operating (or what needs improvement) and why.

Other roles can be used, but are not described here (e.g., Model Builder, Encourager, Checker).

Classroom Management

Group Folder (Collecting and Distributing Materials)

In most POGIL classes each group has a folder. The group folder is used to return graded quizzes or other work at the start of class. Folders are placed on the head table, and a group representative comes and gets the folder before class begins. At the end of class, group managers may be asked to place work to be graded in the group folder before it is returned.

Instructor's Role as Facilitator

The guiding principle in facilitation is that **students gain far more from correcting their own answers** than from being corrected by the instructor. Do not interpret this as meaning that instructors should refuse to answer questions. A successful facilitator will give students the minimum amount of information required for them to correct their own mistake, while being supportive and encouraging. If the instructor gives too much help (e.g. frequently answers the question "Is this correct?") students will quickly get the message that they cannot trust their own self-assessment skills, and must always get expert confirmation of their conclusions. This is poor training for success on quizzes, exams, and especially for the real world, where there is no answer key.

[12] Farrell, J.J.; R.S. Moog, and J.N. Spencer. "A Guided Inquiry General Chemistry Course." *J. Chem. Ed.* Vol. 76, No. 4, April 1999. p 570-574.

If it seems likely a subsequent question will cause a group to correct their own error, the best course of action may be to *not* intervene. In such cases, the facilitator should return to the group to make sure the problem was solved without intervention.

The following table gives examples of some common interventions that can help students self-correct, without developing dependency on instructor confirmation.

Overview of Instructor Facilitation
Observe each group to monitor progress and keep track of which groups are having problems. Focus on the key questions (marked with a key in the margin).
Intervene with individual groups when necessary. For problems affecting more than two groups, a whole-class intervention may be preferable.
Examples of Group Level Interventions: Ask a group to... • Have the manager read the question out loud • Assess confidence in their answer • Check if all group answers match • Read the next question or model • Consult with a neighboring group • Practice using the bullets above *without prompting by the instructor* when stuck or looking for confirmation of an answer
Examples of Whole-class Interventions: • Ask group(s) to write their answer to a key question on the board (or give a group an overhead and marker) • Call on a group to critique or explain an answer • Call on a group to report answer to the whole class • Vote on competing answers • Instructor mini-lecture

Another key principle of facilitation is that **students critically evaluate comments made by other students**; however, many students try to memorize (without reflection) what the *teacher* says. Whenever possible, use student-offered explanations to provide examples or stimulate discussion. An answer offered by a student may be identical to the correct answer offered by the instructor, yet the former will win more scrutiny from students.

A common pitfall of facilitation is for the instructor to fall into a dialogue with just one member of a group. This most often happens when the instructor is responding to a question posed by one member of a group. When the instructor approaches a group with a question, the first thing to do is make sure all members of the group have already considered this question. One good technique is for the instructor to ask a group member (other than the one with his hand up) to state the question on behalf of the group.

It is also tempting for the instructor to engage in a private dialogue with a student who is behind and needs extra help. Involving the rest of the group is worth the time and energy it takes, even if it means slowing down other group members. As the course progresses, the power of learning by teaching will become apparent, and most students will figure out that a less confident student who asks lots of questions is an asset to the group. The instructor may need to jump start group interdependence by, for example, asking a stronger member to explain a concept to a group mate.

Instructor interactions with the group send powerful messages about how the group should interact. **The instructor should model how students should interact, asking for explanations and reasoning.** When two groups have the same problem, the instructor can ask a representative of each group to switch chairs. Or, if they are close enough, the instructor can point to the other group and say: "Those guys are stuck on the same question that you are discussing. I think you will find their answer very interesting." In the future, students may consult a nearby group without being prompted by the instructor.

Summary Lectures and Other Teacher Talk

Every POGIL class typically includes several short periods of teacher talk (lecture). Class often starts with a brief (three-minute) overview of the previous class, and a one-minute preview of the upcoming class. A general rule is that the larger the class, the more time you will spend at the front of the room, speaking to the class as a whole. If several groups are having the same problem, it can be more efficient to stop group work and address the problem as a class.

Whole-class (plenary) teacher talk can also be useful for speeding student work on an activity. If progress is too slow, interrupting group work and giving away an answer shuts down further discussion since the "expert" has ruled in with the "correct" answer. Be aware that instructor input may have this effect, and that each attempt to speed the class along has a cost. The instructor must decide which is more beneficial at that moment: to move students toward the main concept waiting at the end of an activity, or to allow for discussion that will deepen student understanding of this concept.

If the instructor intervenes too frequently this will reinforce the commonly held student belief that no answer is correct until confirmed by the instructor. With that said, however, revealing too little also has a cost. Especially at the start of a course, student frustration thresholds may be low. The number one complaint from POGIL students continues to be "the instructor refuses to answer my questions." Explaining the pedagogy (essentially telling students "it is for your own good when I force you to figure out answers for yourselves") may only intensify anxiety and frustration. The instructor should acknowledge the tricky spot, and perhaps give a hint. This lets the students know he wants to help. Ideally, the instructor gets to know each student's frustration threshold so he can give only the minimum information required for a student to figure out the concept for herself.

According to Piaget, frustration is an important part of learning.[7] The trick is to push students, but not make them snap. Some students have high tolerance for the cognitive discomfort inherent in learning; others do not. It is deeply gratifying for an instructor to know students at this level, and very exciting for everyone involved when this knowledge helps students go beyond their perceived limitations.

Student Self-Management of Class Time

The activities in this workbook are designed to be completed in one class period. For fifty-minute classes, more work will have to be done outside class than for longer classes. The instructor should not let poor progress on an activity cause work on that activity to roll over into the next class period.

The instructor must make it clear that students are <u>responsible for finishing each activity before the start of the next class</u>. Challenging them to complete the activity during or after class will do two things: 1) Encourage students to manage class time effectively and 2) encourage out-of-class group work.

If an individual student or a group is consistently behind the rest of the class, a good strategy is for such students to preview the activity by reading the Model and Questions before class.

Daily Quiz

It is strongly encouraged that each POGIL class start with a one-question (three-minute) quiz. Ideally, the quiz should cover one important concept that was developed during the *previous* class meeting(s), but serves as foundational knowledge for the *upcoming* activity. At the end of the quiz, the instructor should (very briefly) go over the quiz. Otherwise, students will spend the first five minutes of group work time discussing the answer to the quiz.

Group Take-Home Exams

A great way to foster group interdependence is to assign a group take-home exam. For such an exercise, it can be better to let students form their own groups of either three or four. By requiring that each group turn in only one copy of the exam, the instructor forces students to come to consensus. A good rule is to specify that groups can consult any published work, but they can only talk to their group-mates or the instructor about the exam. After such an exercise there is an increase in out-of-class group work. For this reason, it is good to assign a group take-home early in the course.

Use of a Traditional Textbook

These POGIL materials are intended to be used in conjunction with a traditional textbook. Since most textbooks are written at a fairly high level, reading in the text is not usually a good first introduction to a topic; however, students who have already been introduced to a concept via an activity can gain enormously from reading about it in a textbook. For this reason, it is best to assign reading on a topic *after* the students have completed the relevant ChemActivity.

Use of the text as a reference during class is not recommended, especially at the start of the course. As the course progresses, students who have spent significant time using the text outside of class may find it useful as a reference during class.

Most successful students read in the text after each class to try and confirm their answers to the in-class Construct Your Understanding Questions. If you have ever waited for tomorrow's paper to check your crossword answers, you know that "expert" answers are extremely interesting only if you have spent time struggling to come up with your own, possibly incomplete answers.

Assessment: Improving Instructor and Student Performance

Assessment of Course and Instructor by Students

It is strongly recommended that the instructor ask students for feedback about the course sometime in the first half of the course. (So there is time to make adjustments.) One strategy is to give students ten minutes of class time to describe:

- at least one strength of the course, and how this strength helps them learn
- at least one area for improvement, and (if possible) suggest possible changes
- any other insights about teaching and learning.

Another way to collect this information is to have students email responses to a secretary, who can print the emails, cut off the headers and give the instructor the now-anonymous feedback. To ensure full participation, the secretary should keep a list so responders can be awarded bonus points. If no bonus points are offered for an out-of-class feedback exercise, compliance is likely to be low and skewed toward students with extreme viewpoints (positive and negative).

Assessment of Group by Students

Several times, particularly at the start of a semester, it is useful for each group do a self-assessment. This is the same as above, except that students write down one strength of their *group*, and one area for improvement in their *group*. Each group member should take one minute to share their assessment with the rest of the group. Written reports should be turned into the instructor. If time allows, the instructor can ask for volunteers to report key insights to the whole class.

Self-Assessment of Group Participation by Students

Another type of self-assessment is to distribute the table below with the following instructions:

For each row on the table, circle the statement that best describes YOU in terms of participation in your group. (Not to be collected or graded.)

Excellent (4)	Good (3)	Fair (2)	Poor (1)
Lead *and* share the lead without dominating	Lead but dominate a bit	Follow but never lead	Actively resist group goals
Actively pace group so everyone is on the same question & finish on time	Aware of time issues but don't actively work to keep group together & on pace	Don' t think much about group progress or timing	waste lots of group time, fall behind, or work ahead
Stay on task and keep others on task	Keep self on task	Sometimes get group off task	Often get group off task
Actively create environment where everyone feels comfortable participating	Try to engage others in a helpful and friendly way	Rarely initiate interactions, but respond in a friendly way when others initiate	Observe silently, and offer little when others try to engage you
Express disagreement directly and constructively	Usually express disagreement directly	Avoid confrontation even when angry or frustrated	Let negative emotions get in the way of team goals
Enthusiastic and positive	Moderately enthusiastic	Show little enthusiasm	Negative or unenthusiastic
Always come prepared	Usually prepared	Occasionally unprepared	Usually unprepared

Encouraging Students to Answer the Central Meta-Question

On the surface, the students' job is to construct and refine answers to the Construct Your Understanding Questions; however, underlying most every question is another question:

> ***"What is the purpose of this question; what point, distinction or concept is the author trying to convey by asking us this question; what am I supposed to learn about the Model from this question?"***

Answering this *meta*-question requires a student to see the forest from within the trees, and to get inside the author's mind. The instructor may occasionally ask students to explicitly answer the meta-question, but students who get in the habit of doing this will gain a deeper understanding of the material and achieve greater success in the course.

Final Words of Advice

Most students and instructors need at least a few class periods to find their rhythm with POGIL, but by mid-semester it should start to come together. If it does not feel right, seek help early. Instructors can contact the POGIL Project Office (www.pogil.org), and be matched with an experienced POGIL mentor. Students should address their concerns to their instructor during office hours (not during class).

For instructors, the hardest part appears to be figuring out when and how to interrupt. At the beginning err on the side of interrupting more often and saying more. It is easier to step students back from overdependence than to repair their frustration with the instructor.

For students, the hardest part is getting over the anxiety surrounding the question "How do I know if my answers are the right answers." Get to know the feeling of knowing you are right. Keep asking questions, reading and doing problems until you are confident you understand. Now you are ready for the exam.

Frequently Asked Questions *(with answers)*

FAQ For the Instructor (Professor, Peer Leader, or TA)

How do I deal with room and seating issues such as fixed seating?

An ideal POGIL environment has 20 students in five groups of four, each sitting at a square table with four chairs. Few people have the luxury of such a classroom. Most everyone has to make some kind of accommodation to their specific environment.

Two common arrangements are side-arm desks, and rows of tables with chairs facing the front. With side-arm desks, the main problem is that some groups are reluctant to make their circle tight enough for effective group work. Students may require encouragement to scoot closer together. With rows of tables all on the same level, turn half the chairs around to face the back.

Fixed seating in a lecture hall presents special challenges.

- In a fixed seat lecture hall groups of three work best. Groups of four work also, but often function as two loosely affiliated groups of two.
- If your lecture hall is large enough, leave every third row open so the instructor can walk down the empty rows and field questions from above and below without squeezing past student's knees and tripping on their backpacks.
- Students should work in a way that is comfortable to them (without violating fire codes). In a lecture hall some students choose to kneel on their seats to interact with group mates in the row behind. Others groups have preferred to sit on the floor at the front or back.

How are TAs and peer leaders best trained in the use of POGIL?

Many former POGIL students make excellent facilitators, even without additional training. If this is a possibility, it is always a good strategy to select peer leaders from among last year's POGIL students in the same course.

When this is not possible, an experienced teacher (e.g. the lead instructor in the course) with an understanding of POGIL should meet with the TAs prior to each week for the first few weeks of the course and facilitate a "master class" in which TAs do the activity for that week. Put TAs in groups of four and have them discuss each question. Though the content may be easy for many TAs, it is still worth taking the time to go through the whole activity since it allows TAs to address any questions, and the master teacher to model POGIL facilitation techniques.

After a few weeks of modeling POGIL facilitation during TA-meetings, it may make sense to have the TAs take turns running the TA meeting. Having TAs go through the activity on their own is not nearly as beneficial. If a quiz will be given in the recitation section, the TA meeting should begin with the TAs taking a sample quiz.

How do I recruit undergraduate peer instructors for my Peer Led Team Learning sessions?

Recruiting undergrad peer leaders to facilitate *next year's* peer led sessions is as easy as simply asking for volunteers from among your current "A" students. Students who enjoyed being POGIL students are often eager to continue as peer leaders. Other motivating factors include the opportunity to review for the MCAT, PCAT, etc., and the opportunity to receive teacher training and mentoring from the professor. In situations where budgetary constraints preclude payment of peer leaders, these factors can be sufficient motivators with or without course credit. Of course, recruiting from your past POGIL students is impossible the first time you teach a POGIL course at a given institution.

What should I (the instructor) do the first day in class?

A good strategy is to give a three-minute mini-lecture on key expectations (e.g. work together, complete the activity by next class, be prepared for a quiz), then launch immediately into the first activity. Organic chemistry students are generally anxious to focus on the material that is going to be on the exam. Though understanding POGIL will help them do this, the best way for students to learn about POGIL is by doing the activity. Additionally, it is somewhat hypocritical to lecture students on a lecture-less teaching method.

Is it better to let students select their own groups, or for the instructor to assign groups?

There is some evidence that homogeneous groups lead to the most rapid adjustment to group work. On the other side of this ledger is the fact that "learning to work with a diverse range of people" is frequently cited as a key skill that students need for success in today's world. It is up to the instructor to balance these competing goals.

Why do my students seem to have so many misconceptions?

POGIL instructors get to see what their students know and don't know. After lecturing for years, and assuming that certain concepts are straightforward, instructors are surprised at what their students think, especially about topics they always assumed were covered adequately in prerequisite classes. POGIL instructors get to watch students struggle with concepts and grow their understanding through some very sensitive (and even ugly) stages – students' initial conclusions and the misconceptions carried from prior classes can be fascinating and unexpected. Many of these common misconceptions are compiled at the end of each activity in the **Common Points of Confusion** section, and can be useful to both students and instructors.

Does using POGIL take away the fun of being the Sage on the Stage?

No. Though POGIL instructors do not talk as much as lecturers, the talking they do can be very satisfying. When the instructor is not talking to the whole class, she is walking around the room telling stories, giving analogies, and participating in discussions with student groups. I became a professor because I like to explain, profess, and otherwise share my love and understanding of organic chemistry. You do not give that up by using POGIL. In fact, you may be surprised to find your five-minute mini-lecture sets heads nodding (in agreement, not to sleep), and leaves eyes blazing (instead of glazing).

How do I (the instructor) deal with fast vs. slow groups?

There are a number of facilitation techniques to deal with the diversity of group work rates.

- Slow groups are often slowed by one or two students who like to work slowly and think hard about every question. One solution is for these more careful students to preview the activity before class. After trying this once, students usually find it self-evident that this strategy allows them to get so much more out of group discussion and teacher talk.

- Groups that go really fast often miss the point of some key questions. It can help to simply point this out.
- A group of very bright and confident students may work quickly through an activity and get everything they are supposed to get from it.
    - It can be fun to give this group an auxiliary question of a more challenging nature.
    - Alternatively, the instructor can ask such a group to stand up and walk around as co-facilitators, checking the conclusions of other groups and learning by teaching.

What should I (the instructor) do when I see a wrong answer?

Students learn more when they find their own error and correct it with the minimum instructor intervention. The following are techniques you can use to coax students toward understanding.

- Simply having one student in the group read the question out loud can help the group see an error of interpretation.
- Point students to the relevant section of the Model.
- Often one student in a group will have a correct answer. Rather than pointing this out, the instructor can simply point out that there is a difference of opinion within the group.
- If a whole group is wrong, a great technique is to have one member of the wrong group switch chairs (temporarily) with a member from a correct group, not telling them which group is wrong and which is right so both groups critically evaluate both answers.
- If a large number of groups are off target on a question the instructor should lead a whole class discussion. One good format is to have a representative from several (or all) groups put their answer on the board. (If there is limited board space, pass out blank overhead sheets and overhead pens.) Voting usually reveals the strongest answer, and can lead into a useful discussion of why the wrong answers are wrong.

How should the instructor answer the question "Is our answer right?"

Don't be abrupt and say "I can't answer that." Instead, ask students to explain their answer, or why they think they are right or wrong. In every situation, encourage students to talk about what they know and don't know. This helps the instructor hone in on the misconception or bad assumption.

If class is about to end, or the group seems really frustrated, the instructor might consider simply giving the answer, even though this is almost always less beneficial than helping the students come to their own valid conclusion. If a group or individual is so frustrated that they disengage, this can have long term implications that far outweigh damage to the learning process caused by giving away an answer now and then.

How do I (the instructor) encourage internal processing of questions?

When approaching a student with her hand up, a good strategy is to point to another student in the group and say "Do you know what her question is?" At first, the answer might always be no, but students will quickly learn that most questions can be answered within the group without instructor involvement. If the question has not been discussed yet, you can offer to come back (or if there is time) stay and observe while the original questioner explains her question to the group.

What should I (the instructor) do when I make a mistake?

First, and most importantly, try not to feel bad. It happens to even the most experience instructors. Thinking on your feet in a POGIL classroom is in many ways more challenging than lecturing. Because students are thinking deeply about the material, they will invariably ask questions that you have never considered, and make reasonable arguments leading to conclusions you know are contrary to experiment. Occasionally, instructors get confused by one of these and end up confirming an incorrect conclusion or making an incorrect statement that leads students astray.

The climate of the classroom, and the scope of the error should determine how the instructor unwinds an error, but keep any correction simple. There is a strong temptation to carefully detail (in your own defense) the logic of even the most esoteric error. The result is to draw attention to the issue and give students the impression that this issue is more important than it actually is.

What should I (the instructor) do in the last 5-10 minutes of class?

This is a good time for group self-evaluation, especially at the start of the course or on a day when students gained a strong grasp of the material. Such discussion helps groups work through issues that might hinder them going forward.

On other days, it is best to facilitate a whole-class content summation. This can be presented wholly by the instructor or (even better) by students. The more challenging the material, the more important it is to have a content summary at the end.

Are there POGIL activities other than the ones in this book?

Organic Chemistry: A Guided Inquiry *for Recitation* consists of two volumes. Volume 1 covers key Organic 1 topics, Volume 2 covers key Organic 2 topics. A full-course set of materials (called simply, *Organic Chemistry: A Guided Inquiry*) is also available and includes many topics not covered in either of these slim recitation volumes is. The full-course set contains activities on most first year organic topics, enabling an instructor to use POGIL in nearly every class meetings (see next topic).

POGIL activities are available for most chemistry courses (including general, physical, analytical, and biochemistry), and for some courses in biology, mathematics, and engineering. More information is available at www.pogil.org.

Can one do POGIL during every class (instead of just once a week)?

Many POGIL courses consist of group work every day, instead of just in recitation. Such an approach has been employed successfully in both small and large (>350 students) classrooms. In large classrooms (>50 students) POGIL is usually used in conjunction with electronic classroom response devices ("clickers"). More information on the use of POGIL in large classes can be found at www.pogil.org/straumanis.

FAQ for Students and Instructors

Wouldn't it be easier for students to learn the "right" answers if the instructor posted them?

Of course it would be easier if the goal was simply to have students memorize answers to questions. However, the Construct Your Understanding Questions are far too easy to put on an exam. (They are designed to be your introduction to a topic). It turns out that memorizing answers to simple questions does not help you answer harder, more conceptual questions. HOWEVER, wrestling with simple questions and coming up with your own answers is a fantastic entry into a topic, and the best preparation for your further study via homework, lectures, and quiz and exam questions.

Another way of saying this is: "When a student reads a question and doesn't immediately know the answer, the tendency is to immediately look at the answer key if it's available. This hurts the learning process since most learning takes place as you try to figure out the answer to a question by talking to others, reading the textbook, revisiting the activity, etc. Wrestling with the questions is an important part of the learning process in this class, and giving out the answers severely short circuits this process.

If I (a student) understood the activity, do I still need to do the homework and reading?

Some students leave their POGIL class feeling they have learned enough that they do not need to do the homework and reading. No matter how much you feel you understood the activity during class—class is just the start. The homework and reading often contain important extensions of the basic concepts covered in class. The instructor can quickly convince students that skipping homework was an error by pulling the quiz from something that only appeared in the reading or homework. Because POGIL activities are designed to be a student's first introduction to a course, successfully answering the in-class Construct Your Understanding Questions does not, alone, constitute mastery of a topic.

What is a Recorder's Notebook, and how is it used?

Especially if there was no time for a content summary at the end of class, it can be useful to require students or groups to keep a **Recorders Notebook**. In this, students write down the two or three most important concepts from the day's activity. The instructor can spot check these (credit/no credit) during the quiz in the next class.

What if students don't finish the activity in class?

The instructor must make it clear to students that they must finish the whole activity (including homework and reading) before the next class, either on their own, or as a group. This knowledge will help students manage their time, including making a start on the activity before class if necessary.

How does POGIL affect students with learning differences: dyslexia, ADHD, ADD, etc.?

Because POGIL is student centered, it allows students to adapt the classroom to their strengths. Students who are verbal learners can talk it out. Students who learn by listening can listen. Students who learn by reading can read...etc. This same flexibility helps hearing and visually impaired students adapt to a POGIL environment.

Why spend energy improving learning and process skills? (Will they be on the exam?)

A POGIL class is as much about critical thinking, problem solving and group work as it is about content (e.g. organic chemistry). Of course, the two go hand-in-hand. The better a student's learning skills, the more she will learn. It is also the case that in today's world such skills are highly prized. You cannot be a successful scientist or professional without these skills. Alumni of POGIL classes report that the process skills they learned helped them succeed in graduate school, jobs, medical school etc. As an aside, a common comment is that POGIL is a great topic to raise in an interview. POGIL is new enough that many interviewers have never heard of it, and it fits well with the big push at many institutions for creative problem solvers who work well as part of a team.

Will students like POGIL?

Student grumbling about POGIL in the first week of the course is normal, especially the first time POGIL is used for a given course or at a given institution. Any new classroom method is met with resistance by some students. Most students quickly adjust, and by the end of the course only about 10% are negative about POGIL.

Interestingly, at the start of a course the strongest students are the most critical, but this quickly turns around. By the end, the strongest students become the biggest advocates of POGIL. For example, in a recent anonymous survey of 240 POGIL students none (0%) of the 25 students who received an A in the course were negative about POGIL, and only 2 (6%) of the 31 B+/A- students were negative. The majority (80%) of the students who gave POGIL a negative review received a C or lower in the course.

How did the author get involved in POGIL?

When I was a student at Oberlin College I had excellent teachers, but the format was straight lecture. It was my experiences *after* each class that shaped my contributions to what eventually became POGIL.

Every day in my organic chemistry class I wrote down everything the professor put on the board so I could figure it out later. I was lucky to have two other students in the class who were on the soccer team. On the bus going to away games we would sit in the back and work together. Home game weeks we would meet in a study room in the library. (We tried meeting in a dorm room once, but it had a television and we ended up watching Magnum PI.)

In the group sessions we went through each example in our notes and tried to answer the question: "What concept or distinction is the professor trying to convey with this example?" Then we did the homework problems together. If we came across a question we couldn't answer ourselves we asked another group of students who studied in an adjacent room in the library (or, when on the bus, we asked a teammate who had taken the course the year before). If this didn't help, we wrote down our question to ask the professor in the morning.

It struck me that many of the other successful students in the class were doing the same thing as we were, but that there were also many students who were struggling, working on their own.

Later, in graduate school at Stanford University, I encountered the work of P. Uri Treisman and others, which explained why my study group in college was so effective. I had my own recitation sections and I began to experiment with ways to encourage study group behavior in my students during class. A key element of this was the use of guided inquiry: using leading questions to guide students toward understanding of a concept.

About this same time I met Professor Rick Moog, of Franklin & Marshall College, while he was visiting his former graduate research advisor (also at Stanford). He invited me to attend his talk the next day at the American Chemical Society Meeting in San Francisco. It turned out that he had been thinking about similar issues: group learning and guided inquiry. I found Rick was way ahead of me and was about to publish guided inquiry activities for general chemistry. I was so stimulated by his fifteen minute talk that as soon as he finished, my hand shot up and I asked: "Who is writing similar activities for organic chemistry?" He looked me in the eye and said "You are."

Inspired by this, I began work immediately on what eventually became Organic Chemistry: A Guided Inquiry. A few years later, a generous grant from the National Science Foundation allowed us (along with five others) to form the POGIL Project, which has been the focus of our now fourteen-year old collaboration.

My teaching has changed and evolved during the past 20 years, but my primary goal remains the same: to create a classroom environment where students work together in pursuit of a deeper understanding of organic chemistry, and in the process learn how to exceed their own perceived limitations.

Andrei Straumanis, October 2010

Index of Frequently Asked Questions

(answers in the previous section)

Questions for the Instructor (Professor, Peer Leader, or TA)

How do I deal with room and seating issues such as fixed seating?

How are TAs and peer leaders best trained in the use of POGIL?

How do I recruit undergraduate peer instructors for my Peer Led Team Learning sessions?

What should I (the instructor) do the first day in class?

Is it better to let students select their own groups, or for the instructor to assign groups?

Why do my students seem to have so many misconceptions?

Does using POGIL take away the fun of being the Sage on the Stage?

How do I (the instructor) deal with fast vs. slow groups?

What should I (the instructor) do when I see a wrong answer?

How should the instructor answer the question "Is our answer right?"

How do I (the instructor) encourage internal processing of questions?

What should I (the instructor) do when I make a mistake?

What should I (the instructor) do in the last 5-10 minutes of class?

Are there POGIL activities other than the ones in this book?

Can one do POGIL during every class (instead of just once a week)?

Questions for Students and Instructors

Wouldn't it be easier for students to learn the "right" answers if the instructor posted them?

If I (a student) understood the activity, do I still need to do the homework and reading?

What is a Recorder's Notebook, and how is it used?

What if students don't finish the activity in class?

How does POGIL affect students with learning differences: dyslexia, ADHD, ADD, etc.?

Why spend energy improving learning and process skills? (Will they be on the exam?)

Will students like POGIL?

How did the author get involved in POGIL?

Notes

ChemActivity 1: Lewis Structures

(How do I draw a legitimate Lewis structure?)

Read this page once; then answer the Construct Your Understanding Questions on the next page.

Model 1: G. N. Lewis' Octet Rule

In the early part of the last century, a chemist at the University of California at Berkeley named Gilbert N. Lewis devised a system for diagramming atoms and molecules. Though simple, the system is still used today because predictions made from these diagrams often match experimental data.

Lewis proposed the following representations for the first ten elements with their **valence electrons**.

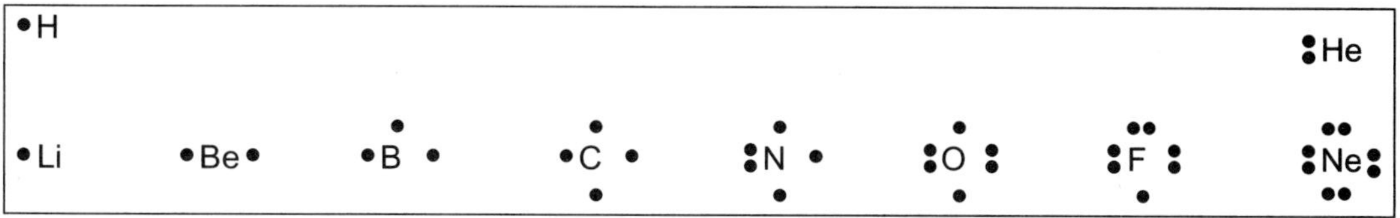

Figure 1.1: Electron Dot Representations of Elements

Only He and Ne are found in nature as shown above. All the other elements are found either as a charged species (**ion**) or as part of a **molecule** that can be represented as a legitimate **Lewis structure**.

CHECKLIST: a Legitimate Lewis Structure is a dot or line bond representation in which…

I. The correct TOTAL number of valence electrons is shown.

II. The sum of the valence electrons around each hydrogen atom is two.

III. The sum of the valence electrons (bonding pairs + lone pairs) around each **carbon, nitrogen, oxygen, or fluorine atom is eight**–an **octet**. (this is the **"octet rule")**

Note that Lewis' rules apply to H, C, N, O and F. We will find that atoms in the next row of the periodic table (e.g., silicon, phosphorus, and sulfur) and beyond commonly violate the octet rule.

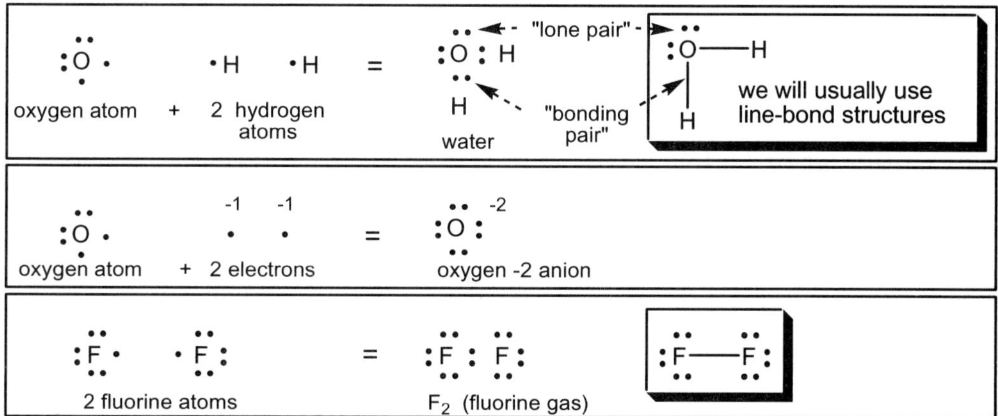

Figure 1.2: Examples of combinations that form legitimate Lewis structures

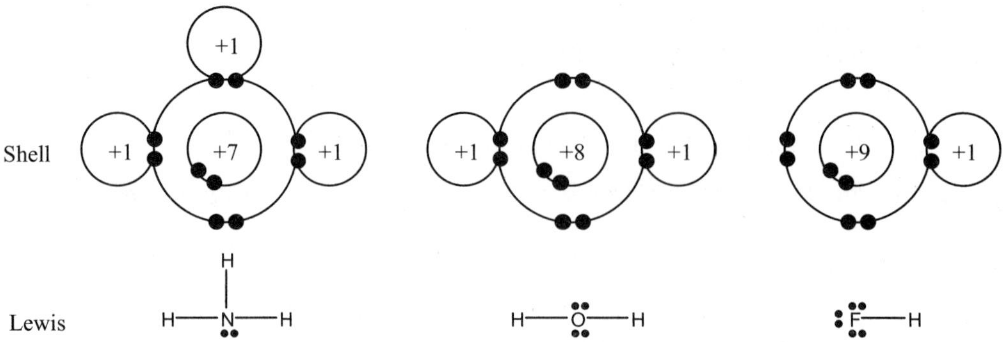

Figure 1.3: Valence Shell and Lewis Representations of Selected Compounds

Construct Your Understanding Questions (to do in class)

Note: Questions marked with an (E) *are meant to be* ***Easy****; a review or direct exploration of the model.*

1. (E)Confirm that each molecule or ion in Figures 1.2 and 1.3 is a legitimate Lewis structure.

2. The **valence shell** of an atom in a legitimate Lewis structure (*see* Figure 1.3) has what in common with the valence shell of a noble gas? (Noble gases are stable elements found in the last column of the periodic table, e.g., He, Ne, Ar, etc.)

3. Draw a <u>shell representation and Lewis structure</u> for the <u>ion</u> of fluorine that you predict is most likely to be stable, and explain your reasoning.

4. Draw a <u>Lewis structure</u> of a neutral molecule that you expect to be a stable and naturally occurring combination of <u>one carbon atom</u> and some number of fluorine atoms.

5. The following structure is <u>NOT</u> a legitimate Lewis structure of a neutral O_2 molecule. :O : O:

   a. Explain why it is not legitimate.

   b. Which item on the legitimate Lewis structure CHECKLIST in Model 1 is violated?

6. It is impossible to draw a legitimate Lewis structure of a neutral NH_4 molecule. Hypothetically, how many valence electrons would such a neutral NH_4 molecule have *if it could exist*?

   a. The +1 cation, NH_4^+, does exist. How many valence electrons does one NH_4^+ ion have?

   b. Draw a Lewis structure of NH_4^+

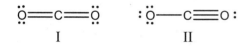

7. Describe how to calculate the total number of valence electrons in a +1 ion… in a -1 ion.

Model 2: Two Lewis Structures for CO_2

$\ddot{O}=C=\ddot{O}$ (I) $:\ddot{O}-C\equiv O:$ (II)

Experiments indicate that both carbon-oxygen bonds of carbon dioxide (CO_2) are identical.

Construct Your Understanding Questions (to do in class)

8. (E)Are both structures of carbon dioxide (CO_2) in Model 2 legitimate Lewis structures?

9. (E)Which Lewis structure best fits experiments indicating that both C to O bonds are identical?

Model 3: Formal Charge

One of the Lewis structures of CO_2 in Model 2 is less favored because it has an imbalance of charge. To find the "hot spots" of + and – charge in a structure we must calculate the formal charge of each atom.

Memorization Task 1.1: Formal Charge = (Group Number) – (# of lines) – (# of dots)

- **Group Number** = Column number on the periodic table (or number of dots on atom in Fig. 1.1)
- **# of lines** = Number of line bonds to the atom in the Lewis structure
- **# of dots** = Number of nonbonded electrons the an atom in the Lewis structure

Construct Your Understanding Questions (to do in class)

10. (E)According to the periodic table at the end of this book, what is the **Group Number** of nitrogen?

    a. (Check your work.) Does this match the number of dots on N in Fig. 1.1?

    b. (E)How many line bonds are attached to N on the structure of NH_3?

    c. (E)How many nonbonded electrons are drawn on N in NH_3?

    d. Calculate the formal charge of each atom in NH_3?

11. (Check your work.) Most atoms in organic molecules (including all atoms of NH_3) have a **zero formal charge**. Confirm that each atom at right has a zero formal charge.

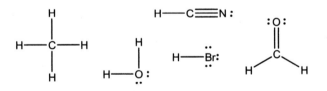

12. In this course we will often encounter **+1 and -1 formal charges**, though rarely will we see formal charges of +2, -2, +3, -3, etc., because they are generally unfavorable. **<u>All nonzero formal charges must be shown on a structure</u>**. +1 and –1 formal charges are often shown as a circled + or – (⊕ /⊖). Add missing formal charges below. (Each has exactly one nonzero formal charge.)

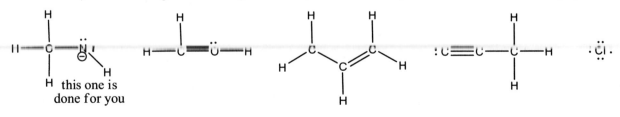

this one is done for you

Extend Your Understanding Questions (to do in or out of class)

13. Complete each of the following Lewis structures by adding any missing formal charges.

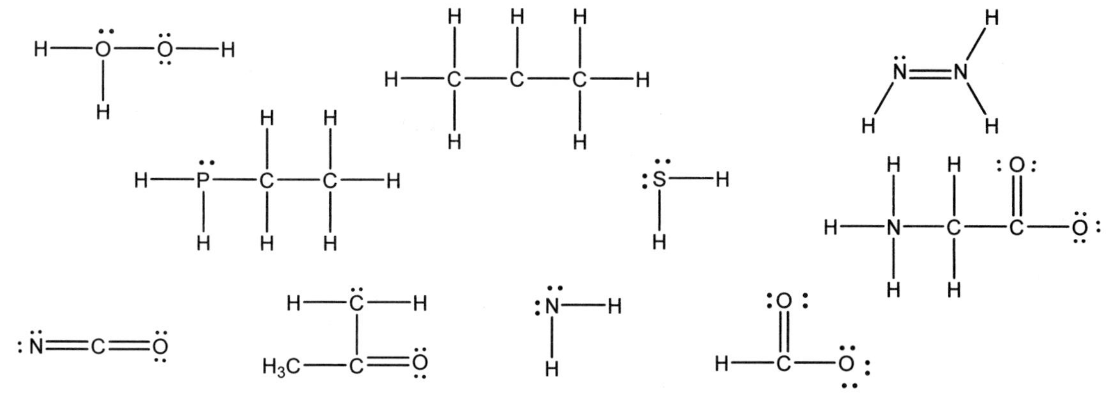

14. (Check your work.) Structures in the top row of the previous question have a **net charge** of +1, structures in the middle row have a **net charge** of zero, and structures in the bottom row have a **net charge** of -1. [**net charge** = total charge on a molecule = sum of all formal charges]

15. T or F: If the net charge on a molecule is zero, the formal charge on every atom in the molecule must equal zero. (If false, give an example from Exercise 1, above, that demonstrates this is false.)

16. Identify the one Lewis structure above that is NOT legitimate, and explain what attribute of a legitimate Lewis structure it is missing.

17. (Check your work.) The top-center Lewis structure above is a **key exception to the octet rule** called a **carbocation.** We will study carbocations extensively in the course. For reasons we will discuss later, a carbocation carbon rarely is involved in a double or triple bond. That is, a carbocation almost always has three single bonds, as shown above and on the next page.

18. Calculating formal charges takes too long. To succeed in this course you must train your eye to immediately recognize a missing formal charge for certain key atoms based on the number of bonds and lone pairs on that atom. Assign a formal charge to each carbon below. What do the three carbons in the left box have in common? What do the carbons in the right box have in common?

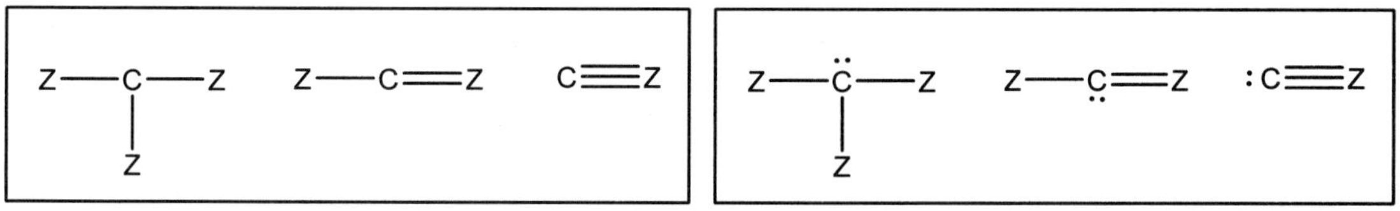

Note that "**Z**" is used as a placeholder in this figure (and on the next page) to represent an unspecified atom.

19. The first row of the table below shows all the common bonding environments of carbon with a +1, 0, or –1 formal charge. Complete the rest of the table. **Make sure each structure you draw has an octet for the central N, O, or X.** Note: X is commonly used to represent the halogens F, Cl, Br, I.

Recognizing Formal Charges for C, N, O, and X

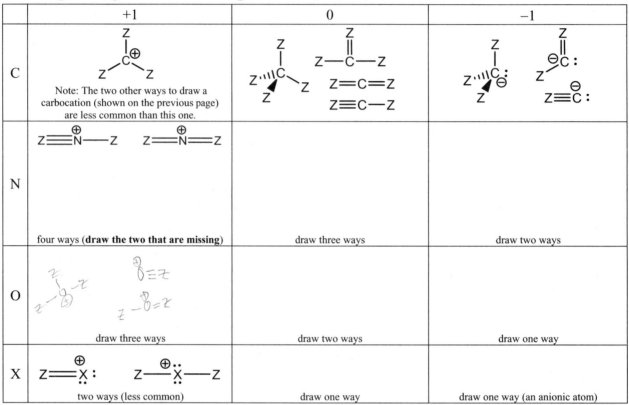

	+1	0	–1
C	Z–C⊕(–Z)–Z (trigonal) Note: The two other ways to draw a carbocation (shown on the previous page) are less common than this one.	Z–C(Z)(Z)–Z (tetrahedral) Z—C(=Z)—Z Z=C=Z Z≡C—Z	Z–C⊖(:)(Z)–Z (tetrahedral) Z–C⊖(=Z): Z≡C⊖:
N	Z≡N⊕—Z Z=N⊕=Z four ways (**draw the two that are missing**)	draw three ways	draw two ways
O	draw three ways	draw two ways	draw one way
X	Z=X⊕: Z—X⊕(:)—Z two ways (less common)	draw one way	draw one way (an anionic atom)

20. It turns out that you can quickly recognize the formal charge on most any N, O and X by counting the number of bonds, without considering the number of nonbonded electrons (dots). Explain this, and why you must count both the bonds and lone pairs to figure out the formal charge on a carbon.

Memorization Task 1.3: Recognizing formal charge for C, N, O, or X

Before the next class: Study the patterns in the table above, and do practice problems until you can QUICKLY recognize the formal charge (+1, 0, or -1) of any C, N, O or X in a structure without counting.

For example, an N with four bonds should look "wrong" without a +1 formal charge; and an N with two bonds should look "wrong" without a -1 formal charge. (Write a similar rule for oxygen!)

21. Make a checklist that can be used to determine if a Lewis structure is correct and that it is the best Lewis structure.

Confirm Your Understanding Questions (to do at home)

22. Turn back to Model 2, and add any missing formal charges to each Lewis structure of CO_2.

    a. Based on the concept of formal charge, which is the better Lewis structure for CO_2 (in Model 2), Lewis structure I <u>or</u> Lewis structure II? Circle one, and explain your reasoning.

    b. Is your choice consistent with the experimental data?

23. Shown below are two possible Lewis structures for the amino acid called glycine.

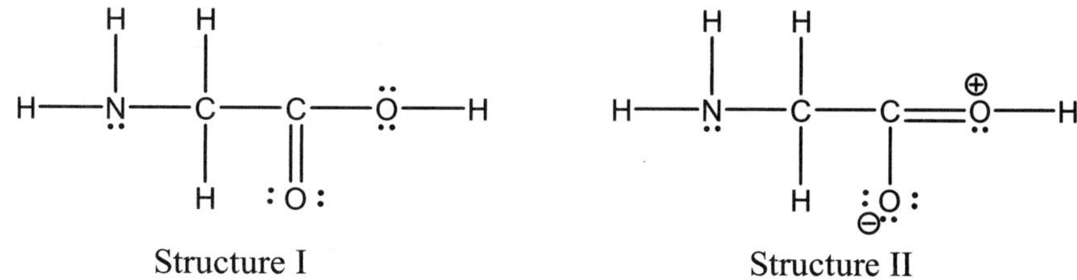

    a. Predict the ∠COH bond angle based on the Lewis structure on the left.

    b. Predict the ∠COH bond angle based on the Lewis structure on the right.

    c. Which prediction do you expect to be more accurate? Explain your reasoning.

24. Draw the Lewis structure of a neutral molecule that is a naturally occurring combination of hydrogen atom<u>s</u> and one sulfur atom. What is the shape of this molecule?

25. Draw legitimate Lewis structures of the following species, and predict the geometry about the central atom (shape).

    a. NH_3
    b. NO_2^+
    c. N_2O (try with N or O as the central atom)
    d. CCl_4
    e. CO_3^{2-}
    f. N_2 (Note: based on the definition, a molecule with only two atoms does not have a shape.)

26. For each element, predict (and draw a Lewis structure of) the most commonly occurring ion (some of these have a formal charge with a magnitude larger than +1 or −1)

    a. sulfur
    b. magnesium
    c. iodine
    d. oxygen

27. Predict which of the following species is least likely to exist.

    CH_2 NO^+ HO^-

28. The molecules BH_3 and SF_6 and the ion SO_4^{2-} exist and are stable. Draw a Lewis structure of each, and comment on whether they are violations of Lewis' octet rule.

29. The structures at right are NOT legitimate Lewis structures (and are missing formal charges). Show (as in the example) where one pair of electrons can be moved to make the Lewis structure legitimate.

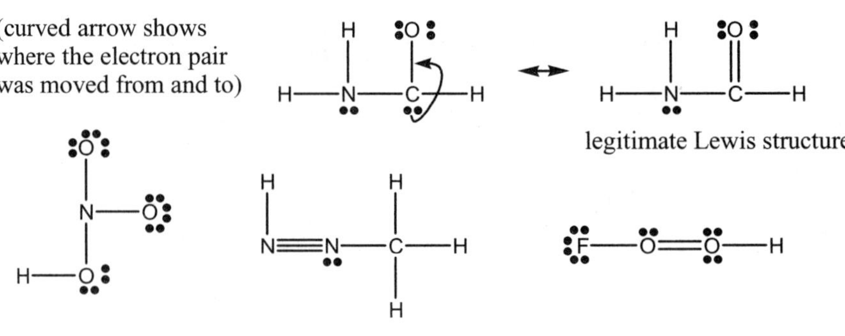

(curved arrow shows
where the electron pair
was moved from and to)
legitimate Lewis structure

30. Fill in missing formal charges where needed (all lone pairs are shown).

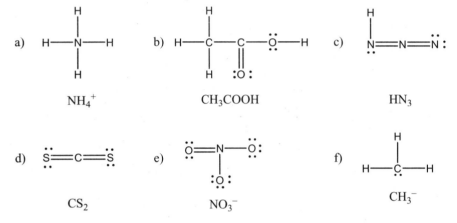

a) NH_4^+

b) CH_3COOH

c) HN_3

d) CS_2

e) NO_3^-

f) CH_3^-

31. Below each structure in the previous question is a "condensed structure" that tells you something about how the atoms are arranged. Draw complete Lewis structures for each of the following condensed structures. (The net charge, if any, on each molecule is given at the end.)

a. $CH_3CH_2^-$

b. CH_2CH_2

c. CH_2CCH_2

d. $C(CH_3)_3^+$

e. BH_4^-

f. NCO^-

g. CH_2OH^+ (two different acceptable answers)

h. $CH_2CHCHCHCH_2^+$ which may also be written as $CH_2(CH)_3CH_2^+$ (more than one acceptable answer)

32. For each structure in the previous two questions, predict the shape of each central atom.

33. Carbon monoxide (CO) is an example of an overall neutral molecule (net charge = 0) that has non-zero formal charges. Draw a Lewis structure of carbon monoxide (CO).

34. The Lewis structure below has no "hot spots" of + or – charge (formal charges), yet it is not as valid as the Lewis structure you drew in the previous question. Explain.

:C=O (with two lone pairs on O)

Read the assigned pages in your text, and do the assigned problems.

Using "The Big Picture" & "Common Points of Confusion" Sections

In this book you are asked to discover your own answers. This can be fun and rewarding, but frustrating when you are not sure if your "discovered" understanding is valid.

Why not just give you the answers to the in-class Critical Thinking Questions?

One answer: If we did, many students would just memorize the answers without thinking about them.

Science is the fun and creative art of constructing your *own* valid explanations for observations. Science is not about *memorizing* answers. Memorization is boring. A practicing scientist cannot "check the answer key" to see if her new theory is correct.

In this course and in real life you build your understanding, then test and improve it by applying it to problems and discussing it with peers. This course is designed to develop your ability to do this. In the process you will learn to recognize the signs when you are correct; and just as importantly, recognize the signs when you are missing something important. This skill is critical for success in school and life. In life (and during the test!) there is no answer key.

After completing a ChemActivity and homework Exercises, check your understanding by reading the Big Picture and Common Points of Confusion sections found at the end of each activity. If the homework or these sections do not make sense, then you are likely missing something important. Go back and study the activity, do more problems, read the textbook more closely, or seek help from a peer, teaching assistant or your instructor.

There is more advice about how to know if you are "**learning the right thing**" in the "To the Student" section that precedes the Table of Contents.)

The Big Picture

Checking for an octet and assigning a formal charge can be done by counting, but this is very slow. These operations must become second nature so that you can quickly determine **formal charge** and use this information to answer higher-level questions. Students who fail to familiarize themselves with the common occurrences of C, N, O, and X with +1, 0, and –1 formal charges quickly find themselves falling behind their classmates at this critical juncture in the course. Things get more complex quickly, so invest some time now and prepare yourself.

Common Points of Confusion

- The number one student error at this point in the course is to confuse the TWO REASONS TO COUNT ELECTRONS on a Lewis structure. One is to check for an octet by looking for eight electrons around an atom (this is very straightforward). The other is to determine an atom's formal charge by counting electrons assigned to that atom (half the bonding electrons plus all the nonbonding electrons). A simple way to do this formal charge accounting is to count the number of lines plus the number of dots; since each line represents two shared electrons, and each dot represents one unshared electron.
- Lewis said that C, N, O, and F must have an octet. (He didn't know about carbocations, and carbocations turn out to be a common exception to the octet rule.) The fact that third-row elements like sulfur and phosphorous can expand their octets is therefore not a violation of Lewis' **octet rule**.
- Carbon is the only atom among C, N, O, or F that can exist without an octet. This can be justified by the fact that carbon is the least electronegative of the four. In other words, N, O and F are so electron-greedy that they will not accept having only six electrons in their valence shell, and will always have an octet.

Notes

ChemActivity 2: Resonance Structures

(Why is baking soda a much *weaker* base than hydroxide, though both have an O with a -1 charge?)

Model 1: Two Lewis Structures for Baking Soda (bicarbonate)

There are two different but equally good ways to draw a legitimate Lewis structure of bicarbonate ion.

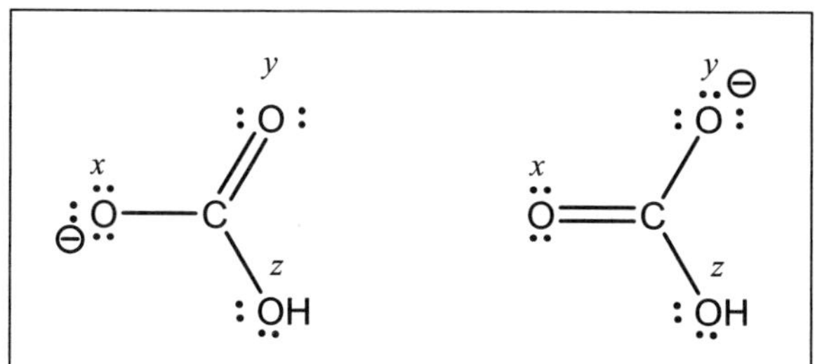

Neither structure by itself tells the whole story.

To get the most accurate picture of bicarbonate ion you must **average the two different structures**.

Figure 2.1: Bicarbonate Ion ($HCO_3^{\ominus}$)

To facilitate the following discussion, the three oxygens of bicarbonate were labeled O_x, O_y, and O_z.

Construct Your Understanding Questions (to do in class)

1. A student looks at Model 1 and says she expects the bond between C and O_x to be halfway between a single bond and a double bond (i.e., shorter and stronger than a single bond, but not as short and strong as a double bond) and that the same will be true of the bond between C and O_y.

   Use the information in Model 1 to construct an explanation for these conclusions.

2. Based on the most accurate picture of bicarbonate ion, what is the charge on O_x? … O_y? … O_z? (Note that it may be a fraction.) Briefly explain how you calculated your answers.

3. Organic chemists say that "the -1 charge on bicarbonate ion is 'delocalized' between O_x and O_y." Come up with at least one synonym for the word "delocalized" in the preceding sentence.

4. Which do you expect to have a more intense and concentrated "hot spot" of negative charge: methoxide ion or bicarbonate ion? Explain.

5. One definition of a strong base is: *"A molecule with a -1 formal charge* localized *on a single H, C, N, or O atom."* Construct an explanation for why bicarbonate ion is NOT a strong base.

Model 2: Resonance Structures

Many ions (e.g., bicarbonate) have charges that are delocalized (shared) among multiple atoms. To obtain an accurate picture, we draw all legitimate Lewis structures of such an ion and average them.

Each of these legitimate Lewis structures is called a **resonance structure** (abbreviated "r.s.").

[Resonate = fit together toward a common purpose or view, e.g., "The candidate's positions really resonate with the voters."]

To generate the full set of resonance structures for an ion:

- Draw any one legitimate Lewis structure of the ion
- CAREFULLY follow the **Rules for Use of Curved Arrows** to generate the other legitimate Lewis structures (resonance structures) of the ion.

Memorization Task 2.1: Rules for Use of Curved Arrows

I. For most **anions**: Use one Type 1 and one Type 2 in tandem to move a ⊖ charge.

II. For most **cations**: Use one Type 3 to move a ⊕ charge.

III. Arrows are only used to move electrons to an adjacent (next-door) atom or bond.

IV. You may not move electrons in a *sigma* bond, move any atoms, change the total number of electrons, or change the total number of formal charges.

V. Each C, N, O and X atom must have an octet (carbocations [$R_3C^{\oplus}$] are acceptable)

Anion Example::

By convention: resonance structures are bracketed together with square brackets and linked with a double-headed arrow.

Cation Example:

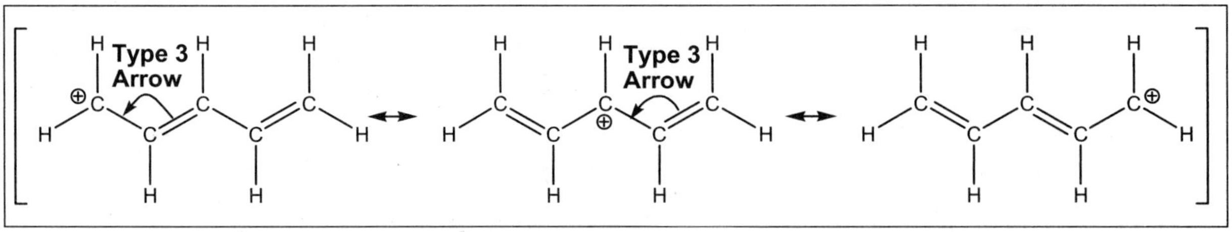

If you cannot legally move any charge (on any resonance structure) to generate a new structure, you are done. There are no more resonance structures in the set for that ion.

Construct Your Understanding Questions (to do in class)

6. [(E)]Match each of the following descriptions to the correct type of curved arrow (1, 2, or 3):

a. A lone pair of electrons becomes a π bond to an adjacent (next-door) atom.

b. A pair of electrons in a π bond slides down to become a lone pair on one atom in that bond.

c. A pair of electrons in a π bond moves to become a π bond to an adjacent (next-door) atom.

7. Complete the set of resonance structures for the following anion.

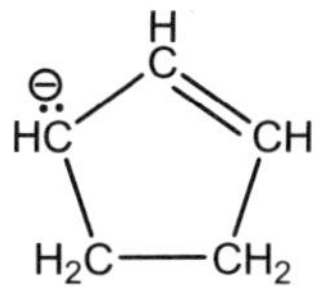

8. The Lewis structure below does NOT belong in the set of resonance structures for the ion in the previous question. What rule must be broken to make it?

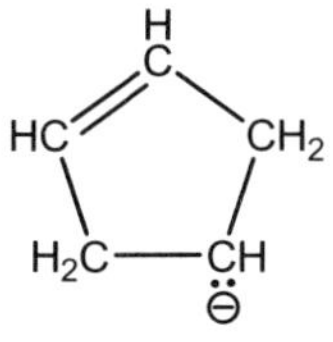

9. Complete the set of resonance structures for the following cation.

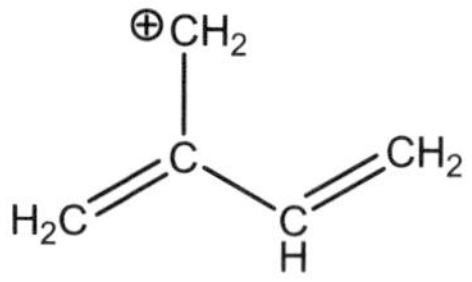

10. (Check your work) The cation in the previous question has a total of **two** resonance structures *including* the one shown above. If you drew more than two, go back and check your arrows.

11. When drawing resonance structures of an anion, the Type 1 curved arrow originates at a lone pair on the atom with a -1 formal charge. Do your best to describe in words where you should start your Type 3 arrow when you are trying to draw a resonance structure of a cation.

12. Only one structure below has a legal set of curved arrows. Each of the other structures has at least one illegal arrow. Cross out any illegal arrow, and explain why it violates a rule in Model 2.

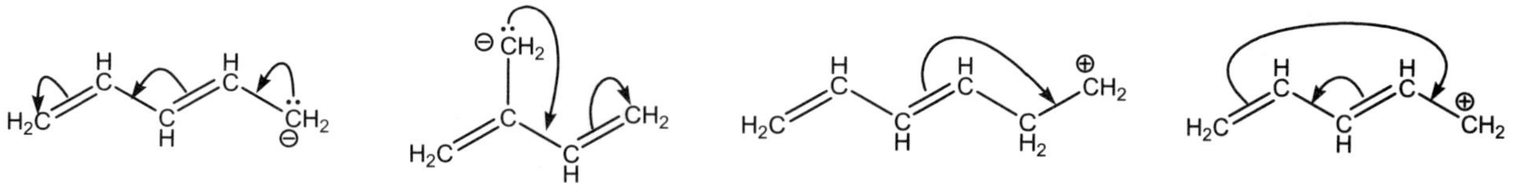

Extend Your Understanding Questions (to do in or out of class)

13. For each proposed set of resonance structures:

    a. Add curved arrows (starting from the left) to show how each successive r.s. was generated.

    b. Cross out the one resonance structure that is not part of each set, and explain your reasoning. (Hint: *See* Rule III in Model 2).

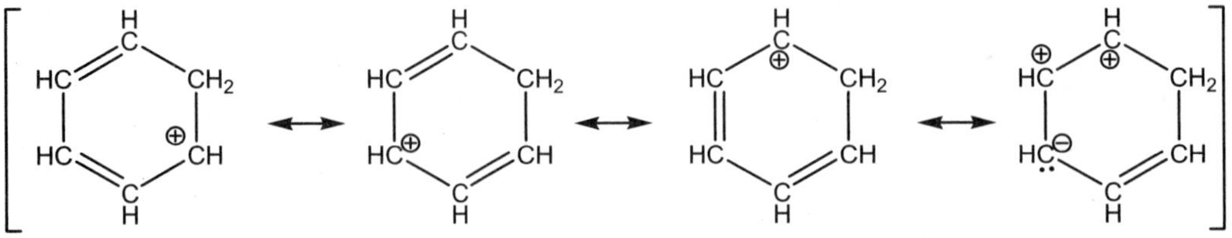

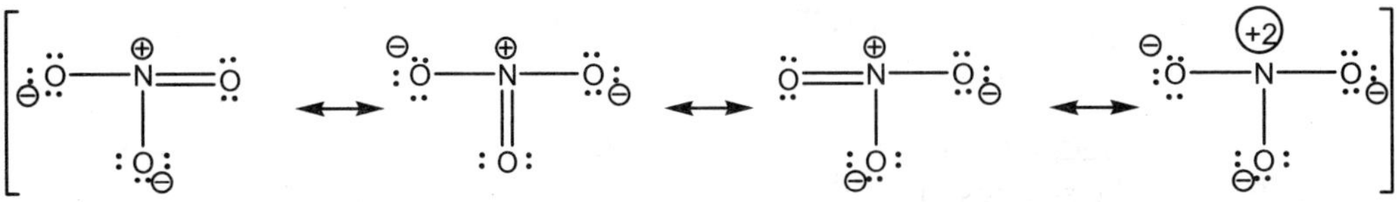

14. Occasionally, you will encounter a neutral molecule that has more than one resonance structure. There are two general types: ***zwitterions*** and **aromatic molecules**.

- For a ***zwitterion*** (an overall neutral molecule with both a ⊕ and ⊖ formal charge) **use a Type 1 and a Type 2 Arrow** (as with **anions**, *see* Model 2).

zwitter = German for "hybrid" or "hermaphrodite"

Ozone

- For an **aromatic molecule** (defined later in the course) **use a Type 3 Arrow** (as with **cations**).

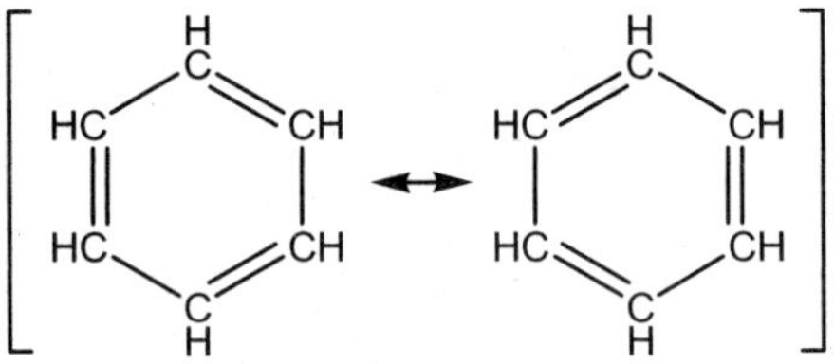

Benzene

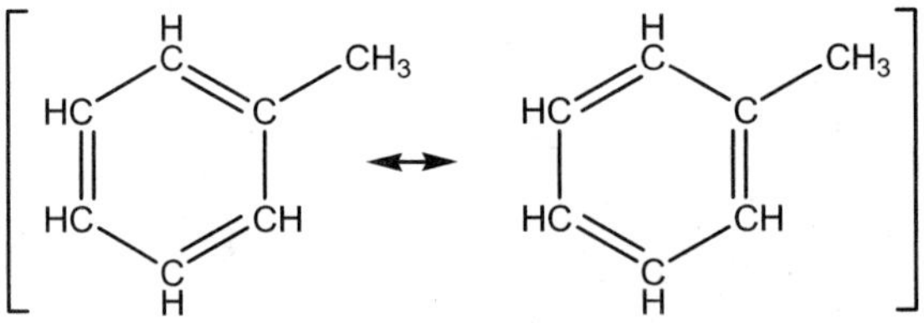

Toluene (also called Methylbenzene)
a common benzene derivative

***Aromatic molecules** are named for the odors of the first members of the family to be discovered. They are frequently encountered, both in nature and the laboratory, and will play a central role in the second half of this book.)*

a. Add curved arrow(s) to one resonance structure of ozone so as to generate the other resonance structure of ozone. What type of curved arrow(s) did you use?

b. Add curved arrow(s) to one resonance structure of benzene and toluene so as to generate the other resonance structure. What type of curved arrow(s) did you use?

c. Confirm that there is no legitimate Lewis structure of ozone with all zero formal charges.

15. The five structures in the top row are considered **identical** representations of the same molecule, whereas the five structures in the bottom row are considered five **different resonance structures**.

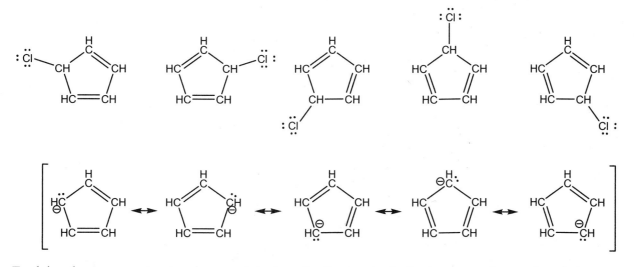

Explain why any one structure is enough to describe the species in the top row, whereas the whole bottom row is needed to describe the species in the bottom row.

16. What is the purpose of the previous question? What concept or distinction is it designed to help you make? What common misconception is it designed to help dispel? (Hint: *See* the "Common Points of Confusion" section at the end of this chapter.)

Confirm Your Understanding Questions (to do at home)

17. Draw all resonance structures of the molecule nitromethane ($H_3C—NO_2$).

18. Below each molecule, draw all other resonance structures (if any).

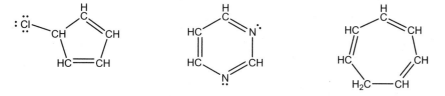

19. The following are -1 anions. Complete each Lewis structure, and draw all important r.s.

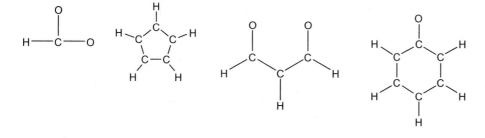

20. For each set, add missing formal charges (all lone pairs shown), and cross out any resonance structures that are NOT important.

a)

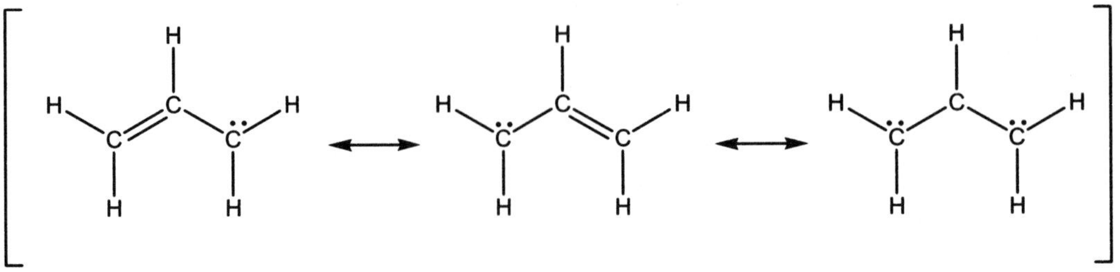

b)

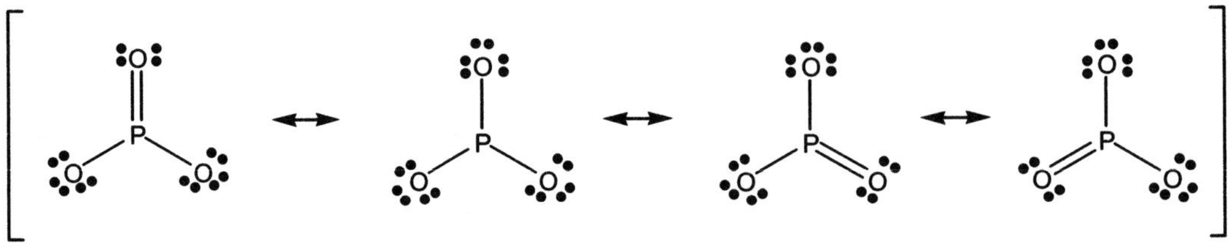

21. Draw all important resonance structures for each ion.

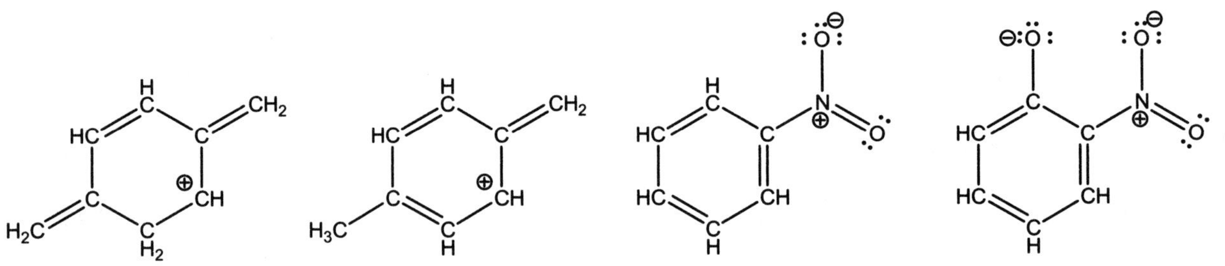

22. Is it possible to draw a resonance structure of nitrate ion ($NO_3^{\ominus}$) that has only one formal charge?

23. The electron movements shown below (left) are "legal."

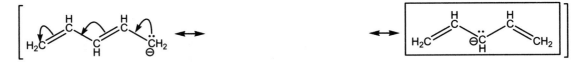

a. Draw the resonance structure that results from these curved arrows.

b. A student comments that "Use of a Type 3 arrow on an anion often causes you to 'skip' an important resonance structure." Construct an explanation for why use of a Type 3 arrow caused us to "skip" the resonance structure in the box.

c. Add curved arrows to the resonance structure you drew in part a that show how the resonance structure in the box can be generated from it.

24. Complete the set of resonance structures for the following anion.

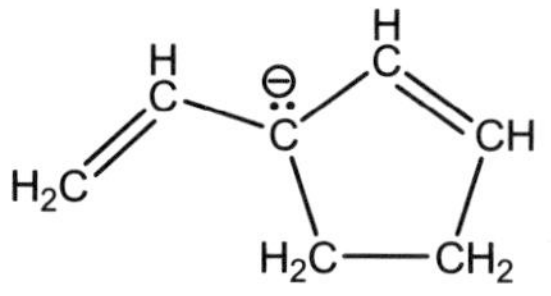

25. "D" is not an element on the periodic table. It stands for deuterium. It is an H with an extra neutron. Think of it as an H with a label. We will frequently use deuterium in this course to track a chemical reaction. For example, when the anion below is treated with D_3O^+ (a strong acid almost identical to H_3O^+), two distinct products are plausible (Products A and B).

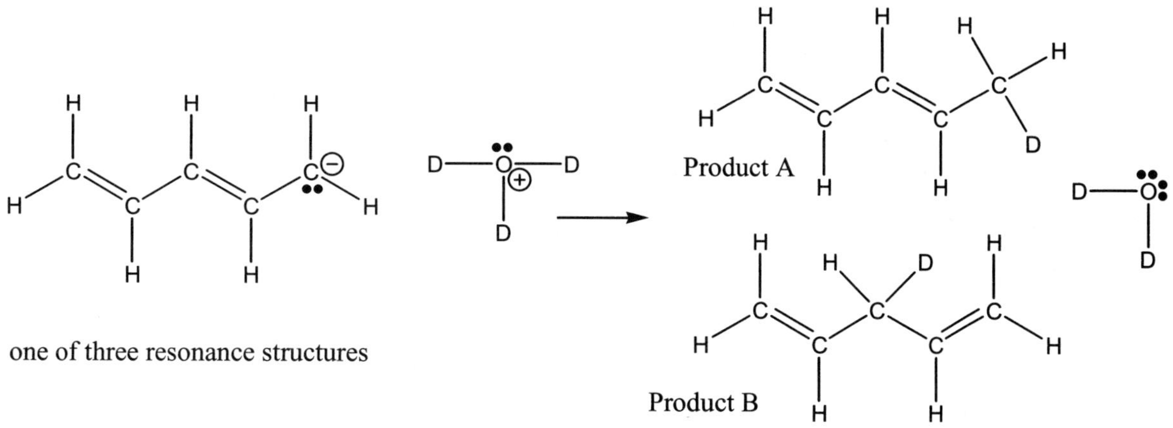

one of three resonance structures

Product A

Product B

a. Use curved arrows to show a reasonable mechanism for formation of Product B. (Hint: draw all three resonance structures of the starting anion.)

b. A student predicts the ratio of A:B in the product mixture will be 2:1. Construct an explanation for this prediction.

Read the assigned sections in your text, and do the assigned problems.

The Big Picture

The main point of this activity is to build your skill at drawing resonance structures. Do as many practice problems as you can to put the rules for drawing resonance structures to the test.

Common Points of Confusion:

- It is critical to recognize that YOU CAN DRAW MULTIPLE RESONANCE STRUCTURES OF A +1 OR -1 ION WHEN THERE IS A PI BOND ADJACENT TO THE CHARGE. Students commonly assume that any ion with a pi bond must have multiple resonance structures.
- Molecules with all zero formal charges rarely have multiple resonance forms. Many students assume that any molecule containing a double bond can experience resonance stabilization—this is incorrect. A possible source of this misconception is the fact that benzene and other aromatic molecules do have multiple (often two) resonance forms.
- A good way to approach drawing resonance structures is to consider that the "purpose" of drawing resonance structures is (usually) to show that a +1 or -1 charge can be "moved" to different atoms in the molecule. (Resonance structures show what atoms share a delocalaized charge.) With this in mind, each set of curved arrows, used to generate the next resonance structure in a series, should move a formal charge to a new location.
- **Tricks for checking your resonance structures**:
    - Count your H's. You cannot break any single bonds, so there must be the same arrangement of H's on ALL your resonance structures. The most common error drawing resonance structures is to lose, gain, or move an H.
    - If you start with exactly one -1 charge on your first r.s., each r.s. must have exactly one -1 charge on it. Same goes for +1 charges. Do not change the total number or type of charges.
- It is very common for students to assume (incorrectly) that a resonance structure that looks like the mirror image (or rotation) of another resonance structure is redundant, and therefore unnecessary. This is incorrect. For example, in Question 15 of this activity the five resonance structures of cyclopentadienyl anion look like the same resonance structure rotated 1/5 of a turn, BUT all five resonance structures are unique and necessary for an accurate picture of the ion. In other words, you need all five resonance structures to demonstrate that the charge is shared among all five carbon atoms.
- The challenges of drawing accurate resonance structures are magnified when using skeletal structures (in which the H's are not drawn but assumed). The number one error that organic students make when drawing resonance structures is to lose track of the number of H's attached to a given atom. As a temporary crutch, it may help to sketch in the H's as you get used to the skeletal structures.

Notes

ChemActivity 3: Constitutional Isomers

WHILE YOU WAIT, BUILD A MODEL OF $CH_3CH_2CH_2CH_2CH_2CH_3$

(Are two molecules with the same molecular formula the same or constitutional isomers?)

Model 1: Representations of Carbon Structures

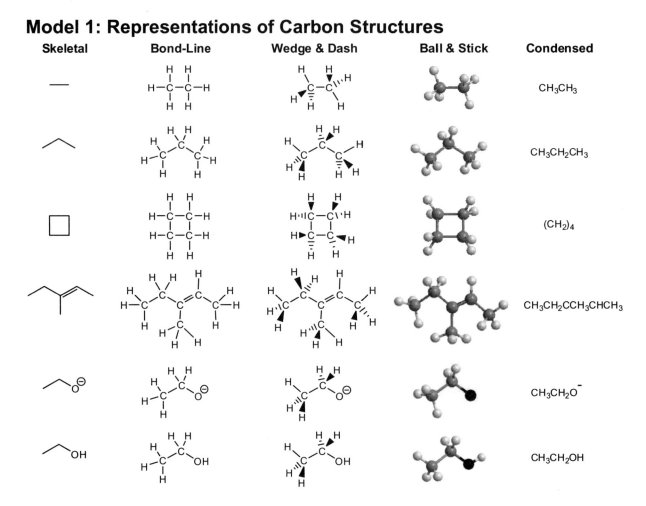

Skeletal	Bond-Line	Wedge & Dash	Ball & Stick	Condensed
				CH_3CH_3
				$CH_3CH_2CH_3$
				$(CH_2)_4$
				$CH_3CH_2CCH_3CHCH_3$
				$CH_3CH_2O^-$
				CH_3CH_2OH

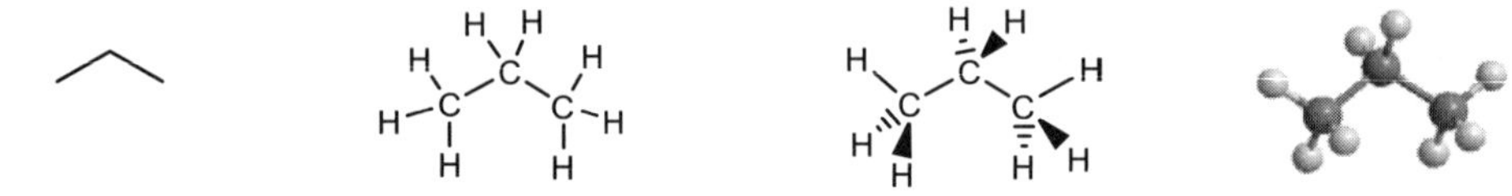

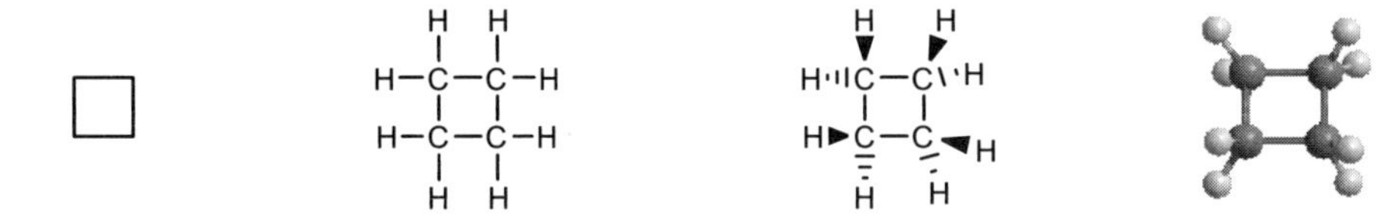

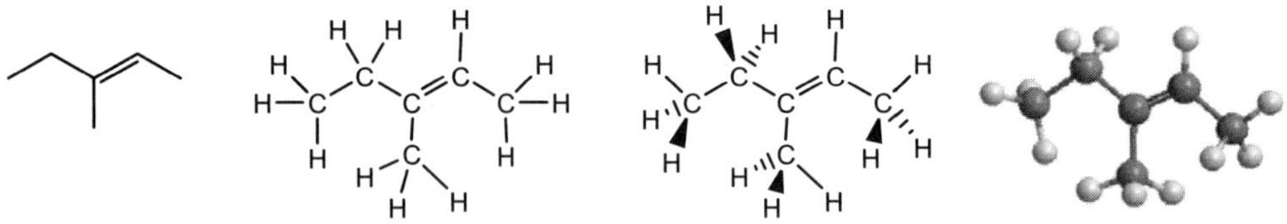

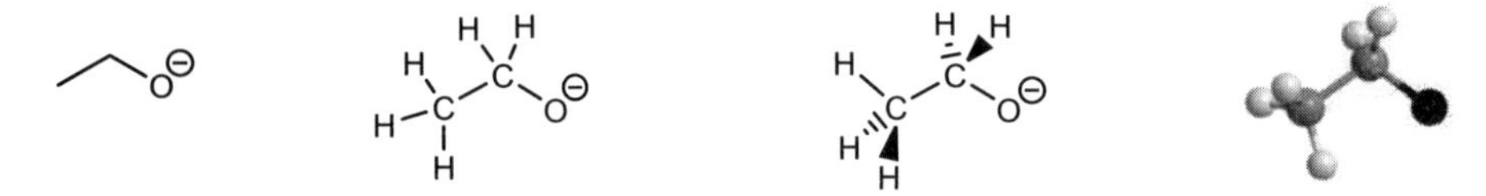

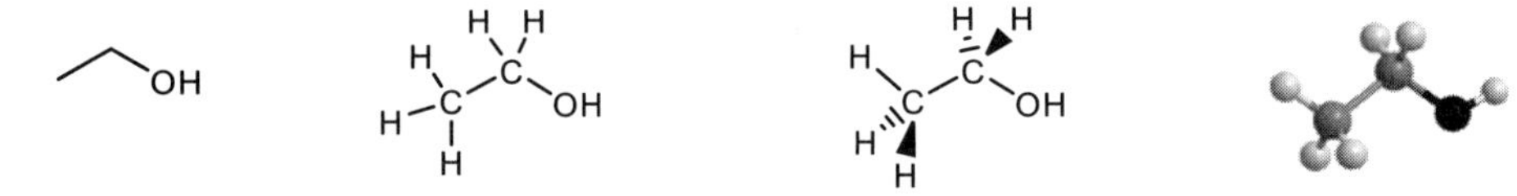

Construct Your Understanding Questions (to do in class)

1. In a skeletal representation the hydrogens attached to carbon are not shown, yet it is still possible to tell how many hydrogens there are on a particular carbon. Explain.

2. Draw a bond-line representation of the molecule shown below as a skeletal representation.

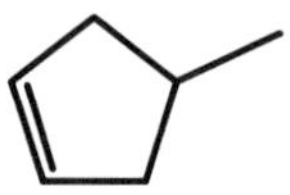

Model 2: Constitutional Isomers

Column 1		Column 2		Column 3	
structure	molecular formula	structure	molecular formula	structure	molecular formula
	C_6H_{14}		C_6H_{12}	Br	C_4H_9Br
	C_6H_{14}		C_6H_{12}	Br	
	C_6H_{14}			Br	
				Br	

Construct Your Understanding Questions (to do in class)

3. [(E)]Complete the table in Model 2 by writing in the missing molecular formulas.

4. What do the molecules in a given column (1 or 2 or 3, above) have in common with the other molecules in that same column?

5. Describe the differences among the four molecules in a given column.

6. All the structures in a given column are **constitutional isomers** of one another, but the structures in Column 1 are not constitutional isomers of structures in the other columns. Based on this information, write a definition for the term **constitutional isomers** that starts: *"Two molecules are constitutional isomers if..."*

7. Without breaking bonds, change your model of hexane ($CH_3CH_2CH_2CH_2CH_2CH_3$) from the conformation shown in Model 2 to the one shown in the bottom left corner of this page.

8. Each structure below is an alternate representation of a structure in Model 2. Under each structure, draw the structure from Model 2 that represents the same molecule. *Recall that:* ***Two molecules are the same if models of each can be interconverted without breaking bonds*** *(i.e., they are conformers).*

> **Memorization Task 3.1: Constitutional Isomers vs. Conformers**
>
> **Constitutional Isomers** = structures with the same molecular formula, & different **atom connectivity**
>
> **Conformers** = structures that can be interconverted via single bond rotation

9. Draw five constitutional isomers missing from Column 2 in Model 2. (There are more than five.)

10. Shown below are only a handful of the many possible six-carbon "skeletons". If the previous question asked you to draw ALL the constitutional isomers missing from Column 2 of Model 2, a good way to start would be to draw all possible six-carbon "skeletons" (taking care that each one had a unique atom connectivity) then add a double bond to those that need one.

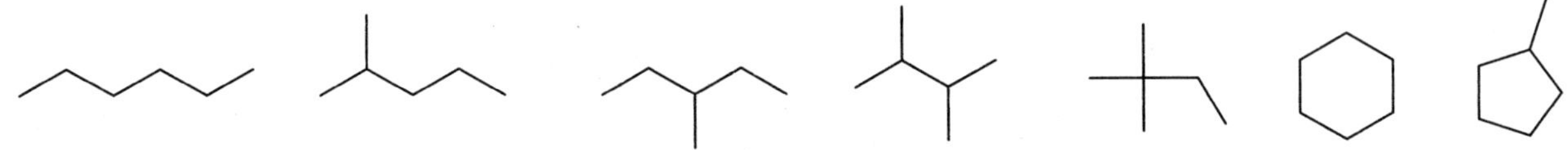

a. Circle the carbon skeletons above that <u>need a double bond</u> to become a constitutional isomer of the molecules in **Column 2** of Model 2.

b. Explain why the following arrangements of six carbons CANNOT form the carbon skeleton of a constitutional isomer of the molecules in Column 2 of Model 2.

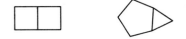

c. T or F: All constitutional isomers of the molecules in Column 2 will have exactly one double bond <u>or</u> one ring.

11. How does the total number of H's in the molecule change if one of the C-C single bonds of pentane is replaced with a C=C double bond (changing the name from pentane to pent**<u>ene</u>**)?

12. How does the total number of H's in the molecule change if the end carbons of pentane are joined to make a ring (changing the name of the molecule from pentane to **<u>cyclo</u>**pentane)?

Memorization Task 3.2: Degree of Unsaturation

Because introducing a π bond or ring has the same effect on the molecular formula of a molecule (to decrease the number of hydrogens by two), the total number of π bonds + rings in a molecule is given a special name...**Degree of Unsaturation** = [Number of π bonds] + [Number of rings]

Extend Your Understanding Questions (to do in or out of class)

13. Write the degree of unsaturation below each structure.

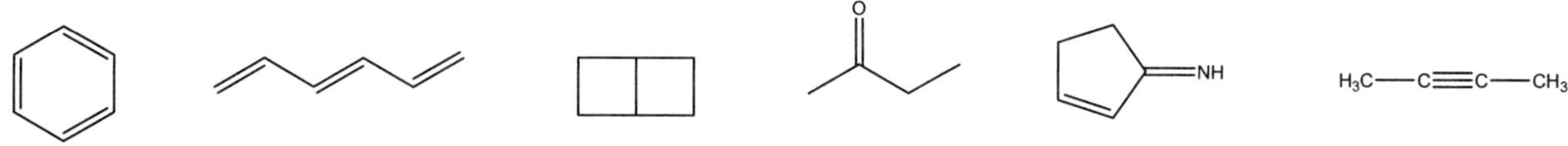

Memorization Task 3.3: Determining Degree of Unsaturation from Molecular Formula

If you are given only the molecular formula of an unknown molecule (e.g., C_7H_8NOBr) and asked to find the molecule's **degree of unsaturation** (or to draw a possible structure of the molecule)...

I. Draw **any** straight-chain structure containing all the atoms but with no π bonds or rings.
(It doesn't matter where you put the groups as long as each atom has the correct number of bonds.)

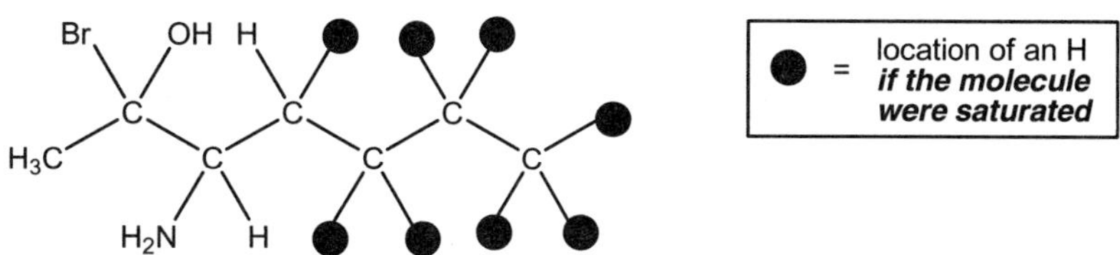

II. Count the number of H's it would take to "saturate" the molecule with H's. (This always will be an even number—if it is an odd number you have made an error.)

$$\textbf{degree of unsaturation} = \frac{\text{No. of "missing" H's}}{2}$$

14. What is the degree of unsaturation for a molecule with molecular formula C_7H_8NOBr?

15. **Without counting hydrogens**, determine which one of the following CANNOT be a molecule with molecular formula C_7H_8NOBr, and explain your reasoning.

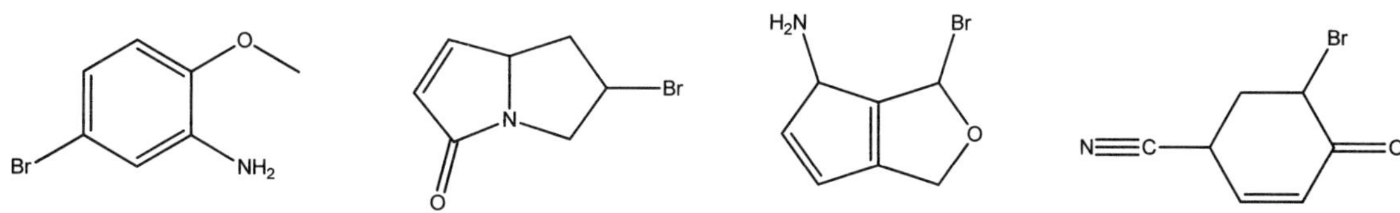

16. Determine the degree of unsaturation (and draw a possible structure) for a molecule with molecular formula $C_6H_4O_2$

Confirm Your Understanding Questions (to do at home)

17. Read Memorization Task 3.4 (below), draw several constitutional isomers of pentane, and construct an explanation for why a branched alkane with n carbons will always have the same molecular formula (C_nH_{2n+2}) as a straight-chain alkane with n carbons (e.g. the molecular formula of all five-carbon alkanes is C_5H_{12}).

Memorization Task 3.4: Molecular formula of an noncyclic alkane = C_nH_{2n+2} *(where n = no. C's)*

noncyclic alkane = molecule with only C & H, and no π bonds or rings

Consider the example of pentane [CH_3-CH_2-CH_2-CH_2-CH_3], which has two H's for each carbon, plus one extra H on each end carbon. Thus, [the number of H's] = 2 x [number of C's] + 2

18. Draw the one constitutional isomer that is missing from column 1 of Model 2.

19. Draw as many of the constitutional isomers missing from column 2 in Model 2 as you can. (Hint: Other than cyclohexane, there are 11 ways to draw a six-carbon skeleton that contains a ring. Plus there are many isomers of cyclohexane that do not contain a ring.)

20. Are any constitutional isomers missing from Column 3 in Model 2? (A good way to answer this and questions is to start by drawing all possible carbon skeletons with unique connectivity. Then figure out how many different ways there are to add the Br atom to each unique skeleton.

21. Draw as many constitutional isomers as you can with the formula $C_5H_{11}F$.

22. Draw the structure of a six-carbon alkene (containing only C and H) with one ring and one double bond.

    d. Draw a constitutional isomer of the structure you drew above with no rings.

    e. Explain the following statement found in many text books: "In terms of molecular formula, a ring is equivalent to a double bond."

23. What is the purpose of Question 8? What concept, distinction, or common misconception is it designed to highlight?

Read the assigned sections in your text, and do the assigned problems.

The Big Picture

This activity is designed to improve your ability to understand how organic chemists define same and different. This course will require you to recognize subtle differences between molecules. To understand the topics around the corner you have to train yourself in the skill of recognizing and drawing constitutional isomers.

It is also critical that you are comfortable with the distinction between conformers (conformational isomers) and constitutional isomers.

The algorithm provided in Memorization Task 3.3 is provided as an alternative to memorizing degree of unsaturation formulas found in most textbooks.

Models were not absolutely critical for this activity, but they are very useful for the next activity, and critical for the following week. If you do not have a model set, you must borrow or purchase one immediately.

Common Points of Confusion

- Many students incorrectly identify some pairs of conformers as being constitutional isomers. Learning how to name molecules is the best solution to this problem (see Nomenclature Worksheets 1 and 2). Two molecules will have the same chemical name if they are conformers, but different chemical names if they are constitutional isomers.
- The previous bullet hints at the difficulty of defining the difference between the term conformers and the term identical. The terms **conformational stereoisomer**, **conformer**, **rotational stereoisomer, rotomer,** and **conformational isomer** are all used interchangeably. Difficulty arises in defining the relationship between the terms above and the term **identical** (or you will sometimes see the terms **chemically identical**, or **same molecule)**. When chemists say two compounds are the same, picture going to the stockroom, pulling a bottle of each from the shelf, and confirming that the contents of each bottle are the same. Every possible conformation is represented simultaneously in any large sample of a molecule (i.e. the bottle). Only if you freeze the sample toward absolute zero will you begin to see one or a handful of conformations (those with the lowest energy). Occasionally you will encounter a molecule with groups that are so large that they hinder free rotation of a single bond at room temperature. The bottom line is that, for most molecules, different conformers are just different representations of the same molecule. For this reason you will sometimes see the terms conformers and identical lumped together as though they are synonyms.

Notes

ChemActivity 4: Cycloalkane Stereoisomers

BUILD MODELS: 1,2-DIMETHYLCYCLOPENTANE & CYCLOHEXANE

(as described in Model 4)

(How can you tell if two groups on a ring are *cis* or *trans* to each other?)

Model 1: *Cis* and *Trans* Rings

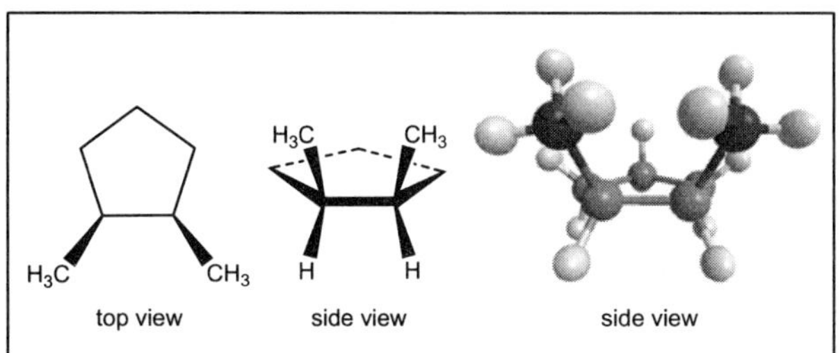

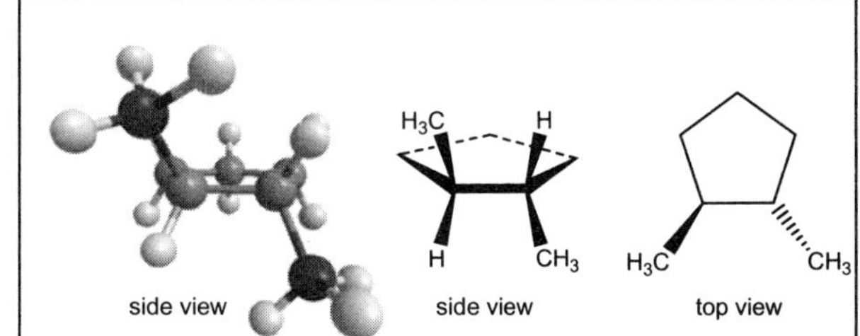

Construct Your Understanding Questions (to do in class)

1. Use the rules in Memorization Task 4.1 to label each box in Model 1 with the term ***cis*** or ***trans.***

Memorization Task 4.1: Determining *cis/trans*

When the ring carbons lie in the plane of the paper (as in each "top view" above), determine if each (non-H) group is attached to the ring with a wedge (coming out at you) or dash (going into the paper).

- If both groups are attached to the **same carbon** → groups are **neither** *cis* nor *trans* to one another
- If both groups lie on the **same side** of the plane of the paper → groups are ***cis*** to one another
- If the groups lie on **opposite sides** of the plane of the paper → groups are ***trans*** to one another

2. Use models of the structures in Model 1 to confirm that the molecule in the left box is *cis* and different from the molecule in the right box (which is *trans*). Recall that *two molecules are the same/conformers ONLY if models of each can be interconverted without breaking bonds.*
3. Write the name of each molecule including *cis* or *trans*, if appropriate. (Notes: Each name ends in "dimethylcyclobutane;" the first one is done for you.)

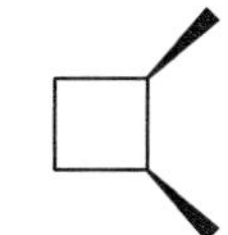

cis-1,2-dimethylcyclobutane

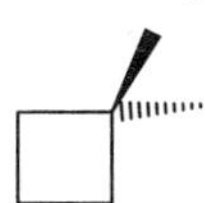

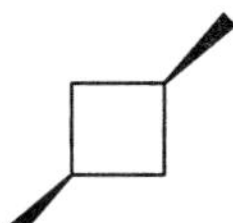

Model 2: Cyclohexane

← Flat cyclohexane would have bond angles very far from 109.5°

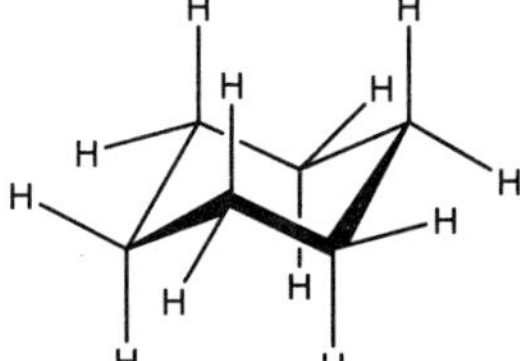

Cyclohexne is often represented as a hexagon, but it is not flat. At right is a more realistic representation of cyclohexane that shows 109.5° bond angles →

Memorization Task 4.2: Determining *cis* and *trans* for 3D drawings of cyclohexane

Draw a dotted line to define the plane of the average carbon position—which is often the horizontal.

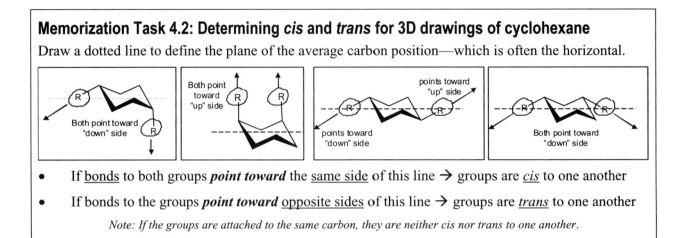

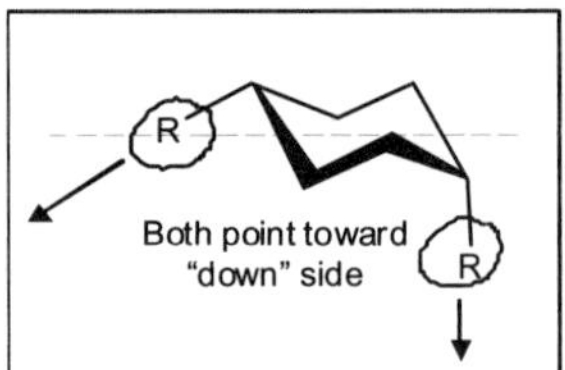

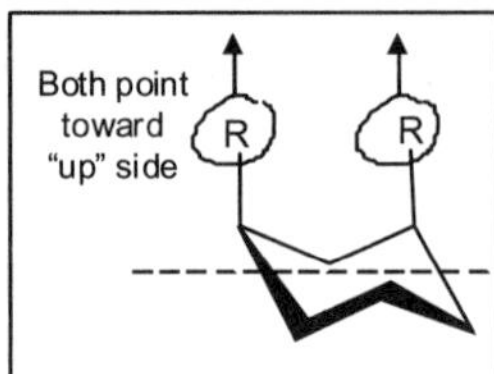

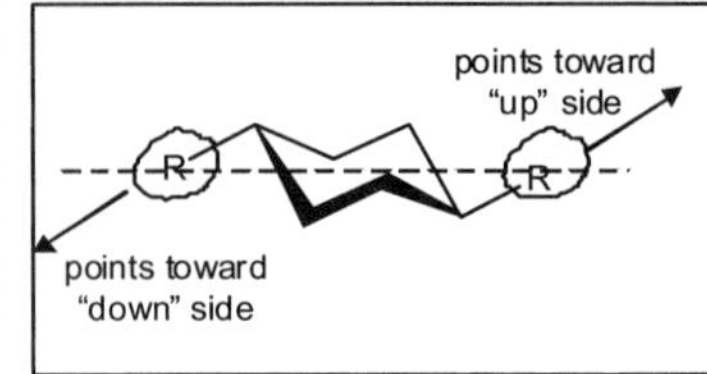

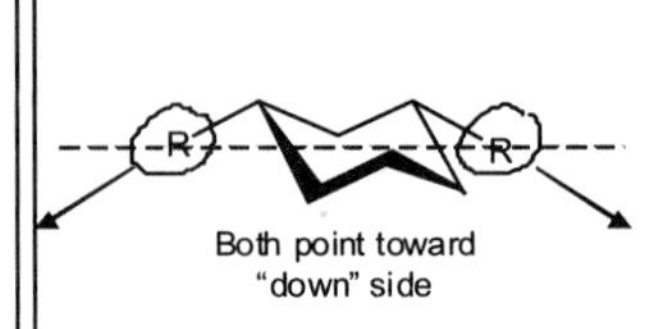

- If bonds to both groups ***point toward*** the same side of this line → groups are *cis* to one another
- If bonds to the groups ***point toward*** opposite sides of this line → groups are *trans* to one another

Note: If the groups are attached to the same carbon, they are neither cis nor trans to one another.

Construct Your Understanding Questions (to do in class)

4. Label each box in Memorization Task 4.2 *cis* or *trans*.

5. Label each of the following disubstituted cyclohexane rings as *cis, trans*, or neither.

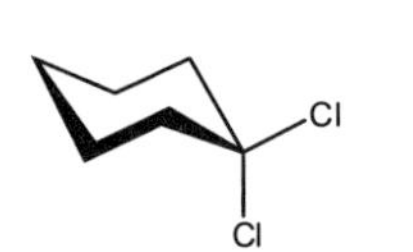

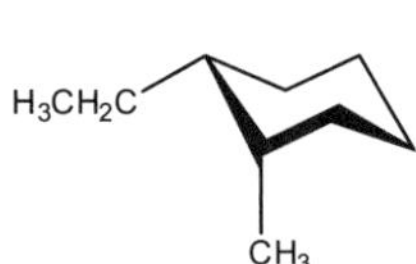

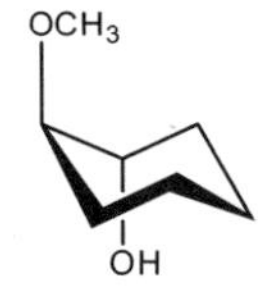

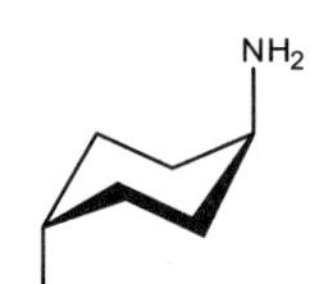

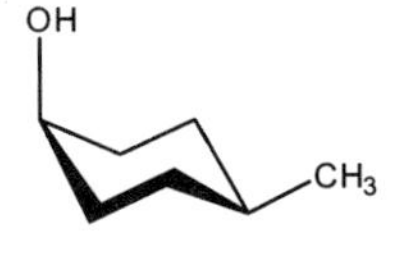

Model 3: Cyclohexane Chair Conformation

The lowest potential energy conformation of cyclohexane is called the "chair" conformation because (with a little imagination) it looks like a side view of a recliner with a footrest.

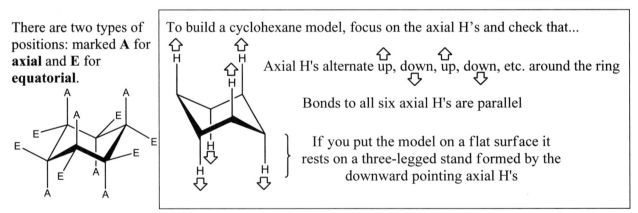

There are two types of positions: marked **A** for **axial** and **E** for **equatorial**.

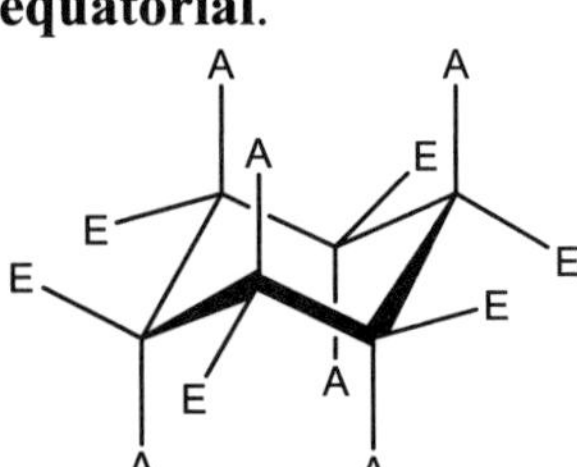

To build a cyclohexane model, focus on the axial H's and check that...

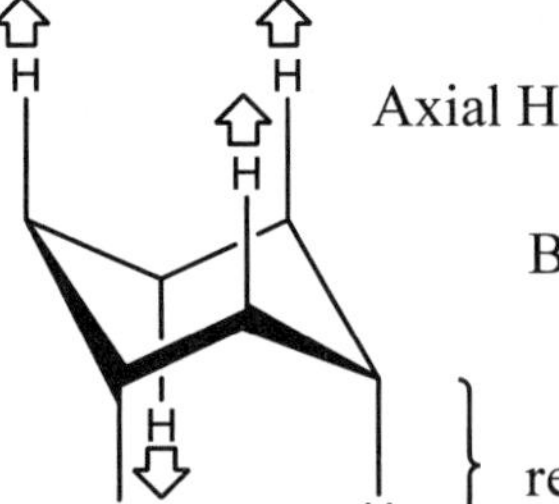

Axial H's alternate up, down, up, down, etc. around the ring

Bonds to all six axial H's are parallel

If you put the model on a flat surface it rests on a three-legged stand formed by the downward pointing axial H's

Construct Your Understanding Questions (to do in class)

6. Label each non-hydrogen group in Question 5 with an A (for axial) or E (for equatorial).

7. Make a model of cyclohexane in the chair conformation, and construct an explanation for why the names "axial" and "equatorial" are used. (Hint: How are these words applied to a globe of Earth?)

Memorization Task 4.3: Groups Attached to a Cyclohexane "Chair"

(A) axial position = attached to a chair via a bond that **points straight up or down** *like an Axis.*

(E) equatorial position = attached to a chair via a bond that **points slightly up or down** *(On a model you can see that equatorial groups point slightly up or down, but mostly out toward an imaginary "equator" circling the cyclohexane molecule.)*

Model 4: Chair "Flip"

Let us start by stating what a chair flip is NOT.

It is NOT flipping a model like a pancake →

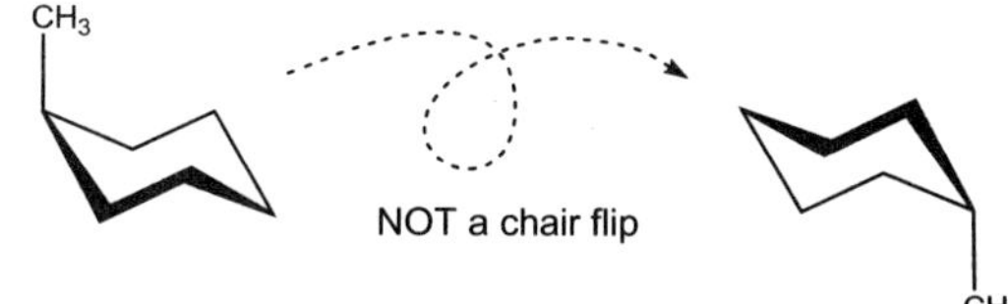

A **chair flip** is a conformation change (no bonds are broken) in which the "headrest" becomes the "footrest" and vice versa, and **all axial groups become equatorial groups and vice versa.**

The best way to see this with your model is to replace each axial H with a colored ball (black, below). When you successfully perform the chair flip the colored balls will all be in equatorial positions.

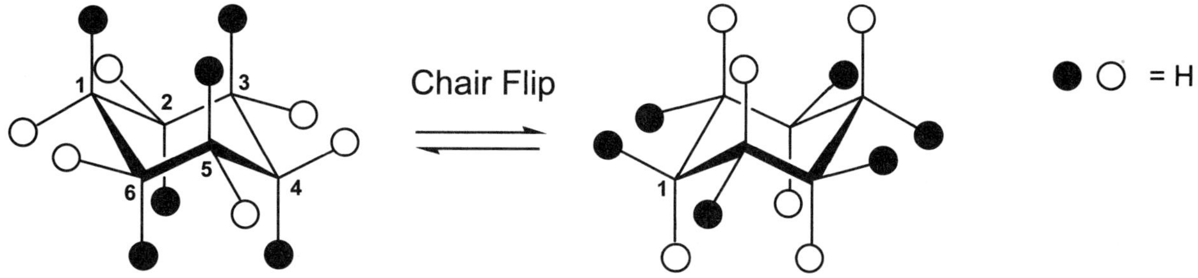

Construct Your Understanding Questions (to do in class)

8. Use your model of cyclohexane in a chair conformation to finish numbering the carbons on the ring on the right side of the figure above (C_1 is marked for you). Careful! This is harder than it looks.
9. Label each black ball on both structures as "up" or "down." ***Hint*:** To determine if an equatorial group is up or down, look at the *axial* group. If the axial group is up, the equatorial group is down.
10. Label each methyl (Me) group on the structure below left as axial or equatorial and up or down. (To determine up/down, it may help to sketch in the axial H's on the carbons of interest.)

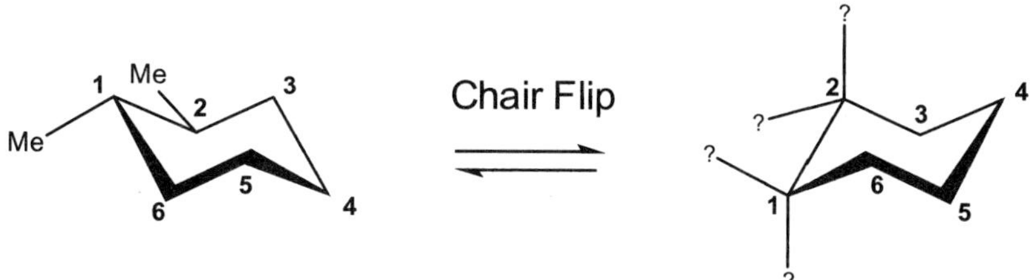

a. Label the structure on the left with its name including *cis* or *trans* if necessary.

b. Complete the drawing on the right by replacing each ? with an H or Me to show the correct positions of the methyl groups after a chair flip has occurred.

c. Label the structure on the right with its name including *cis* or *trans* if necessary.

11. (Check your work) Is your answer to the previous question consistent with the fact that a chair flip does not break any bonds, and so does not change the name of a molecule?

Memorization Task 4.4: It is more favorable to have a large group in an equatorial position

A group in an equatorial position points away from the ring where there is more space. A lower potential energy conformation is usually achieved when the largest group is equatorial (not axial).

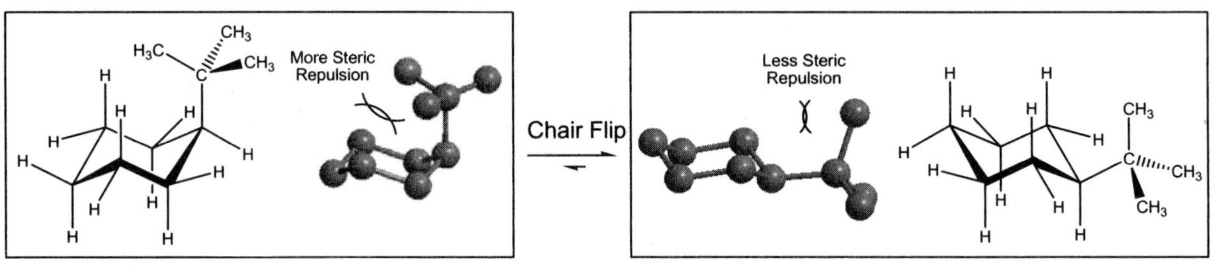

12. Draw chair representations of ***cis*-1-*tert*-butyl-4-methylcyclohexane** and ***trans*-1-*tert*-butyl-4-methylcyclohexane** in their lowest potential energy conformations.

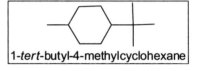

1-*tert*-butyl-4-methylcyclohexane

13. What does the pair on the left have in common that the pair on the right does not?

and

and

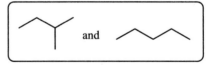

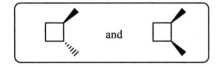

Memorization Task 4.5: Definitions	*Examples...*
Conformers (conformational stereoisomers) can be interconverted via single bond rotation (without breaking bonds). *Two conformers are usually considered the same molecule.	**anti & gauche** or **equatorial & axial**
Constitutional Isomers have the same molecular formula but **different atom connectivity**	**butane and 2-methylpropane**
Configurational Stereoisomers have the same molecular formula AND the same atom connectivity, but **CANNOT be inter-converted via single bond rotation.**	***cis* & *trans*** rings

**You cannot buy a bottle of "anti" butane because in any sample of butane at normal temperatures the molecules are rapidly interconverting among all possible conformations. You can buy a bottle of* cis-1,3-dimethylcyclohexane.

Extend Your Understanding Questions (to do in or out of class)

14. What term best describes the relationship between each of the following pairs? Choose from identical, conformers, configurational stereoisomers, constitutional isomers, different molecular formula.

(a) 1 and 6	(f) 2 and 7	(k) 4 and 14
(b) 2 and 3	(g) 8 and 9	(l) 11 and 2
(c) 2 and 4	(h) 7 and 10	(m) 11 and 3
(d) 4 and 5	(i) 11 and 12	(n) 12 and 2
(e) 5 and 8	(j) 11 and 13	(o) 7 and 14

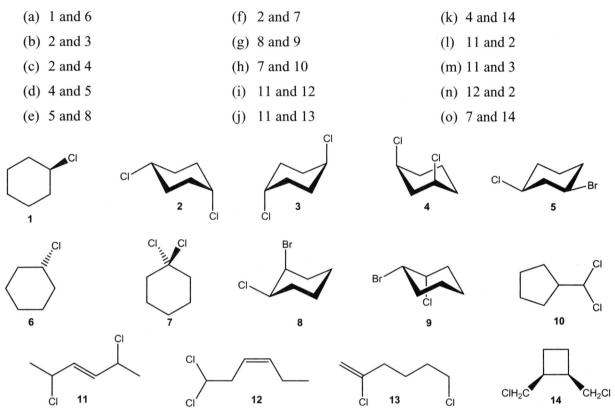

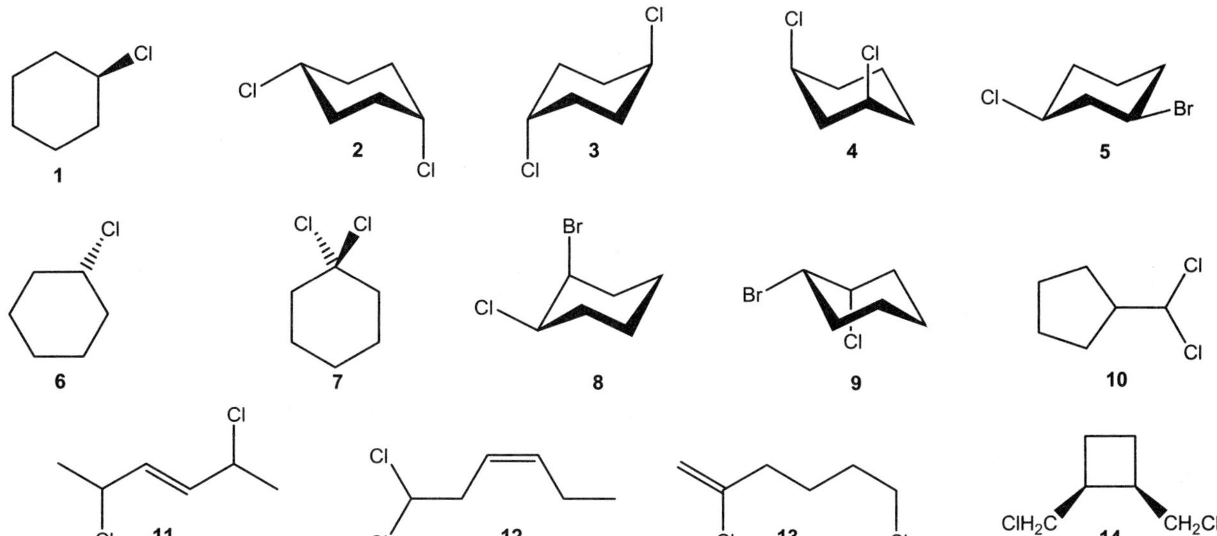

15. Fill in the table by drawing a representation of a *cis* and a *trans* configurational stereoisomer for each row. **Show each molecule in its <u>lowest potential energy</u> chair conformation.**

Generalized name	*cis* configurational stereoisomer	*trans* configurational stereoisomer
1,2-dimethyl-cyclohexane		
1,3-dimethyl-cyclohexane		
1,4-dimethyl-cyclohexane		

16. For each <u>row</u> in the table above, if one of the two stereoisomers is <u>lower</u> in potential energy, circle it and explain your reasoning. If the two are equivalent in energy, circle neither.

Confirm Your Understanding Questions (to do at home)

17. A student names the two molecules on the left *cis* and the two molecules on the right *trans* and defends these names by citing that the methyl groups are on the <u>same side</u> for the first two and <u>across from each other</u> for the second two. What misconception does this student appear to hold?

18. State what information is conveyed by each part of the name:

<u>*cis*</u> <u>1,2-</u> <u>di</u> <u>methyl</u> <u>cyclo</u> <u>pentane</u>

19. Name each molecule in Question 14.

20. Label each of the following as *cis*, *trans* or neither. Below each structure that is *cis* draw a *trans* configurational stereoisomer; below each structure that is *trans* draw the *cis* configurational stereoisomer.

Br Cl

Cl Cl

OH CH_2CH_3

NH_2 H_2N

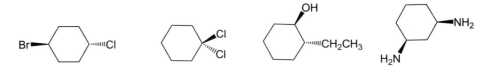

21. True or False: If you perform a chair flip on *cis*-1,4-dichlorocyclohexane, the result is *still* called *cis*-1,4-dichlorocyclohexane.

22. Each group on a cyclohexane ring is either <u>axial or equatorial</u> and either <u>up or down</u>.
    a. During a chair flip, only ONE of these changes. Which changes? Which stays the same?
    b. Consider a group that is <u>up</u> and in an <u>axial</u> position on a given cyclohexane ring. Use the words up/down/axial/equatorial to describe this group after a chair flip has occurred.

23. Fill in the blanks: *cis*-1,3-dibromocyclohexane has two different chair conformations: one with both Br groups in ________ positions and one with both Br groups in ________ positions.

24. Which is lower in potential energy, structure 4 or 5 in Question 14? Explain your reasoning.

25. Construct an explanation for why *trans*-1-*tert*-butyl-4-methylcyclohexane is lower in potential energy than *cis*-1-*tert*-butyl-4-methylcyclohexane.

26. Draw the following molecule in its lowest potential energy chair conformation when…

a. $R_{1\text{-}3}$ = Me
b. $R_{1,3}$ = Me, R_2 = *t*-butyl
c. R_1 = Me, R_2 = Et, R_3 = Pr

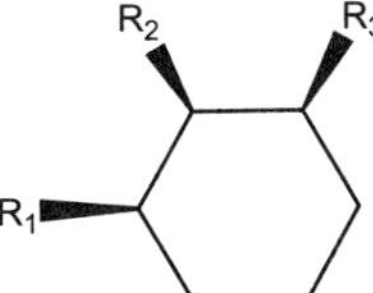

27. Label each molecule drawn below as *cis*, *trans*, or neither.

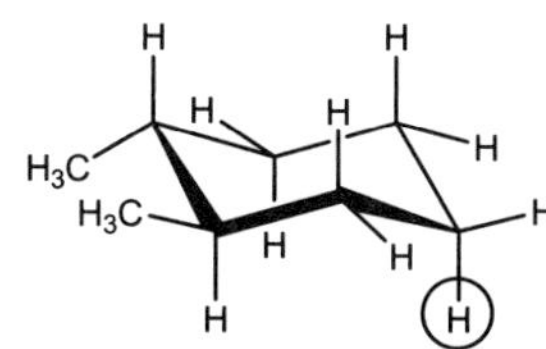

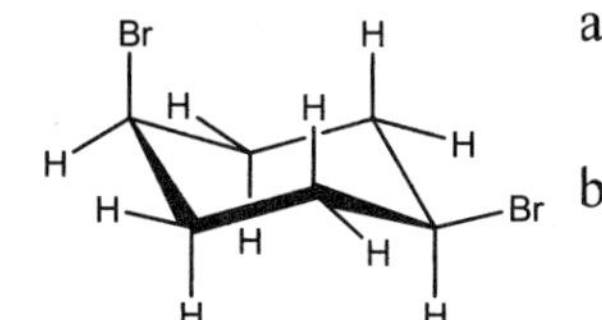

a. What word is used to describe the type of position held by the circled H in the first drawing.

b. DRAW a picture of what each molecule will look like after it has undergone a chair flip. **<u>Do not include any hydrogens in your drawings.</u>**

c. Consider all four drawings (the two above, and the two you drew). For each molecule, circle the conformation, if any, that is lower in potential energy. If it is a tie, circle neither.

d. Name all four molecules. Be sure to include *cis* or *trans* in the name, if appropriate.

28. Describe what is wrong with each of the following chair representations?

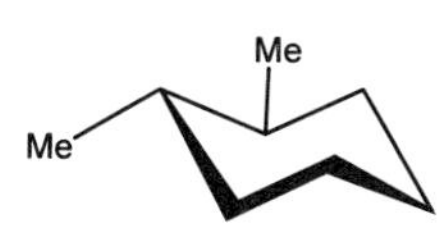

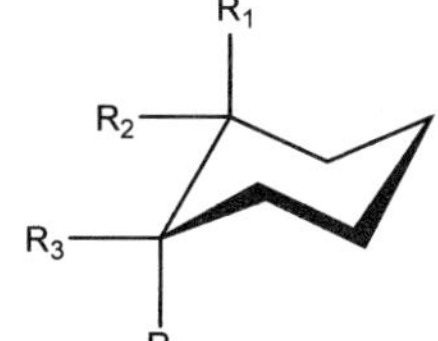

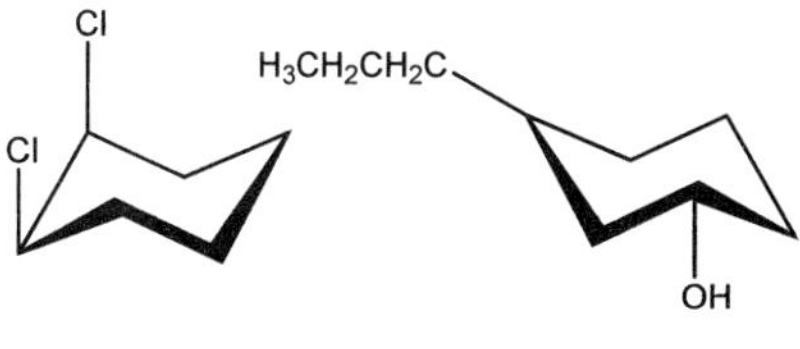

29. The words *cis* and *trans* can be used to name a ring with exactly two non-hydrogen groups. A ring with more than two groups on it cannot be *cis* nor *trans* as a whole, but these terms (*cis* or *trans*) *can* be used to specify the relationship between any two groups (not on the same carbon) on a ring.

State the relationship (*cis, trans*, or neither) between each pair…

a. X and H_1
b. X and Z
c. Y and Z
d. H_2 and H_3
e. Z and H_3
f. Z and H_4
g. Y and H_3
h. H_3 and H_4

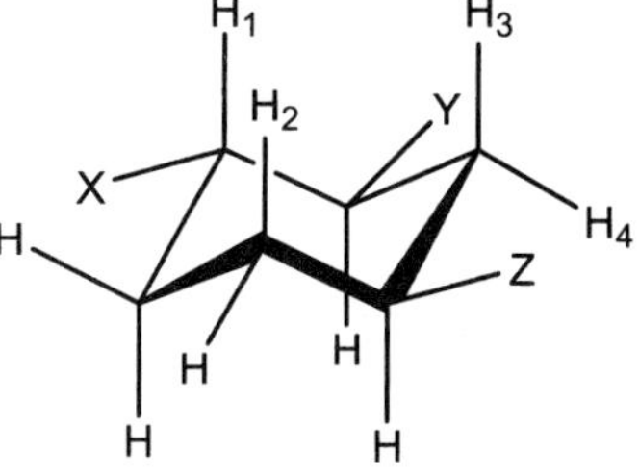

30. Draw *trans*-1- *tert*-butyl-3-methylcyclohexane in its most favorable chair conformation, and explain your reasoning.

31. Build a model of methylcyclohexane, and use the model to complete the following Newman projections of methylcyclohexane in the chair conformation:

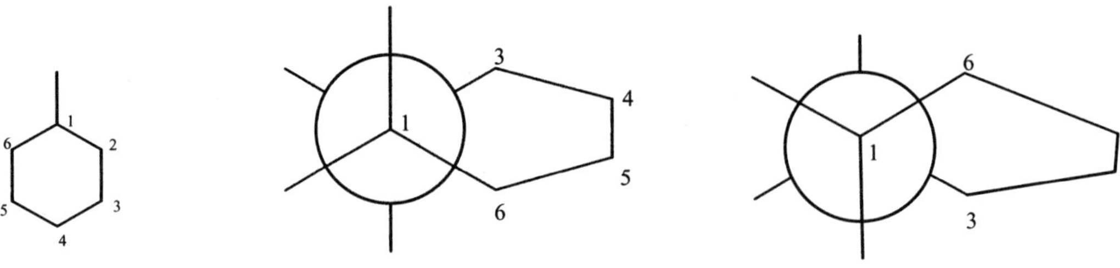

methylcyclohexane

methylcyclohexane
(show CH_3 in axial position)

methylcyclohexane
(show CH_3 in equatorial position)

a. When the methyl group is in an **axial** or **equatorial** (circle one) position, the molecule is in its lowest potential energy conformation.

b. Label one Newman projection above **anti** and the other **gauche** to describe the relationship between the methyl group and C_3 of the ring.

c. In general, which is a lower PE conformation, anti or gauche?

d. Explain how your answer to b and c provide an explanation for why it is more favorable for a large group to be in an equatorial than an axial position.

Read the assigned pages in your text, and do the assigned problems.

The Big Picture

Cyclohexane is typically the only ring system whose conformations are studied in detail in an introductory organic chemistry course. This is partly because cyclohexane rings are so common in biology and chemistry (e.g. many sugars including glucose adopt a chair conformation in their cyclic form). Your investigations in this activity also provide a great opportunity for you to develop your ability to think and problem solve in three dimensions. The terms *cis, trans*, conformers, configurational stereoisomers, up, down, axial, equatorial, etc., will return again and again in future activities. Take the time now to develop a solid grasp of these concepts or you may struggle going forward.

The best way to improve your understanding of these terms and concepts is to spend some quality time with a model set and a group mate who feels more comfortable with these terms.

The next activity extends the concepts of configurational stereoisomers to alkenes. You will again need your model sets, so bring them to the next group work session.

Common Points of Confusion

- *Trans* means "across" and *cis* means "same side." The hard part is figuring out across from what or on the same side as what? A common misconception with respect to this is highlighted by Question 17.
- Spend some time practicing drawing a cyclohexane chair; then draw the same chair after a flip. Do this with a model so that you can "program" your brain to see your drawings in 3D.
- The key to making a model or drawing of chair cyclohexane is to pay close attention to the axial groups. The student who draws an axial group up when it is supposed to be down will likely lose big points on cyclohexane questions. Remember that each axial group extends straight up or down from the point formed by the ring at each carbon and that axial groups alternate, up, down, up, down, etc., around the ring.
- It can be hard to decide if an equatorial group is "up" or "down". A good trick is to draw in the axial group, since it always points directly up or down and follows the bend in the ring at that carbon. With this information you can confidently state which way the equatorial group on this same carbon points (since it is the opposite of the axial group).
- Many students incorrectly believe that performing a chair flip will change a *cis* cyclohexane into a *trans* cyclohexane or vice versa. Flipping a cyclohexane chair does not break any bonds, and so does not change the name of a molecule. As you discovered in Model 1, **<u>you must break bonds to change from *cis* to *trans* or vice versa</u>**.
- Another thing that does NOT change during a chair flip is the direction of the group ("up" or "down"). A group that is up before a chair flip, will be up after the flip. A group that is down before a chair flip will be down after the chair flip.
- It helps some students to focus on what DOES change during a **chair flip**: all **axial positions** become **equatorial positions** and vice versa. The words axial and equatorial refer to conformational relationships, which, like *anti* and *gauche*, are constantly changing as the molecule undergoes single-bond rotation, and (for most cyclohexanes) rapid chair flips, at room temperature.

Notes

ChemActivity 5: Alkene Stereoisomers

BUILD A MODEL OF $H_3C—CH=CH—CH_3$

(How can you tell which **stereoisomer** of an *E*/*Z* pair is *E* and which is *Z*?)

Model 1: But-2-ene (or 2-Butene)

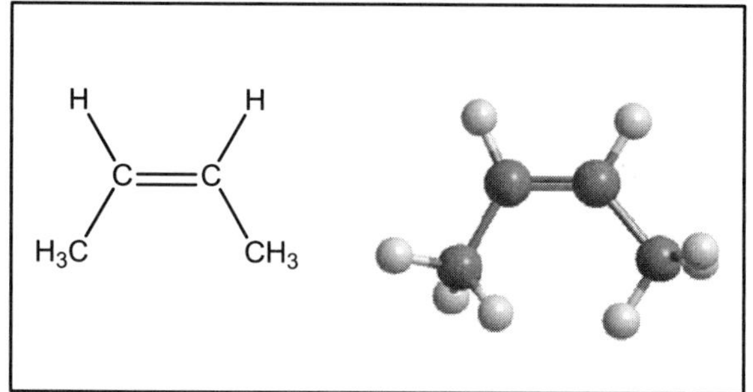

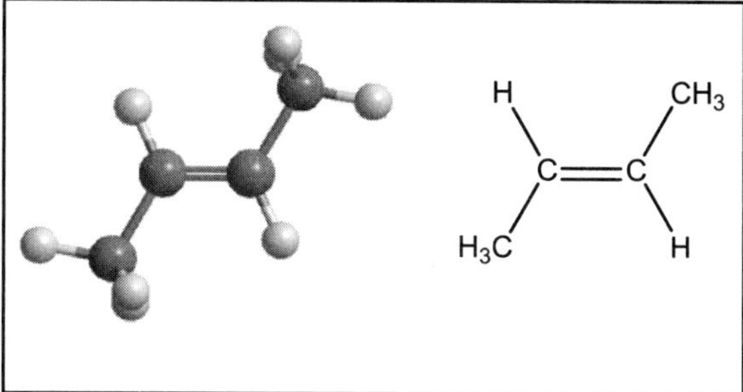

Figure 5.1: Representations of but-2-ene molecules

Construct Your Understanding Questions (to do in class)

1. Make a model of each molecule shown above: Is the molecule in the left box the **<u>same molecule</u>** as the molecule in the right box? <u>Use your models to answer the question</u>, and recall that *two molecules are the <u>same</u> ONLY if models of each can be interconverted without breaking bonds.*

Model 2: *E*, *Z*, or Neither?

A single bond can rotate freely, but the π portion of a double bond prevents rotation. As a result, many double bonds have two different orientations that give rise to <u>two different molecules</u>.

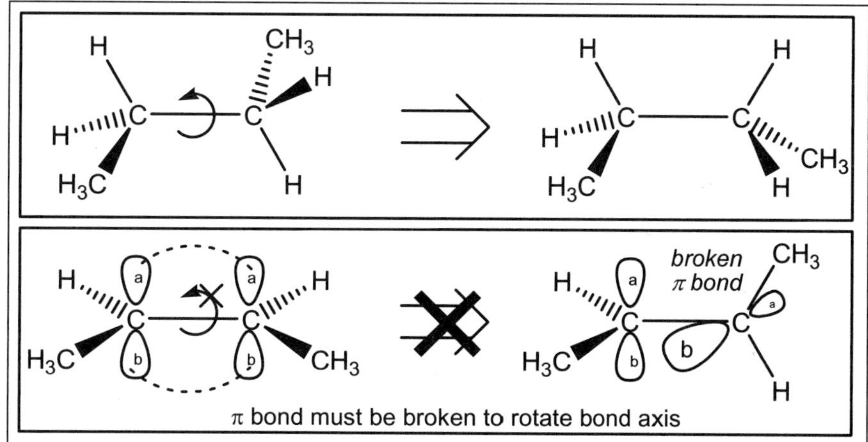

Figure 5.2: Single bonds rotate freely, but double bonds do not.

The letters *E* and *Z* (or the terms *trans* and *cis*) are used to name the two possibilities that result.

Rules for determining E and Z are on the next page.

Memorization Task 5.1: Memorize the definitions of *E* and *Z* (and *trans* and *cis*)

To determine if a double bond is ***E*, *Z*, or neither**…

I. Put a box around each carbon involved in the double bond (as shown below)

II. Are there <u>two identical groups</u> on *either* boxed C? → If *YES* double bond is <u>neither</u> *E* nor *Z*

*(If a double bond is **neither** there is only one way to draw the molecule, and its name will include neither E nor Z.)*

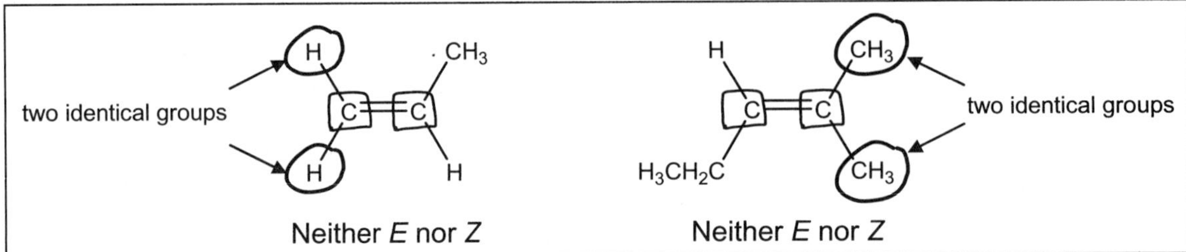

STOP here if neither E nor Z

III. Otherwise, draw a dotted line along the double bond, and **<u>circle the atom with the highest atomic number on each boxed C</u>**. *(Atomic numbers from the periodic table: H = 1, C = 6, N = 7, O = 8, etc.).*

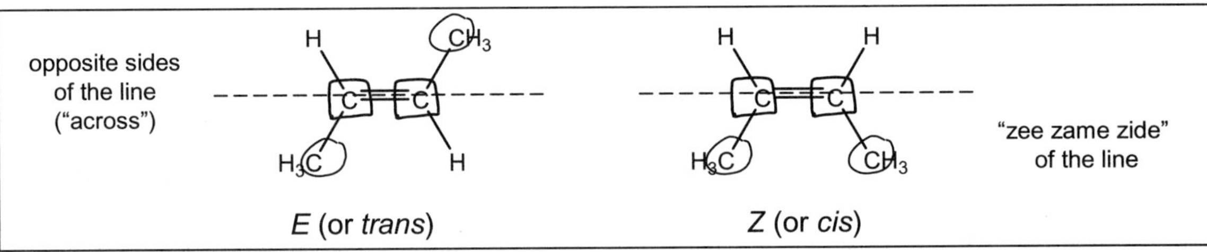

- If circled groups are on <u>opposite sides</u> of this line → double bond is ***E***
  E stands for *entgegen*, which is German for "opposite."
- If circled groups are on the **<u>same side</u>** of this line → double bond is ***Z***
  Z stand for *zusammen*, German for "together." (Say in a German accent: "*zee zame zide*")

Trans and *Cis*: alternates for *E* and *Z*

The two circled groups in the *E* molecule above are said to be *trans* to each other. (*trans* = "across")

The two circled groups in the *Z* molecule are *cis* to each other. (Say *cis* with a German accent: "***Zis***")

trans or ***cis*** is often used in place of *E* or *Z* when the molecule has one (non-H) group on each boxed carbon. Use *E*/*Z* when there are three or more groups, as in the last example at the top of the next page.

Construct Your Understanding Questions (to do in class)

2. Label each but-2-ene molecule in Model 1 with the appropriate two names. Choose from: *Z*-but-2-ene, *E*-but-2-ene, *cis*-but-2-ene, and *trans*-but-2-ene.

3. Draw skeletal representations of *Z*-hex-3-ene and *E*-hex-3-ene.
(The "3" tells you the π bond starts on the 3rd carbon, i.e. $CH_3CH_2CH{=}CHCH_2CH_3$)

4. Label <u>each</u> double bond *E, Z*, or neither. (It may help to draw in some missing H's.)

F

Model 3: Rules for Assigning *E/Z* Priorities

The rules for assigning priority outlined in this activity are called the Cahn-Ingold-Prelog rules after their three co-inventors.

In the structure at right, the atoms labeled C_A and C_B are tied since both are carbons (atomic number = 6).

To determine which gets circled, look at the 3 atoms (other than the boxed carbon) bonded to C_A and C_B.

Think of these atoms as "cards."

- C_A's "cards" = H H H
- C_B's "cards" = **C** H H

Compare the highest "card" (based on atomic number) in each hand: C_B beats C_A, and so the molecule is *Z*.

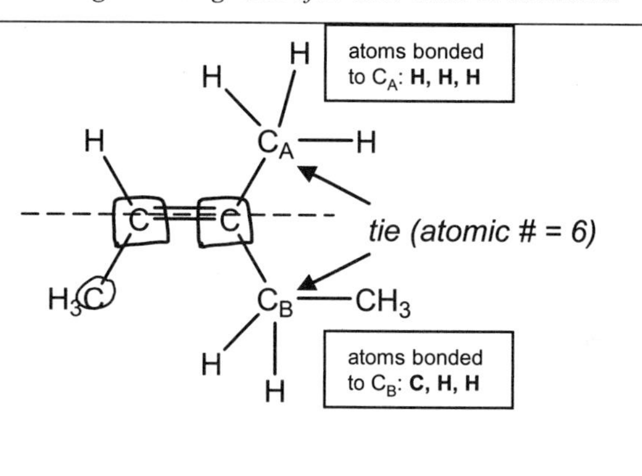

Additional Rules:

isotopes → the heavier atom wins: ^2H (also called "D" for Deuterium) beats ^1H, ^{13}C beats ^{12}C, etc.

multiple bonds → a double bond counts double [e.g., **C**=O same as C bound to two O's (i.e., O-**C**-O)]

If a tie results after the first round of the card game, play a 2nd round using the atoms attached to the highest "card" from each hand.

Keep playing until an assignment of *E* or *Z* is made.

Construct Your Understanding Questions

5. Explain why the molecule at right is *E* even though CH_2CH_2OH has more atoms than CH_2COH.

6. Label each molecule below *E, Z*, or neither.

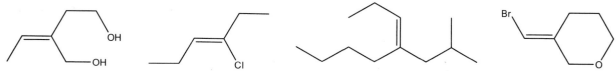

OH OH Cl Br O

Model 4: Configurational Stereoisomers

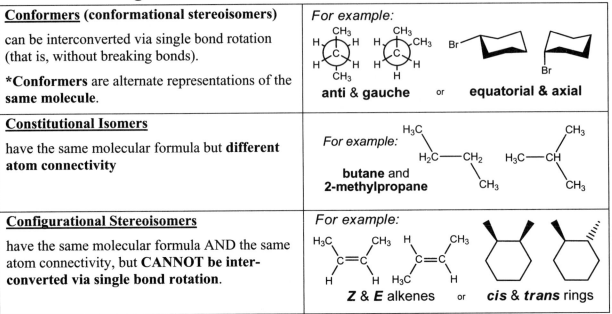

Conformers (conformational stereoisomers) can be interconverted via single bond rotation (that is, without breaking bonds). ***Conformers** are alternate representations of the **same molecule**.	*For example:* 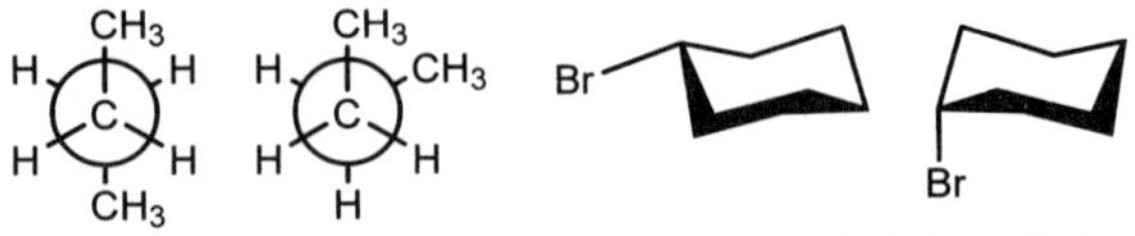 **anti & gauche** or **equatorial & axial**
Constitutional Isomers have the same molecular formula but **different atom connectivity**	*For example:* 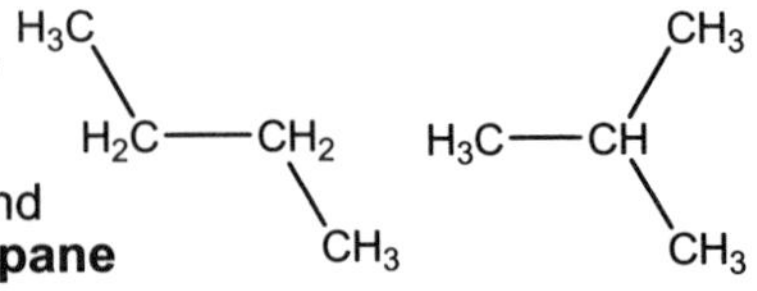 **butane** and **2-methylpropane**
Configurational Stereoisomers have the same molecular formula AND the same atom connectivity, but **CANNOT be inter-converted via single bond rotation.**	*For example:* 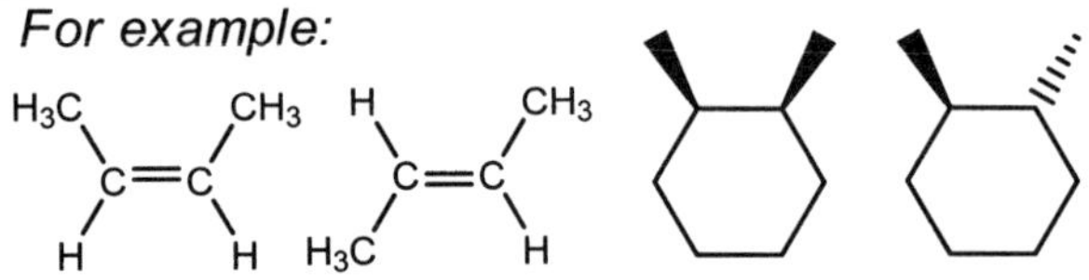 ***Z* & *E*** alkenes or ***cis* & *trans*** rings

**You cannot buy a bottle of "anti" butane or "gauche" butane because in any sample of butane at normal temperatures the molecules are rapidly interconverting among all possible conformations. You can buy a bottle of E-butene or Z-butene.*

Extend Your Understanding Questions (to do in or out of class)

7. Indicate the relationship between each pair. Choose from: **configurational stereoisomers, conformers or identical, constitutional isomers,** or **different formulas** (use *each* at least twice).

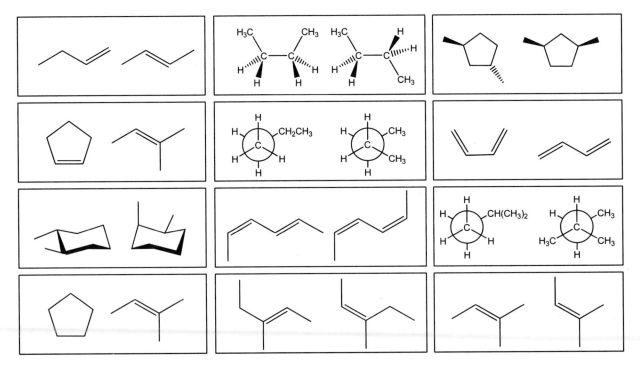

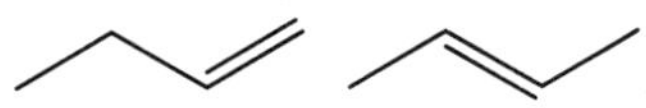

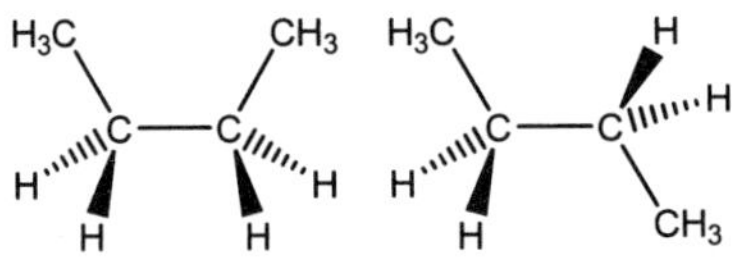

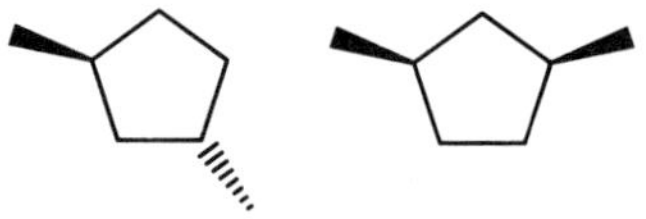

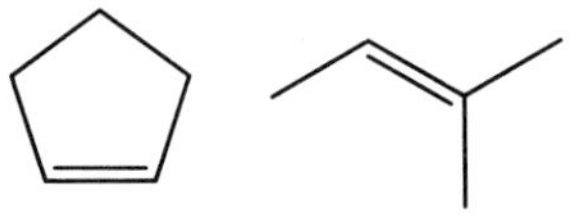

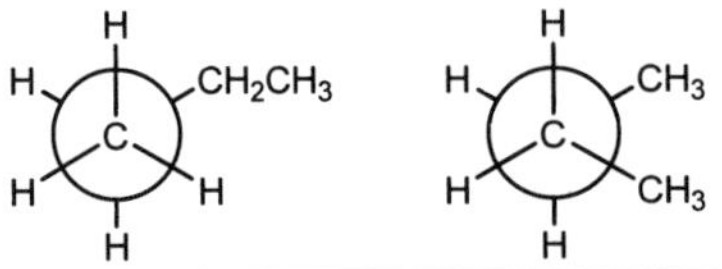

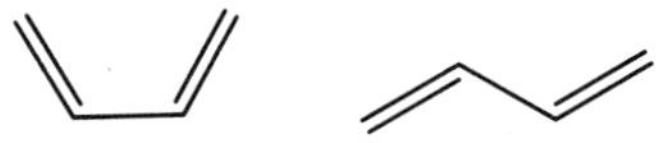

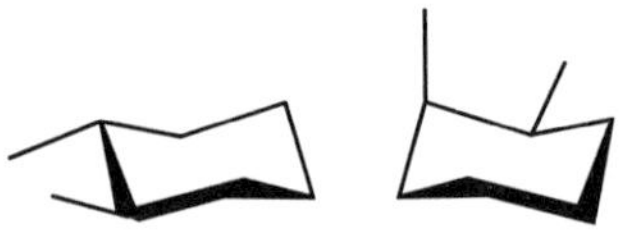

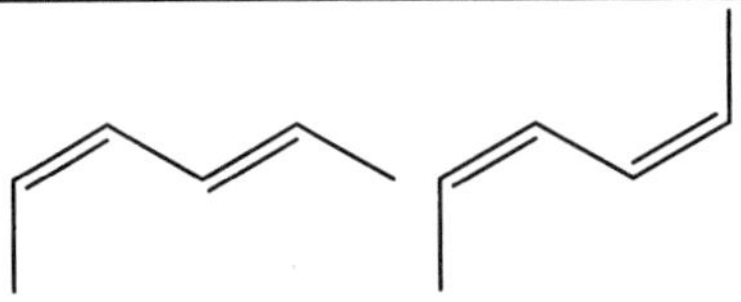

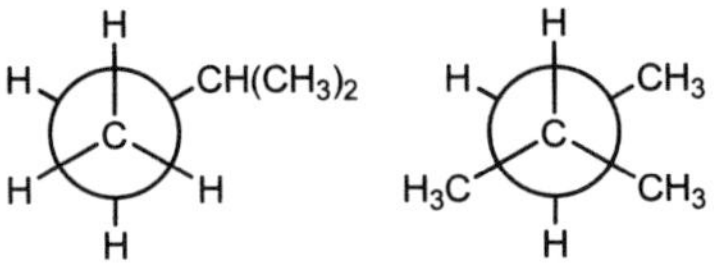

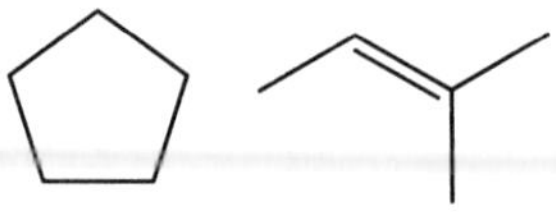

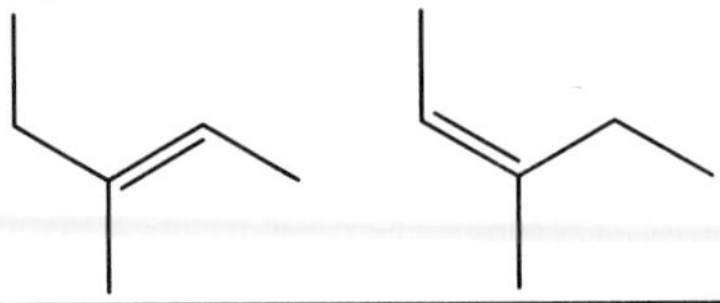

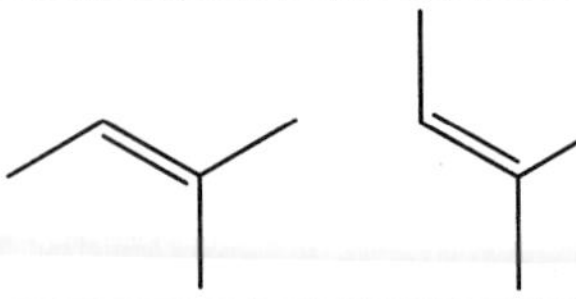

Model 5: Alkene Nomenclature

Naming can be a great way to figure out the relationship between two molecules. For example: If two molecules have the same name, they are the same or conformers. If two names differ only in the prefixes *Z* vs. *E* (or *cis* vs. *trans*), the molecules are configurational stereoisomers.

Memorization Task 5.2: Rules and Conventions for Naming Alkenes

- Use a number to designate where a double bond is located on a carbon skeleton. For example, but-1-ene (shown in the first box in the previous question).
- To name a molecule with one double bond that is *Z* or *E*, include the appropriate prefix before the name (e.g. *E*-but-2-ene). Omitting the prefix implies a mixture of *Z* and *E*.
    - Exception: If not specified, a double bond in a ring is assumed to be *Z*. e.g., "**cyclohexene**", not *Z*-cyclohexene.

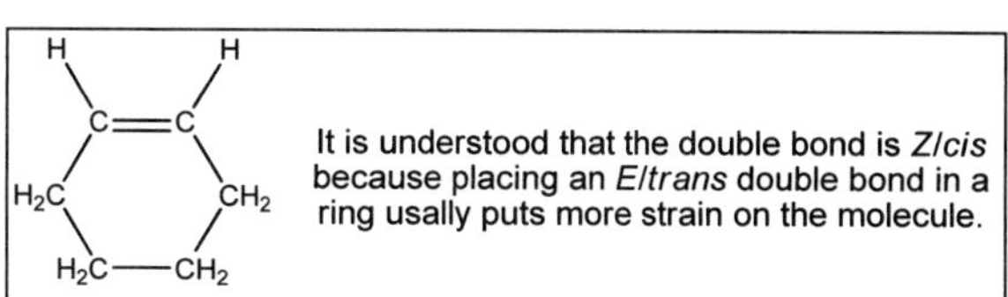

H
H
C=C
H2C
CH2
H2C—CH2
It is understood that the double bond is Z/cis because placing an E/trans double bond in a ring usally puts more strain on the molecule.

- If there is more than one double bond that can be *Z* or *E*, place multiple prefixes and numbers at the start to designate which bonds are *Z* and which are *E*.
    - Molecules with two, three, or four double bonds have the base names "diene," "triene," and "tetraene," respectively.

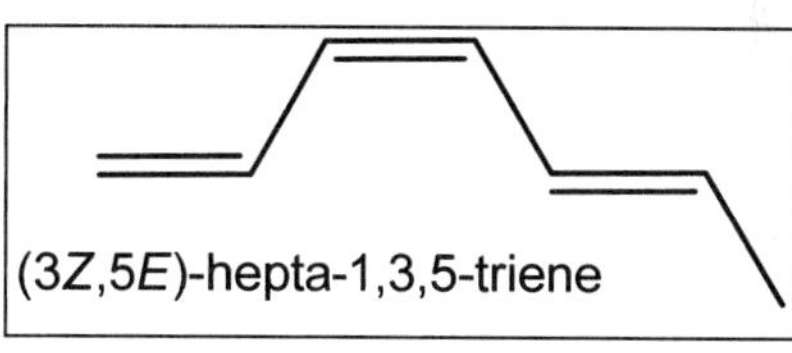

Extend Your Understanding Questions (to do in or out of class)

8. Name each molecule on the previous page using information found in Model 5 and Nomenclature Worksheet 1.

Confirm Your Understanding Questions (to do at home)

9. Draw a skeletal representation of *Z*-2-hexene and *E*-2-hexene.

10. Draw but-1-ene. Why does it not make sense to specify either *Z*- or *E*-but-1-ene while you must specify *Z*- or *E*-but-2-ene to draw the correct molecule?

11. Label each double bond *E (trans)*, *Z (cis)*, or neither. (It may help to draw in critical H's.)

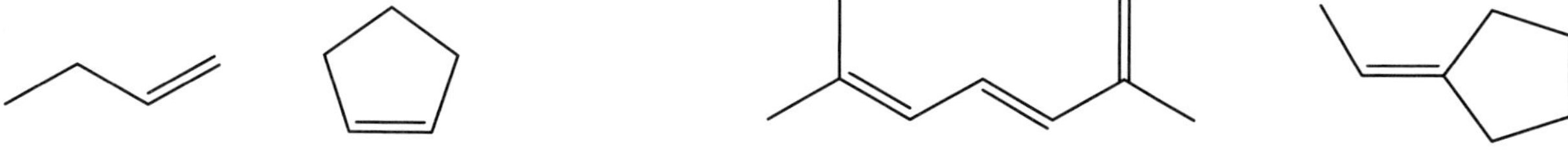

12. Label each double bond *Z*, *E*, or neither.

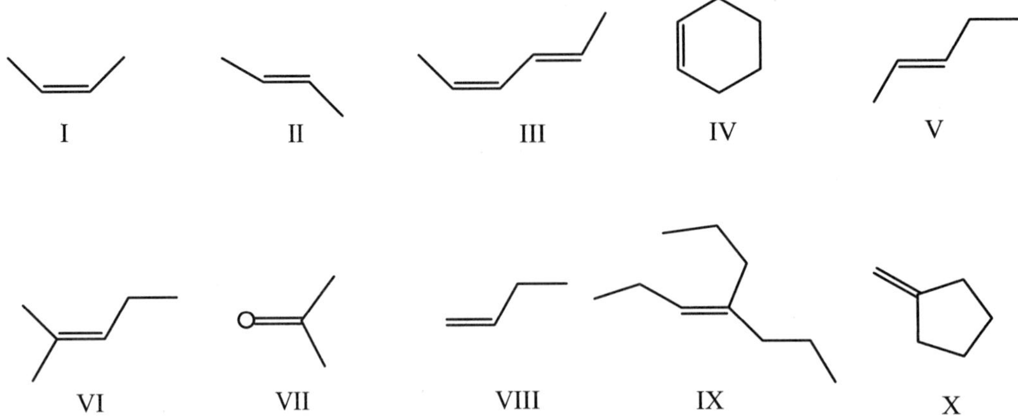

a. For each structure draw one constitutional isomer and all possible configurational stereoisomers.

b. A "terminal" double bond is a double bond found at the end of a carbon chain (e.g., VIII and X). What generalization can you make about all "terminal" double bonds in terms of *Z/E* (or neither)?

c. What is the relationship between Compounds III and IV above?

13. Draw a skeletal structure of *E*-3,4-dimethyl-3-heptene.

a. Following the algorithm in Memorization Task 5.1 results in two circled groups attached to the carbons involved in the double bond that are not identical. Nevertheless, the molecule still qualifies as *trans* (having two identical groups *trans* to one another). Explain.

b. Draw and name the configurational stereoisomer of *E*-3,4-dimethyl-3-heptene

14. Name this molecule, draw and name its configurational stereoisomer, and construct an explanation for why naming this molecule using *cis* or *trans* (instead of E or Z) could lead to ambiguity.

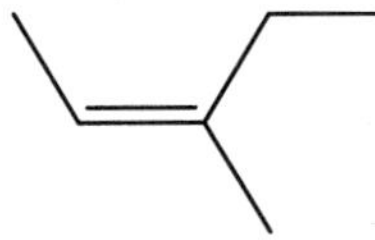

15. Draw another example of an alkene that cannot be named using the *cis/trans* nomenclature. (Such a molecule is not *cis, trans* or neither, but instead requires the *E/Z* nomenclature.)

a. Draw the configurational stereoisomer of the molecule you drew above. (If it has no configurational stereoisomer, your original structure is not correct.)

b. Describe the characteristics of a molecule that falls outside the *cis, trans* or neither naming scheme and requires the *E/Z* naming scheme.

16. Double bonds do not rotate freely under normal conditions. The change from Z to E requires a reaction. This can occur in the presence of a catalyst or with the addition of an appropriate amount of energy (e.g., at high temperature).

    One such reaction is diagramed below:

    (1) Add enough potential energy to break the double bond (E_{act}),

    (2) free rotation occurs at high energy transition state, then

    (3) reforming the double bond as a mixture of Z and E.

    Draw E-but-2-ene in one box and Z-but-2-ene in the other box, and explain your reasoning.

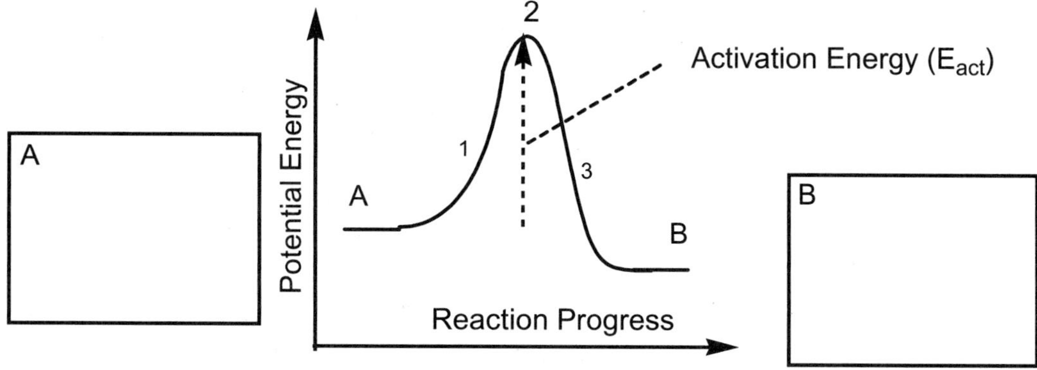

17. Make up an example (not appearing in this ChemActivity) of a pair of molecules that are a) constitutional isomers, b) conformers, c) configurational stereoisomers.

18. Construct an explanation for why Question 7 in this activity has the choice two terms "conformers or identical" lumped together as one choice, rather than offered as separate choices (i.e. "conformers" or "identical"). Hint: see Common Points of Confusion.

19. In the previous activity you were asked to draw all possible constitutional isomers with molecular formula C_6H_{12}. At the time, you might not have considered all possible configurational stereoisomers. For each double bond with the possibility of E/Z isomerism, you must draw both. For example, a six-carbon straight chain can accommodate a double bond in the five ways shown below.

    Draw the five constitutional isomers with molecular formula C_6H_{12} that utilize the backbone at right.

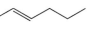

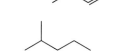

Read the assigned pages in the text, and do the assigned problems.

The Big Picture

In a subsequent activity we will break down the term configurational stereoisomers into finer categories, so it is critical that you are comfortable with the distinction between conformers, constitutional isomers, and configurational stereoisomers before you move on.

If you are having trouble with the terms *Z/cis* and *E/trans* as applied to alkenes, work with some models, and a team member who understands these terms. If you do not have a model set, borrow or purchase one immediately.

Common Points of Confusion

- When assigning priorities using the Cahn-Ingold-Prelog rules, the number one error that students make is to sum atomic numbers instead of considering the atomic number of each atom, one at a time, along a chain. A good illustration of this is the example below, left. Students assume that the longer chain ending in OH must get higher priority because it has more atoms, but this is not correct. In this case, using the tie breaker rules gives the CH_2OH chain priority because the higher atomic number oxygen is "encountered" sooner in the progression along each chain. A similar effect means that when considering two carbon chains, the one with an earlier branch will get priority (*see* example below, right).

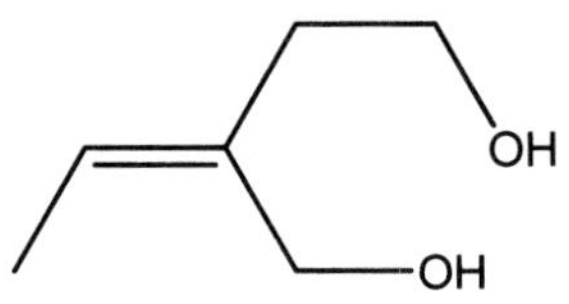

OH

OH

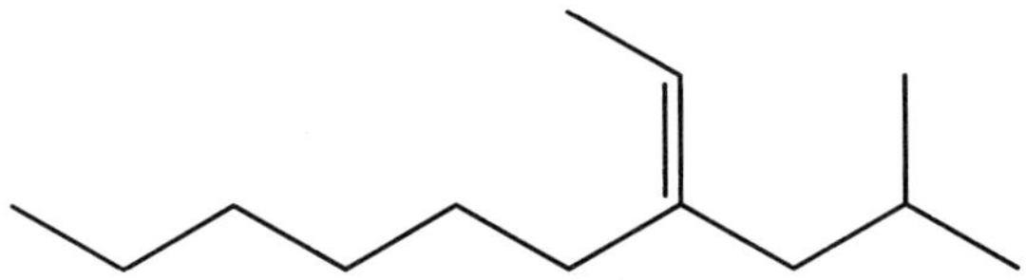

- Watch out for the term "**stereoisomer**" used by itself. It is sometimes used in place of the term "**configurational stereoisomer**" though this use is technically incorrect since stereoisomer (by itself) implies inclusion of all stereoisomers, both conformational stereoisomers (conformers) and configurational stereoisomers.

- What is the difference between the term *E* and the term *trans* (or *Z* vs. *cis*)? For alkenes, the terms *cis/trans* are not as useful as the terms *Z/E*. That is, all *cis* alkenes could be named using *Z*, and all *trans* alkenes could be named using *E*, but the opposite is not the case. For example, tri- or tetra-substituted alkenes such as *E*-3-methylhept-3-ene cannot be named using the *cis/trans* nomenclature. However, since the *cis/trans* nomenclature is used, you need to be familiar with it. Additionally, these terms are useful for talking about the relative relationship of two groups on an alkene. For example, you can say that on the molecule *E*-3-methylhept-3-ene the H on C_4 is *trans* to the methyl group on C_3.

- Some students wonder why the terms *E* and *Z* are not applied to rings. So far we only know how to describe/name the stereochemistry of disubstituted rings (rings with exactly two non-hydrogen groups), so the terms *cis* and *trans* are adequate. Later in the course we will learn a different naming strategy for rings with three or more groups.

- *Z/E* configurational stereoisomers drive a few students toward the exit door each semester, but a few minutes with a model and a knowledgeable peer, TA, or instructor can always bring them back. If you are having trouble recognizing *Z* and *E*, **make some models**, and talk to someone who is comfortable with these terms.

Notes

Notes

ChemActivity 6: Addition to Alkenes

(What is the mechanism of the addition of HBr to an alkene?)

<u>If you have already completed ChemActivity 12: Two-Step Elimination (E1), skip to Model 2.</u>

Model 1: Carbocation Stability

Recall that a **carbocation** is a cation with a +1 formal charge on a carbon (the **carbocation carbon**). The **carbocation carbon** lacks an octet. This is unfavorable and drives up the overall potential energy.

An <u>adjacent</u> **electron-donating group** will <u>help complete the octet of a carbocation carbon</u> by donating some electron density to its electron-deficient neighbor.

Memorization Task 6.1: Alkyl groups (-CH_3, -CH_2CH_3, etc.) are more electron-donating than H

Construct Your Understanding Questions

1. Which carbocation carbon in Figure 6.1 is <u>farthest from having an octet</u>? Explain your reasoning.

2. Is your answer to the previous question consistent with the fact that highest PE/least stable carbocation in Fig. 6.1 is at the top?

3. What stabilizing feature do the boxed carbocations in Figure 6.1 have that the others do not?

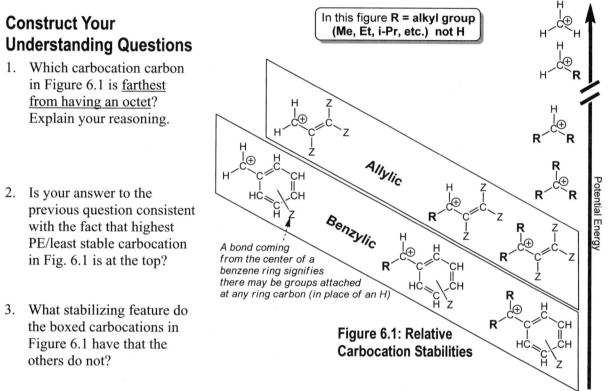

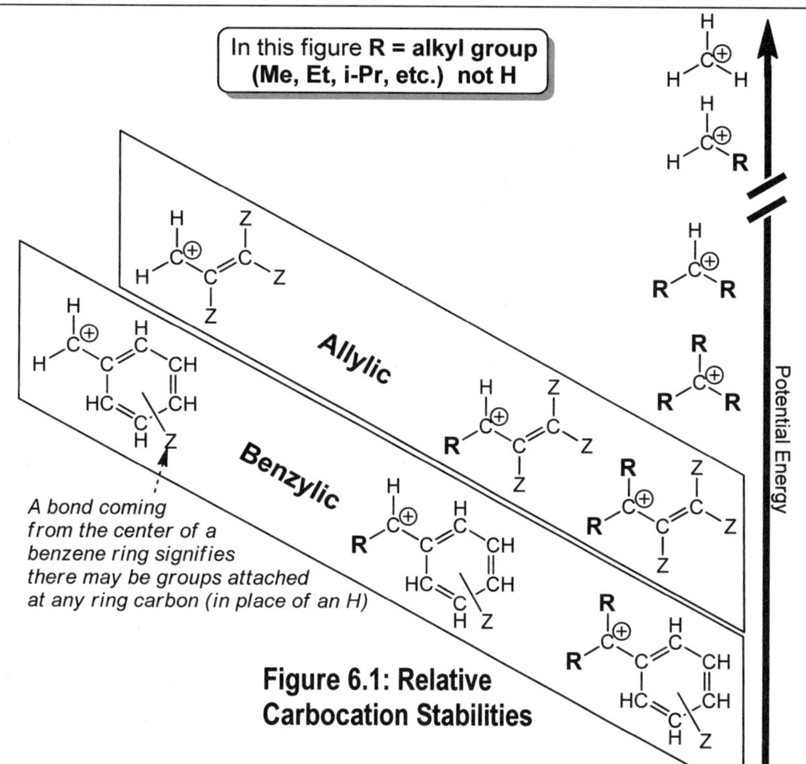

Figure 6.1: Relative Carbocation Stabilities

Memorization Task 6.2: Types of Carbocations *[In Figure 6.1 and below, assume R = alkyl (R ≠ H)]*

- **tertiary (3°)** carbocation → **three** R groups attached to C (**zero H's** attached to C)
- **secondary (2°)** carbocation → **two** R groups (**one H**)
- **primary (1°)** carbocation → **one** R group (**two H's**)
- **methyl (0°)** carbocation → **zero** R groups (**three H's**)

Appropriate proximity to π bond(s) leads to two special categories of resonance-stabilized carbocations:

- **benzylic** carbocation → + charge next to benzene ring (and in resonance with benzene π bonds)
- **allylic** carbocation → + charge next to (and in resonance with) π bond(s) (not benzene ring)

4. Label each carbocation in Figure 6.1 with **3°**, **2°**, **1°**, or **0°** as described in Mem Task 6.2.

5. More than one term from Memorization Task 6.2 can simultaneously apply to some carbocations. That is, a given resonance structure of an allylic or benzylic carbocation will also be **3°**, **2°**, or **1°**. Find and label the **tertiary allylic carbocation** resonance structure in Figure 6.1 (previous page).

6. It turns out that **methyl carbocations** and **ordinary (non-resonance-stabilized) primary carbocations** are so unstable that they essentially do not form. Cross out these two carbocations in Figure 6.1, and write **DO NOT FORM** next to them.

Memorization Task 6.3: Assume methyl and *ordinary* primary carbocations do not form.

7. Label each carbocation below with one (or more) of the terms in Memorization Task 6.2.

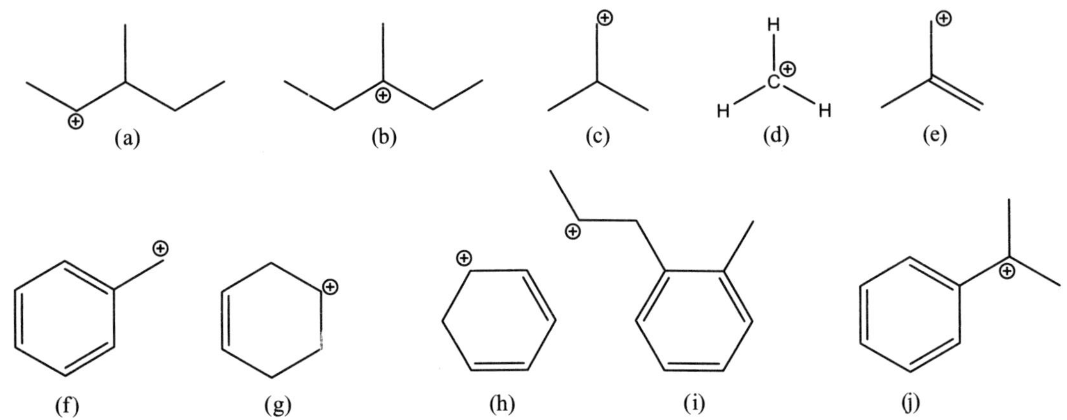

8. (Check your work) In the previous question, only structures (f) and (j) are benzylic carbocations. Speculate: What feature makes these two structures benzylic carbocations, while the other three rings (g, h, & i) are not benzylic.

9. (Check your work) Structures (e) and (h) are allylic, but structure (g) is not. Explain. [Note: (i) is neither benzylic or allylic.]

Model 2: π Bond Acting as a Base with a Strong Acid

H_3CH_2C CH_3
C=C
H CH_2CH_3
H—Br
solvent = diethyl ether

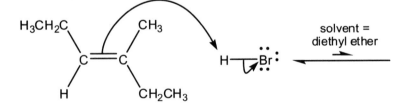

Construct Your Understanding Questions (to do In class)

10. Complete the following sentence describing what the first curved arrow in Model 2 is saying. Then write a sentence describing what the other curved arrow is saying.

"The π portion of the C=C bond breaks and the two (formerly π) electrons form a new bond to...

11. You may have noticed that the curved arrow showing bond formation in Model 2 (previous page) does not specify which carbon atom makes the bond to the H. The special *bouncing curved arrow below adds this information. Draw the result of the curved arrows below (same as in Model 2).

J. Chem. Educ. **2009**, 86 (12), p 1389.

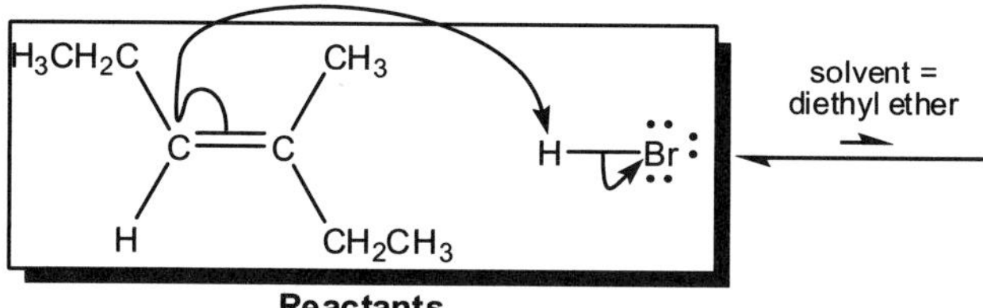

Draw the Products

Figure 6.2: Bouncing Curved Arrow Illustrating a π Bond Acting as a Base with a Strong Acid

12. (Check your work) Did you draw a tertiary carbocation in the previous question?

13. The reaction shown below (in Figure 6.3) does not happen; however, for the purpose of discussion, draw the products that *would* result if the arrow "bounced" toward the other carbon.

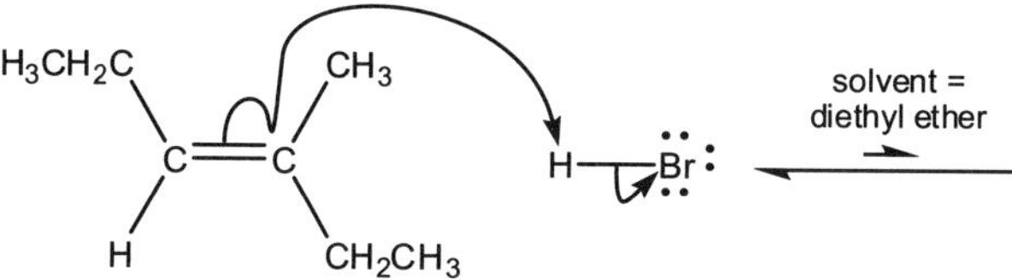

Figure 6.3: Curved arrow bouncing the other way

14. Construct an explanation for why the reaction takes place as shown in Figure 6.2 (not Figure 6.3).

Model 3: Markovnikov's Rule

The reaction in Model 2 is the first half of a **two-step** addition reaction, so the products of the first step are called intermediate products or simply "**intermediates.**" The last product is the **final product**.

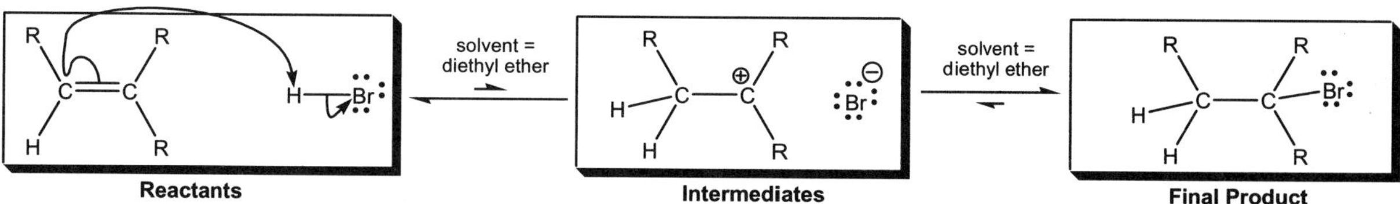

In 1869 Russian chemist Vladimir Markovnikov noticed a pattern in the products of additions to asymmetrical alkenes that has come to be called **Markovnikov's Rule**. It states that…

Markovnikov's Rule:

In addition of HX to a π bond, the X ends up on the carbon with more alkyl (R) groups and fewer H's.

Construct Your Understanding Questions (to do in class)

15. Add a curved arrow to Model 3 to show bond formation in the second step of the addition reaction.

16. Markovnikov did not know why his rule worked. (Carbocations had not been discovered yet.) Complete the following modern restatement of Markovnikov's Rule:

In an addition of HX to a π bond, the X ends up on the carbon that is more likely to form a carbocation. This is the carbocation that is **lower <u>or</u> higher** [circle the correct one] in potential energy.

Memorization Task 6.4: π bonds on a benzene ring do NOT undergo addition.

For reasons we will discuss later, π bonds on a benzene ring are almost always inert (do not react).

17. The **reactants, intermediates, final products**, and all **curved arrows** showing bonds forming and breaking are collectively referred to as the **mechanism** of a reaction. For the following reactants:

   a. Explain why the original statement of Markovnikov's rule does not help in this case, but the **modern restatement of Markovnikov's rule** tells you which carbon will bond to X (Cl).

   b. Show the complete mechanism of the most likely addition reaction between these reactants.

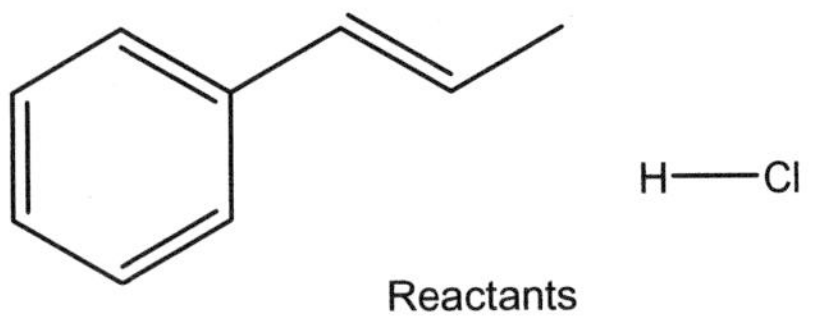

H—Cl

Reactants

Model 4: Carbocation Rearrangements

In a polar solvent, a carbocation will rearrange if the result is a lower potential energy carbocation. *(Rearrangements are the other legal use—besides a π bond acting as a base—for bouncing arrows.)*

There are two types: (the electrons that are shifting are highlighted in **bold**)

Hydride (H) Shift	Alkyl (R) Shift
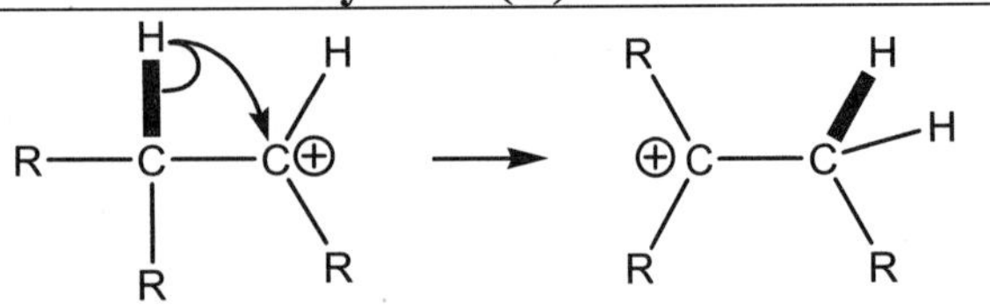 H, H, R—C—C⊕, R, R → R, H, ⊕C—C—H, R, R	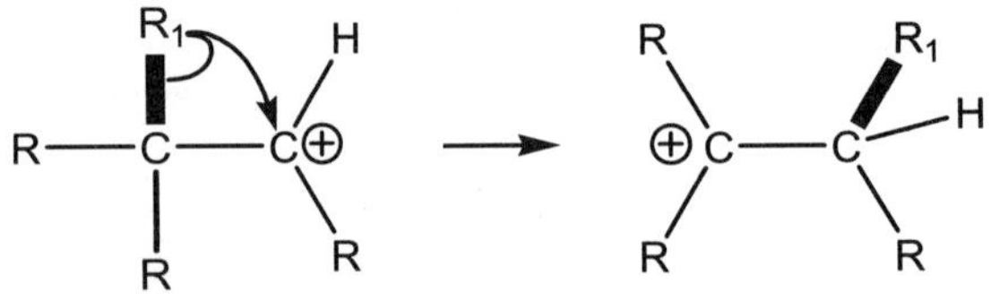 R_1, H, R—C—C⊕, R, R → R, R_1, ⊕C—C—H, R, R

For our purposes, we will assume that an **H or R can shift a total of one bond to an adjacent carbon,** though, given optimal conditions, a variety of exotic carbocation rearrangements can occur.

Extend Your Understanding Questions (to do in or out of class)

18. For each example in Model 4, explain why the rearrangement shown is downhill in terms of PE.

19. Write a sentence that describes what one of the **curved bouncing arrows** in Model 4 says about the movement of the electrons.

20. Show a mechanism that accounts for the product shown. Include all important resonance structures of carbocation 2. (*Hint*: A rearrangement is part of the mechanism.)

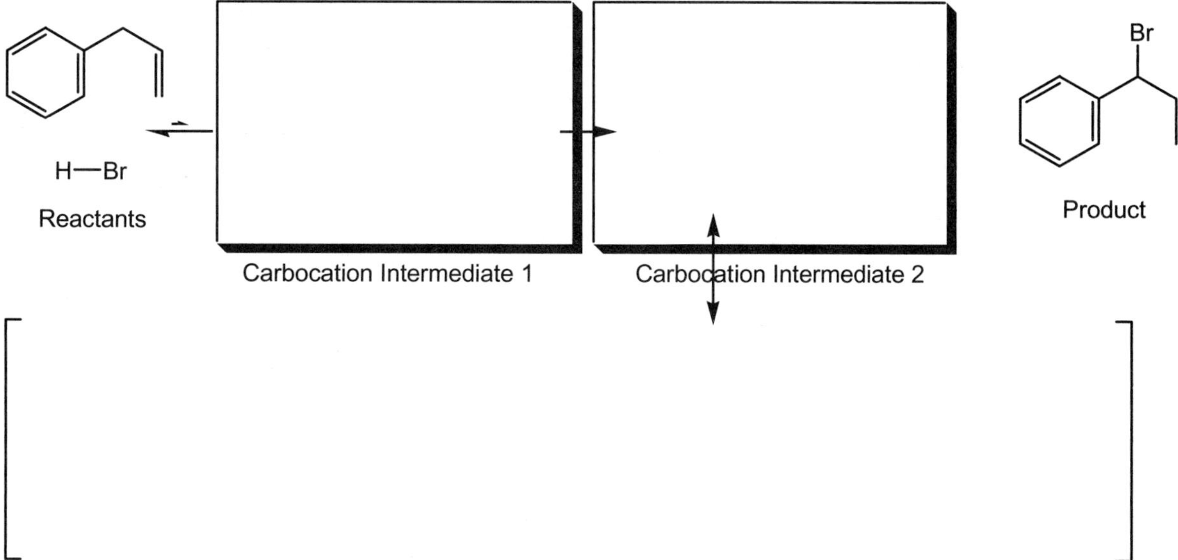

Important Resonance Structures of Carbocation 2

Model 5: Synthetic Transformations

The Markovnikov Addition of HX to an Alkene is an example of a **synthetic transformation**. It is called this because an organic chemist can use this reaction to *transform* alkenes (which are often cheap and readily available from petroleum sources) into alkyl halides, which, we will see, are very useful for other reactions. You must memorize all synthetic transformations covered in this course!

Synthetic Transformation 6.1: Markovnikov Addition of HX to an Alkene (X = Cl, Br, or I)

$$R_2C{=}CHR \xrightarrow[\text{cold, dark, no peroxides}]{HX} X{-}CR_2{-}CH_2R$$

As you learn more synthetic transformations, you will be asked to design a way to make a target molecule from a given starting material using these transformations.

This type of problem is very intimidating IF you have not memorized the synthetic transformations as we have encountered them. For some transformations, like 6.1 (above), you will learn the full mechanism (the "curved arrows"). For others you may learn only a partial mechanism, or no mechanism if it is unknown, very complex, or poorly defined. These are typically harder to memorize.

You might wonder, since there is a reaction called Markovnikov Addition of HX to an Alkene, is there a comparable reaction that is NOT Markovnikov? Indeed, there is, and we will learn its mechanism later in the course. For now, please be familiar with the reagents and overall transformation below.

Synthetic Transformation 6.2: Anti-Markovnikov Addition of HBr to an Alkene

$R_2C{=}CHR \xrightarrow[\text{heat or light}]{\text{HBr, peroxides (ROOR)}} R_2CH{-}CHR(Br)$

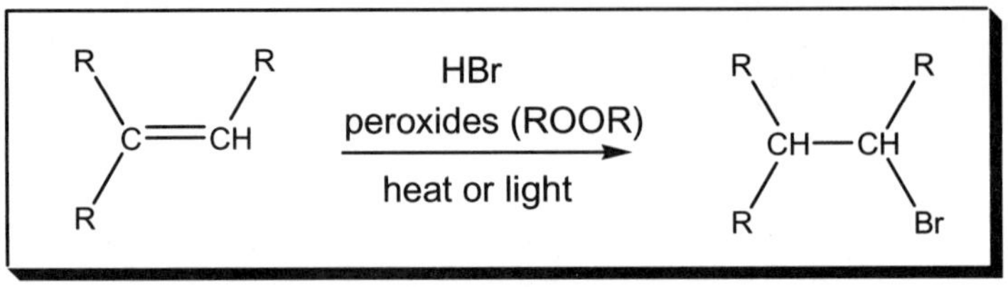

Extend Your Understanding Questions (to do in or out of class)

21. Describe the differences between Anti-Markovnikov addition of HBr to an alkene (Synth. Transf. 6.2) and Markovnikov addition of HX to an alkene (Synth. Transf. 6.1).

Confirm Your Understanding Questions (to do at home)

22. Draw the complete mechanism including the intermediate and most likely product for each reaction.

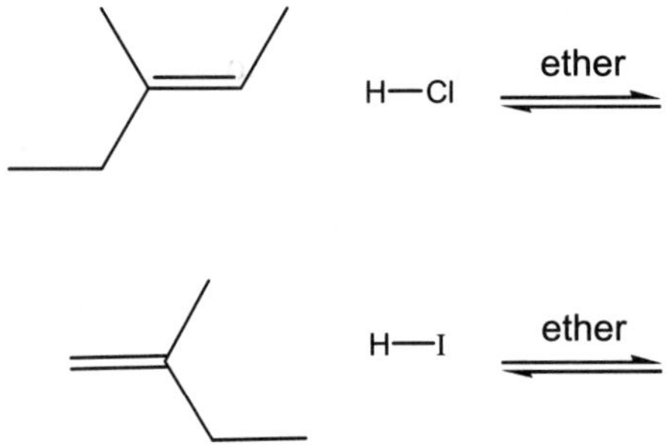

H—Cl, ether

H—I, ether

23. Explain why the following products are NOT observed in the reactions above.

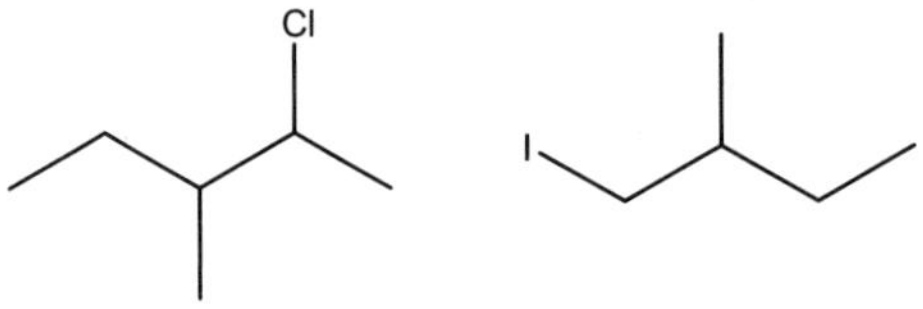

Cl

I

24. Construct an explanation for why ethene does not react with HX (X = Cl, Br or I) under normal conditions.

25. Draw the complete mechanism of each pair of reactants including favorable rearrangements and all important resonance structures of all intermediates.

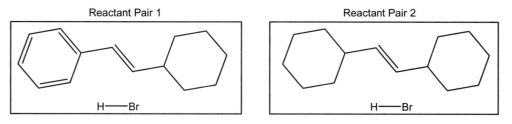

Reactant Pair 1

H—Br

Reactant Pair 2

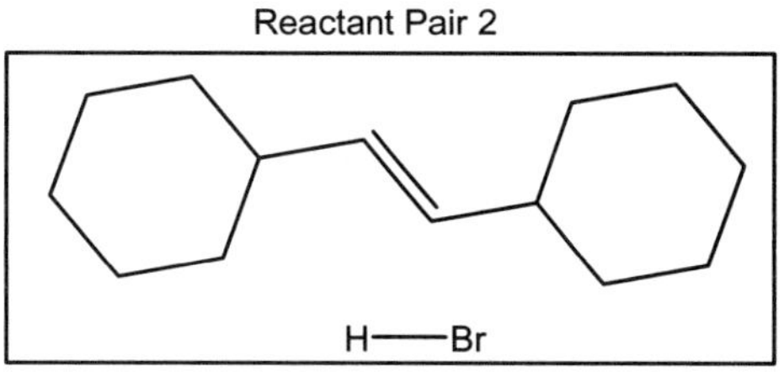

a. Which reaction has a lower PE carbocation intermediate?

b. Draw an energy diagram showing the reaction profiles of both reactions in the previous question. Use a dotted line for the first pair of reactants and a solid line for the second pair of reactants. (Assume the energy of the starting materials and products are the same for both pairs and the reactions are neither uphill nor downhill on net.

c. Mark points on the energy diagram corresponding to each carbocation in your mechanisms.

26. Use a curved bouncing arrow to depict the hydride shift that will likely occur for each carbocation below, and **explain why the new carbocation is lower in potential energy** than the original. (Draw any important resonance structures of each!)

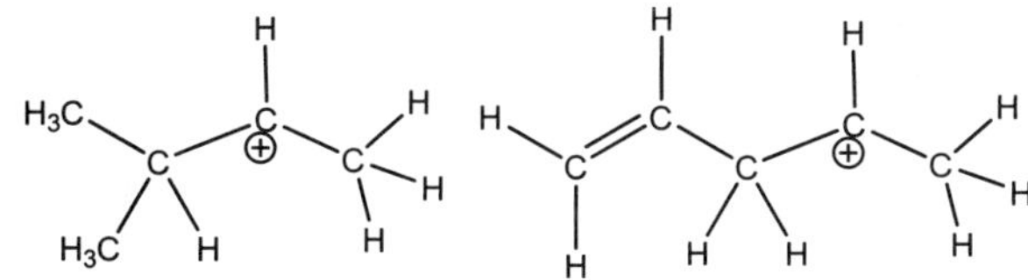

27. Use a curved bouncing arrow to depict the alkyl shift that will likely occur for each carbocation below, and **explain why the new carbocation is lower in potential energy** than the original.

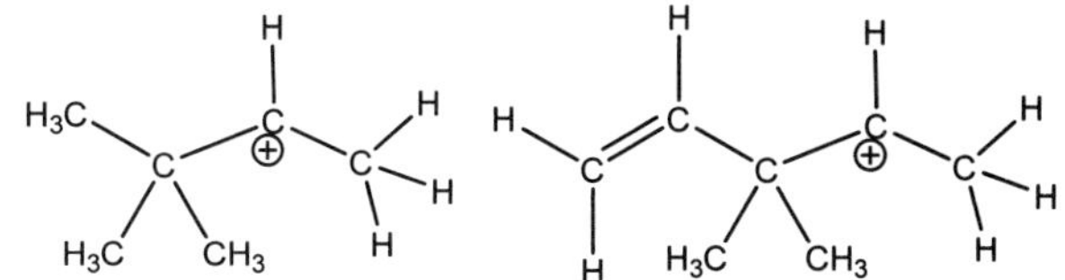

28. For each alkene draw all the likely products of electrophilic addition of one equivalent of strong acid, HX (where X = Cl, Br or I), and use curved arrows to show the mechanism of each reaction. (Some of the reactants are likely to undergo a carbocation rearrangement.)

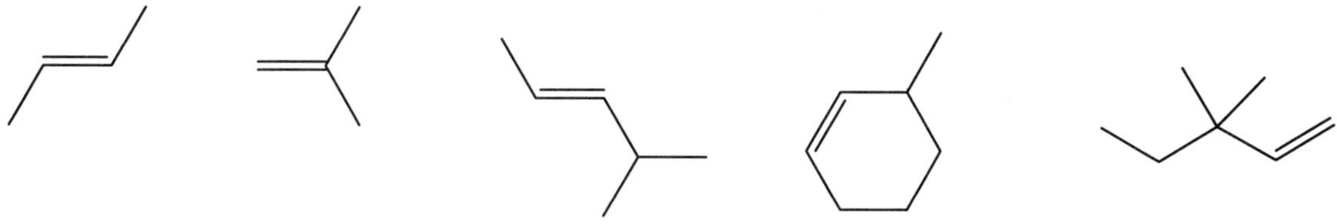

29. Some special cyclic carbocations undergo ring expansion with an alkyl shift (as shown below).

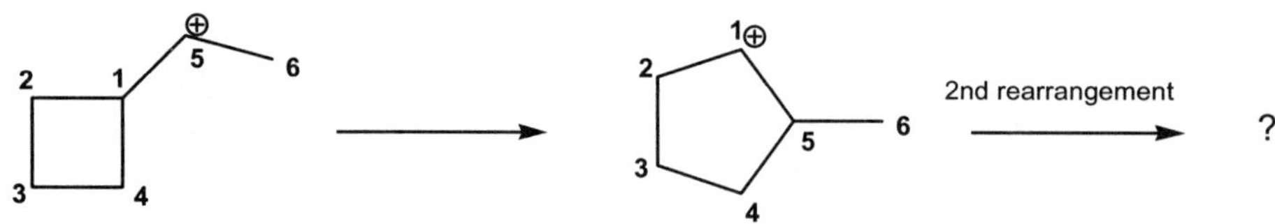

a. Use a curved bouncing arrow to show electron movement in the first carbocation rearrangement above. (Carbons are numbered to assist you.)

b. Use a curved bouncing arrow to show electron movement in a second carbocation rearrangement that this carbocation is likely to undergo, and draw the resulting carbocation.

c. What is the driving force for each of these rearrangements (It is different for the first rearrangement and the second rearrangement.)?

30. Use a curved bouncing arrow to show a likely rearrangement for each of the following carbocations, and draw the resulting lower-energy carbocation.

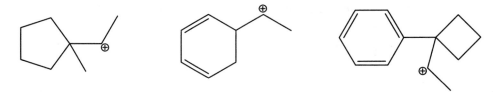

31. The product of the following electrophilic addition reaction, due to symmetry, has only two chemically distinct kinds of H's. (For those who have done the NMR activity, the proton NMR spectrum consists of only two peak clusters.) Propose a structure for this product, and use curved arrows to show a reasonable mechanism by which it would form.

H—I

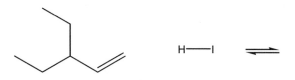

32. Electrophilic addition in water yields an alcohol product (in addition to the expected alkyl halide.

H—Br H_2O

alkyl halide alcohol

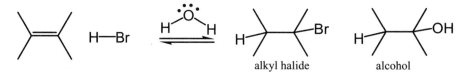

a. Devise a mechanism for formation of the alcohol product.

b. Draw the product you would expect if the same reaction were run using methanol as the solvent instead of water.

Read the assigned pages in the text and do the assigned problems.

The Big Picture

This activity introduces the first few synthetic reactions. These synthetic transformations pile up fast. They are the tools in your toolbox for transforming one molecule into another or building a complex molecule from simple starting materials. Synthesis problems bring together a complex set of skills and are often challenging to students, but they are IMPOSSIBLY FRUSTRATING if you have not memorized the tools in the toolbox. Make note cards and memorize these reactions as we encounter them.

Common Points of Confusion

- You will not find bouncing curved arrows in most textbooks. Electrophilic addition (e.g. addition of HX to an alkene) is most often represented as in Model 2. The bouncing curved arrows are a useful tool for making Markovnikov's rule explicit in the curved arrow formalism. As this rule becomes second nature, you may not need this tool; however, unlike simple electrophilic addition to an alkene, most students find that complex carbocation rearrangements such as ring expansions are always made easier by the use of a bouncing curved arrow.
- Please do not use a bouncing curved arrow when it is not needed. The ONLY two reactions we will learn this term that make use of bouncing curved arrows are electrophilic additions, and carbocation rearrangements.
- For more information about bouncing curved arrows see *J. Chem. Educ.* 2009, 86 (12), p 1389.
- Students sometimes reason that a higher-potential-energy (less stable) carbocation should lead to the major product because a higher-energy carbocation will be more reactive. The flaw in this reasoning is that (though the higher-PE carbocation *would* produce a more exothermic reaction) it never has a chance to form! The lower-PE carbocation forms preferentially over the higher-PE carbocation, because the lower-PE carbocation takes less energy to make. Since the first step in the reaction has a higher activation energy than the second, the first step is the hard step or slow step (or rate-limiting step). The result is the modern restatement of Markovnikov's rule. For the addition of HX to an asymmetrical alkene, the X will end up on the carbon that is more likely to form a carbocation.

Notes

ChemActivity 7: Radical Halogenation

(How can we add a functional group to an un-functionalized alkane?)

Model 1: Radical Halogenation of Alkanes

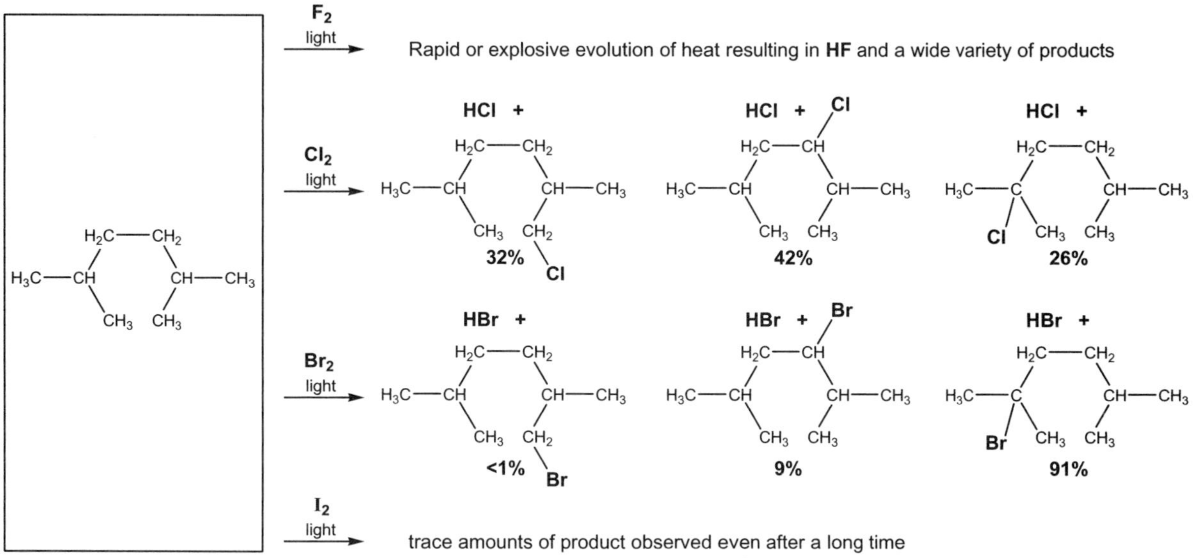

Starting material: $H_3C-CH(CH_3)-CH_2-CH_2-CH(CH_3)-CH_3$

F_2, light → Rapid or explosive evolution of heat resulting in **HF** and a wide variety of products

Cl_2, light → **HCl +** 1-chloro product (Cl on CH_2 of a methyl group) **32%**; **HCl +** Cl on ring CH (secondary) **42%**; **HCl +** Cl on tertiary C **26%**

Br_2, light → **HBr +** Br on CH_2 (primary) **<1%**; **HBr +** Br on CH (secondary) **9%**; **HBr +** Br on tertiary C **91%**

I_2, light → trace amounts of product observed even after a long time

Construct Your Understanding Questions (to do in class)

1. Describe the transformation shown in Model 1. (What is replaced by what?)

2. Which replacement of an H with a halogen (F, Cl, Br, or I) is fastest? …which is slowest?

Memorization Task 7.1: Relative rates of halogenation: $F_2 >> Cl_2 > Br_2 >> I_2$

Fluorination of an alkane is dangerously fast. Iodination is so slow as to be impractical. Only chlorination and bromination are commonly used in laboratories.

3. Each reaction in Model 1 gives a mixture of products. Based on the percentage of each product, what type of H is most likely to be replaced with Br? Choose from: [1° H (primary)], [2° H (secondary)], or [3° H (tertiary)].)

4. Because of symmetry, the starting material in Model 1 (2,5-dimethylhexane) has only one type of 1° H, one type of 2° H, and 1 type of 3° H. Report below the number of…

   i. 1° H's

   ii. 2° H's

   iii. 3° H's

Memorization Task 7.2: Selectivities of Cl_2 and Br_2 in Radical Halogenation

For reasons we will explore later, a halogen (Cl or Br) is more likely to replace a tertiary (3°), allylic, or benzylic H. Approximate selectivities are show below.

Halogen (X)	X sub. for H in a **1°** Position	X sub. for a H in a **2°** Position	X sub. for a H in a **3°** Position	X sub. for a H in **allylic** Position	X sub. for a H in **benzylic** Position
Cl	1	4	5	similar to 3°	
Br	1	100	2000	>> 3°	>> allylic

Examples:

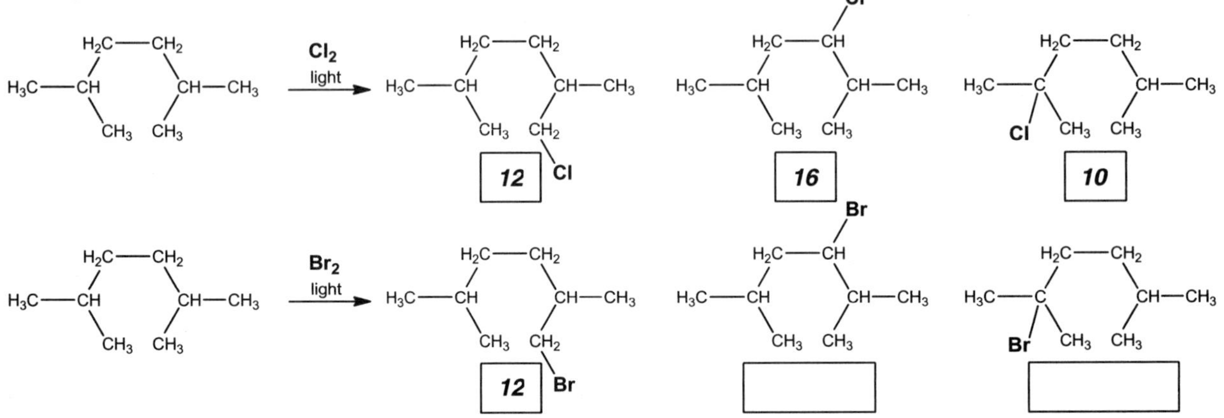

Cl_2, light → products: Cl on CH_2 (primary) **12**; Cl on CH (tertiary, ring-type position) **16**; Cl on C (tertiary) **10**

Br_2, light → products: Br on CH_2 (primary) **12**; Br on CH: [ ]; Br on C: [ ]

5. Confirm that the relative amounts of the three chlorinated products, given by the numbers in the boxes above, were generated using the following formula…

   Relative Amount of Product = [Selectivity from Mem. Task 7.2] x [Number of Equivalent H's]

6. Use this formula to calculate the relative amounts of the secondary and tertiarty brominated products and write these numbers in the empty boxes in Memorization Task 7.2.

7. (Check your work.) Are your answers to the previous question consistent with the data in Model 1?

8. Radical <u>chlorination</u> of the following molecule gives four different mono-chlorinated products (ignore stereoisomerism). Draw them and calculate the relative amounts using the formula above. Put a box around the **major product** (the one formed in the largest amount).

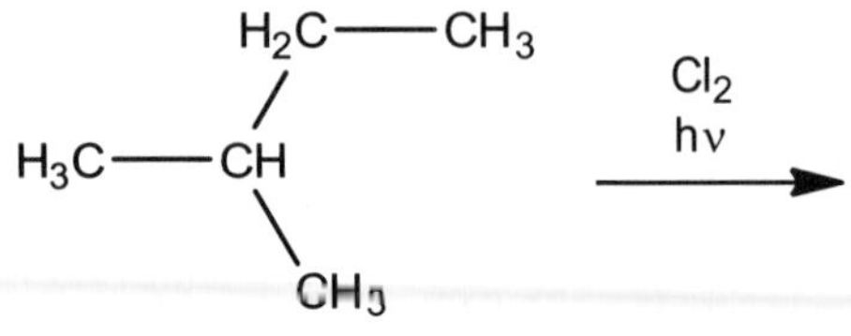

   H_3C—CH(CH_3)—H_2C—CH_3 $\xrightarrow{Cl_2,\ h\nu}$

9. Without doing any calculations (by inspection) draw the one mono-brominated product that will account for nearly 90% of the product resulting from radical bromination of this same alkane.

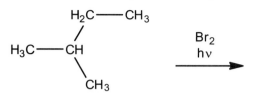

H_2C—CH_3 / H_3C—CH \ CH_3 $\xrightarrow[h\nu]{Br_2}$

10. Construct an explanation for why organic chemists say that **radical bromination is more selective** than radical chlorination. (The reason *why* bromination is more selective than chlorination has to do with the slower rate of the radical bromination reaction.)

Model 2: Polar and NonPolar Bond Breakage

Most mechanisms in this course <u>outside of this chapter</u> involve **polar bond breakage** in which the two electrons comprising a bond move together as a pair when the bond breaks, generating + and – charges.

radical = species with an <u>unpaired electron</u> (Radicals result from nonpolar bond breakage.)

⟶ = "double barbed" arrow (or **full arrow**)	⇀ = "single barbed" arrow (or **half arrow**)
Polar (heterolytic) Bond Breakage	Non-polar (homolytic) Bond Breakage
Y—Z ⟶ $Y^{\oplus}$ $:Z^{\ominus}$	Y—Z $\xrightarrow{\text{light } (h\nu)}$ Y• •Z
for example:	*for example:*
:Br—Br: ⟶ :BrH2⊕ $:Br:^{\ominus}$	:Cl—Cl: $\xrightarrow{\text{light } (h\nu)}$:Cl• •Cl:
H—Br: ⟶ $H^{\oplus}$ $:Br:^{\ominus}$	:Br—Br: $\xrightarrow{\text{light } (h\nu)}$:Br• •Br:
H—$\overset{\oplus}{O}H_2$ ⟶ $H^{\oplus}$ OH_2	RO—OR $\xrightarrow{\text{light } (h\nu)}$ R—O• •O—R

Construct Your Understanding Questions (to do in class)

11. Identify the radical species in Model 2.

12. What energy source is shown generating two halogen radicals (X·) from a molecule of X_2?

13. A **double-barb arrow** depicts the movement of a <u>pair</u> of electrons. What does a **single-barb arrow** depict?

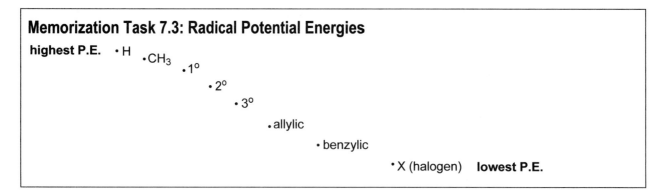

Memorization Task 7.3: Radical Potential Energies

highest P.E. •H •CH_3 •1° •2° •3° •allylic •benzylic •X (halogen) **lowest P.E.**

14. [(E)]Which is lower in potential energy: a tertiary radical or a primary radical?

15. Complete the sentence: Radicals follow approximately the same potential energy trends as **carbocations** or **carbanions** [circle one].

Model 3: "Red-Rover Red-Rover, send Radical right over!"

We will use a variation of the school-yard game "Red-Rover" to model radical bond breakage. In this version of Red-Rover the radical bonds to one of the two atoms it splits apart.

Recall from oxidation reactions that a C—H bond is often easier to break than a C—C bond.

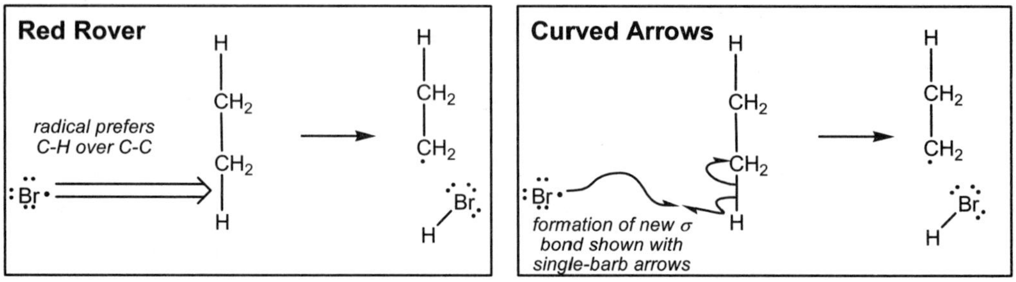

Red Rover

radical prefers C-H over C-C

:Br• ⟹ H–CH_2–CH_2–H ⟶ H–CH_2–$\dot{C}H_2$ + H–Br

Curved Arrows

:Br• + H–CH_2–CH_2–H ⟶ H–CH_2–$\dot{C}H_2$ + H–Br

formation of new σ bond shown with single-barb arrows

Construct Your Understanding Questions (to do in class)

16. Construct an explanation for why the Br radical in Model 3 sticks to the H instead of the C. (*Hint*: Draw the radical that would be produced if the Br radical had stuck to the C instead?)

17. In the first step in a radical halogenation (shown in Model 2), light is used to generate two halogen radicals (X·). The next step is shown below. Use curved single-barb arrows to show the mechanism.

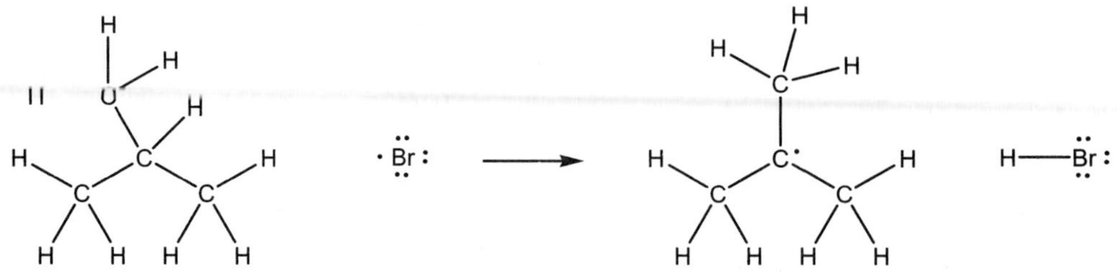

H H H H H H H H H H (C C C C) + ·Br: ⟶ H H H H H H H H H (C C Ċ C) + H—Br:

Extend Your Understanding Questions (to do in or out of class)

18. Use curved single-barb arrows to show a **multi-step** mechanism for the following reaction

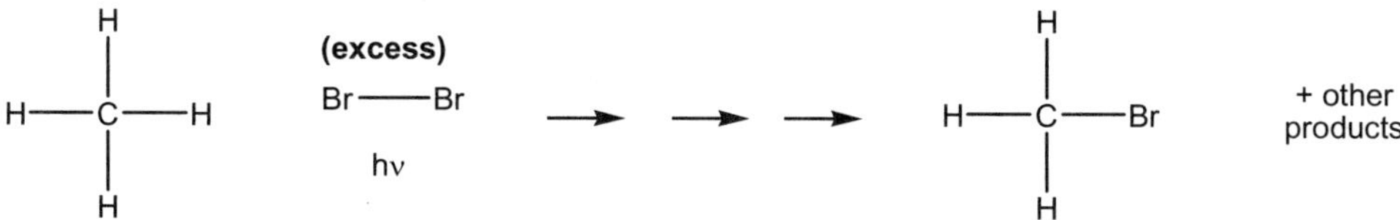

Model 4: Collision Statistics

A reaction between two radicals may seem very likely from an energy standpoint because, in such a reaction, two high-potential-energy species combine to form one low-potential-energy species.

for example:

$$H_3C\cdot \;\; \cdot Br \xrightarrow[\textbf{but statistically unlikely!!}]{\text{rxn is very very downhill}} H_3C\text{—}Br$$

However, radicals are **so** high in energy that they react with anything they "bump into." This means a radical has a short life-span and usually does not have time to "find" another radical.

Extend Your Understanding Questions (to do in or out of class)

19. Picture the old Veterans Stadium in Philadelphia filled to capacity with 62,382 people. Eagles fans are *normally* peaceful folks, but on this day there are five violently reactive fans in the stadium that will start a fight with the first person they bump into (getting themselves thrown into Veterans' in-house jail). Assume the violent fans are dispersed randomly in the crowd. What are the chances that two of the five "violently reactive fans" get in a fight with one another?

20. Based on the information in Model 4, circle any step in <u>your</u> mechanism at the top of the page that has a low probability of occurring, and explain your reasoning.

21. The first two steps of the mechanism from the previous page are drawn for you below.

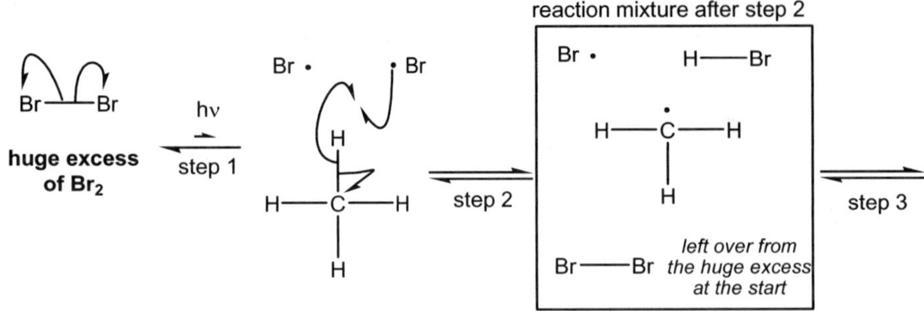

a. The radical ($\cdot CH_3$) is like the "fighter" in the stadium analogy. It will react with the first bond it bumps into. What is the most prevalent species in the reaction mixture after step 2?

b. Show the statistically most likely Step 3 in the mechanism above. **Note: This step leads to formation of a CH_3Br, but there are also another product.**

Model 5: The Three Parts of a Domino (or Radical) Chain Reaction

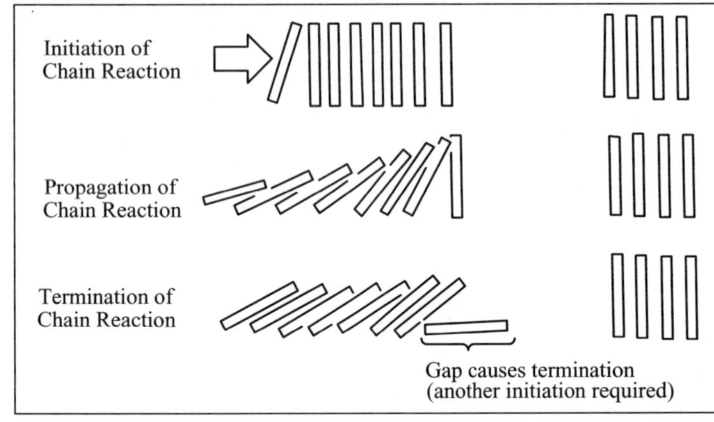

A **radical chain reaction** is a reaction with the following parts:

I. **Initiation** = net generation of radicals

II. **Propagation** = one radical consumed; one radical produced

III. **Termination** = net consumption of radicals

Extend Your Understanding Questions (to do in or out of class)

22. The mechanism at the top of the page is part of a radical chain reaction.

a. Identify the **initiation step.**

b. Identify the two **propagation steps.**

c. A collision between two radicals is unlikely, but it does happen every once in awhile. Eventually such reactions will stop the chain reaction. List at least two possible **chain termination steps** (they don't have to lead to H_3C-Br).

Synthetic Transformation 7.1: Radical Chlorination of an Alkane

R(CH$_3$)CH—R $\xrightarrow[\text{"hv" = common symbol for light}]{Cl_2 \text{, light (hv)}}$ R(Cl—CH$_2$)CH—R + R(CH$_3$)C(Cl)—R + **HCl**

Synthetic Transformation 7.2: Radical Bromination of an Alkane

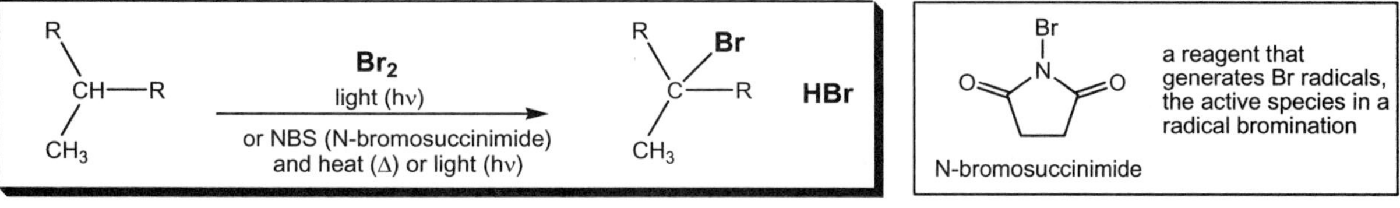

Confirm Your Understanding Questions (to do at home)

23. Along with the methyl radical, there is a Br radical in the box in Question 21. What happens if the Br radical reacts with the most abundant species in the reaction mixture (Br-Br)? Draw this reaction, and explain why it is not very interesting.

24. Construct an explanation for why the solvents on the left are suitable for radical halogenation reactions, but the solvents on the right are not.

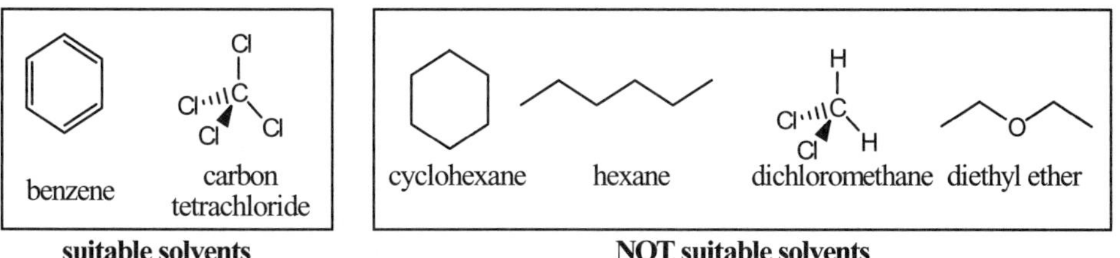

25. Give an example of each of the following: primary radical, secondary radical, tertiary radical, methyl radical, benzylic radical, allylic radical.

26. Consider the partial resonance representation of a benzylic radical species.

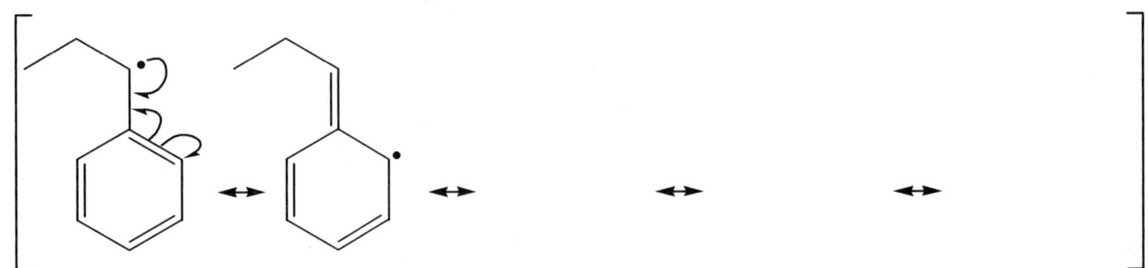

a. Use curved half-arrows to generate the other three resonance structures of the radical above.

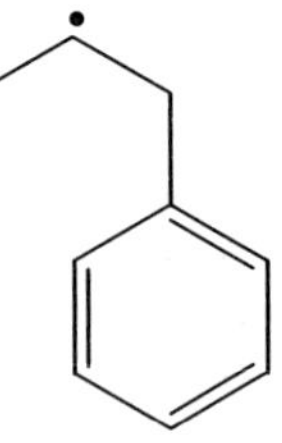

b. Explain why a benzylic radical is much lower in potential energy than the radical drawn at right. (For reasons we will not discuss here, radicals do not rearrange—even when, as in this case, there is a driving force for a rearrangement.

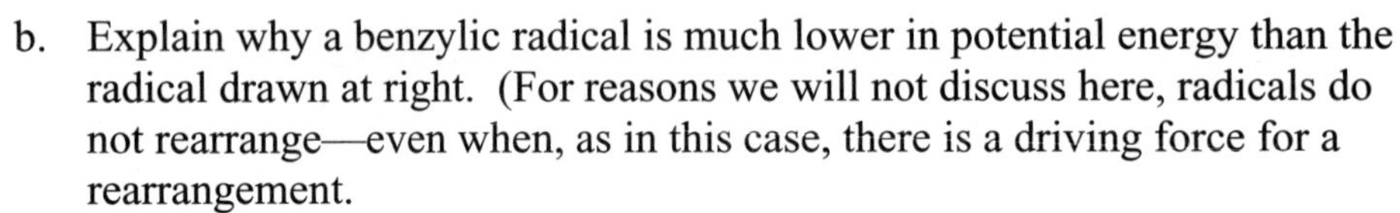

27. Circle the species below that is lower in potential energy.

a. Construct an explanation for your choice including drawing at least one other structure.

b. Which one of these radicals is an allylic radical?

c. Draw an allylic carbocation, and summarize the similarities and differences between an allylic radical and an allylic carbocation.

28. Which of the following diagrams best tracks the products formed from a single initiation event (breaking apart of one Br_2 molecule) in a **radical chain reaction**?

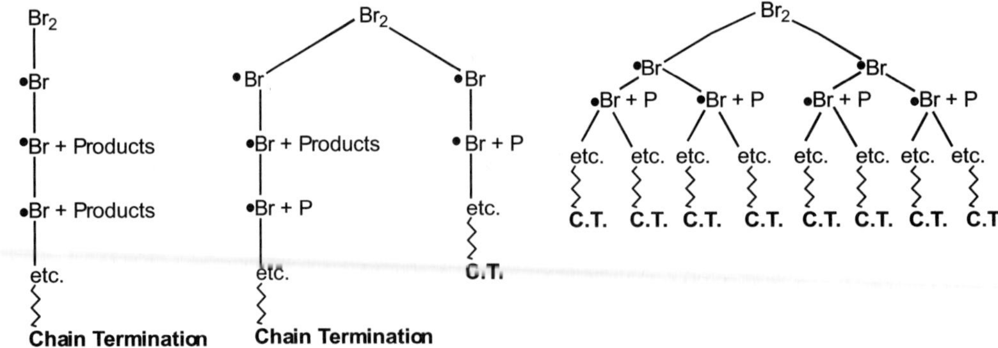

29. Use curved single-barb arrows to show an initiation step and propagation steps leading to each of the following products.

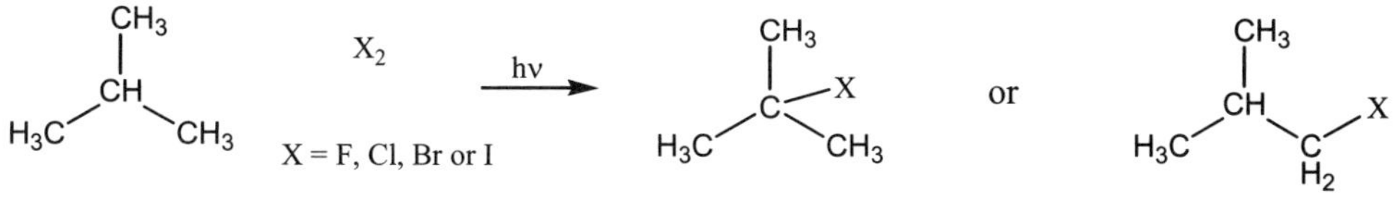

$$(CH_3)_2CH{-}CH_3 \xrightarrow[h\nu]{X_2} (CH_3)_2CX{-}CH_3 \text{ or } (CH_3)_2CH{-}CH_2X$$

X = F, Cl, Br or I

30. Summarize the effect of each factor on the ratio of the two products shown above.

a. The type of H (benzylic, allylic, 3°, 2°, 1° or methyl).

b. The number of H's of a given type.

c. The identity of X (F, Cl, Br, or I).

31. Fill in the blank with F_2, Cl_2, Br_2 or I_2

a. Radical halogenation with _____ is unselective, violent and dangerous, limiting the usefulness of this reaction in organic synthesis.

b. Radical halogenation with _____is so slow as to be useless in organic synthesis

c. Radical halogenation with _____is just right to be useful in organic synthesis because the rate is manageable, and the reactions are very selective.

d. Radical halogenation with _____ is often used in the laboratory to halogenate molecules with only one kind of H (e.g., cyclohexane).

32. Consider the following synthesis of the alkyl halide shown below.

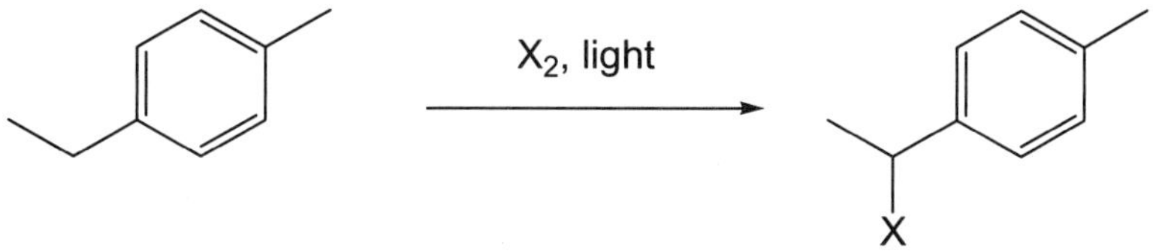

X_2, light

X

a. A student chooses Cl_2 as the halogen and gets a mixture of three different products. Draw them, and estimate the approximate product ratio.

b. Explain why replacing Cl_2 with a different halogen (specify which one) would give a better yield of the desired product.

33. Consider the following reaction.

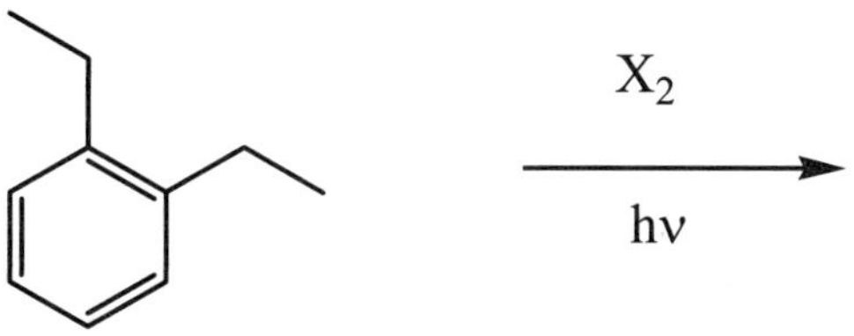

$$\xrightarrow[h\nu]{X_2}$$

a. There are two different H's in this molecule that can be replaced by X. Add them to the drawing above, and label them H_a and H_b.

b. There are two different H's on this molecule that will not react with X radical. Label them H_c and H_d.

c. Specify whether H_a and H_b are primary, secondary, tertiary, allylic or benzylic.

d. Calculate the relative amounts of each of the two different products when the reaction is run with Cl_2.

e. Calculate the relative amounts of each of the two different products when the reaction is run with Br_2.

f. Which reaction is more selective: the reaction with **Cl_2** or **Br_2**? (Circle one, and explain your reasoning).

34. Draw all possible radical halogenation products using X_2 and light (ignoring stereoisomerism).

a. Estimate the ratio of the products if X = Cl_2

b. Estimate the ratio of the products if X = Br_2

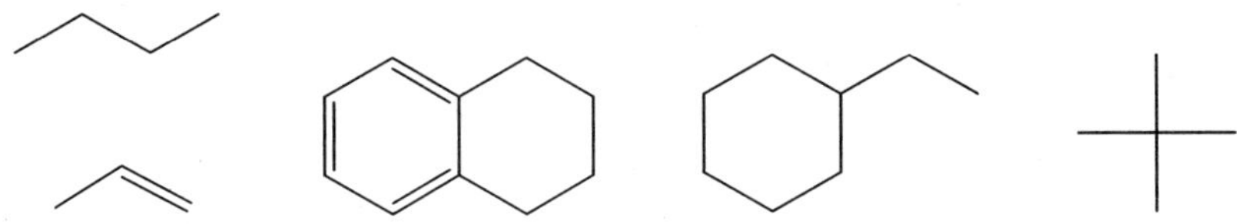

35. Design a synthesis of each of the following target molecules starting from cyclohexane.

OH D D O

OH OH Br Br OH O OH

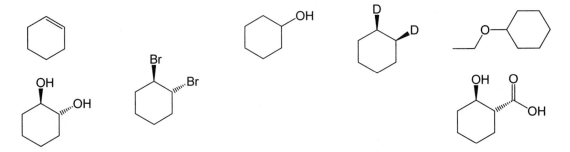

36. Design a synthesis of each of the following target molecules starting from any alkane.

O Br OH OCH_3 *cis* only CN O H

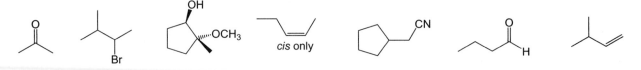

Read the assigned pages in the text, and do the assigned problems.

The Big Picture

This course focuses on polar reactions because these are better understood and easier to diagram using curved arrows. Radical reactions are just as common, but are not emphasized in organic chemistry courses because they are complicated, poorly understood, and often lead to products via a wide array of competing mechanisms. This chapter only scratches the surface when it explores the mechanism of one well-behaved radical reaction: radical halogenation. You will encounter radical reactions throughout your course (e.g., oxidation and reduction reactions), but may not discuss their mechanisms. In fact, in a typical organic chemistry course, most of the reactions that are presented without a mechanism are radical reactions.

Common Points of Confusion

- In Question 18 students frequently show a mechanism that leads to product, but does so via a termination step. This certainly happens, but since termination steps are hundreds or thousands of times less likely than propagation steps, it is unreasonable to claim this is the primary mechanism of product formation.
- You may see a reagent such as… "Br_2, cold, dark, no peroxides" for the dibromination of an alkene. The purpose of stipulating the temperature and conditions is simply to rule out the possibility of a radical reaction (replacement of an allylic H with Br). Prior to this activity, the reagents for this reaction were listed simply as Br_2.

Notes

ChemActivity 8: Chiral Centers

BUILD MODELS: H | Cl—C—I | Br *and 2-chlorobutane*

(How do you name two molecules that are non-identical mirror images of each other?)

Model 1: Same or Not the Same

Recall that two molecules are the **same** if models of them can be **superimposed** without breaking bonds.

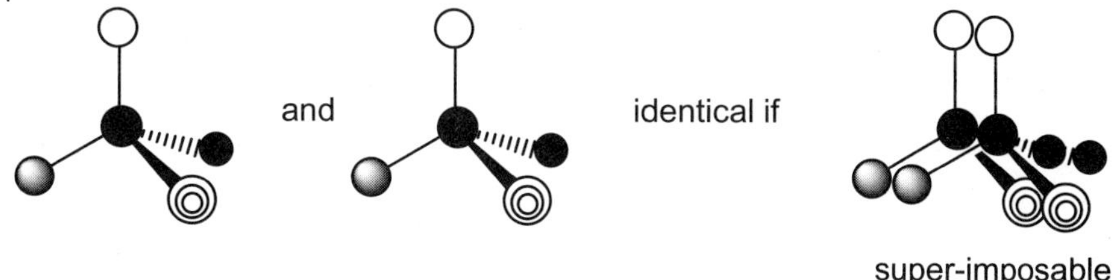

Construct Your Understanding Questions (to do in class)

1. [(E)]Make two identical models of the following molecule, and **confirm that they can be superimposed on one another as shown above**. Use any four different colors to represent the four attached atoms. If your set has green, orange and purple, use the following color code.

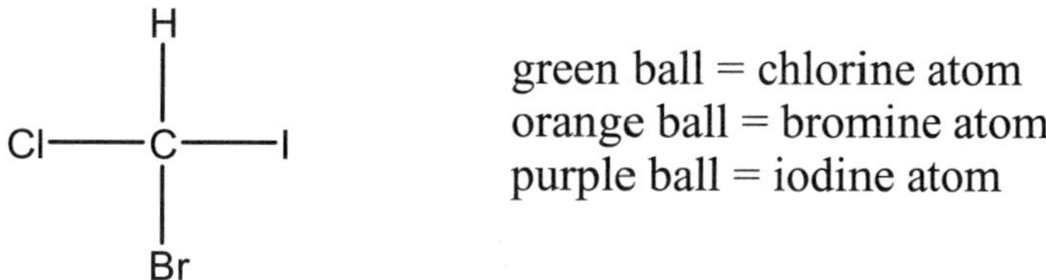

2. Modify <u>one</u> of your models by switching any two balls. (**Leave the other model unchanged.**) Is the modified model **identical** to the unmodified model?

3. Which of the following words describe(s) the relationship between the modified model 6nd the unmodified model? (Circle more than one choice, if appropriate.)

   a. **identical** (or **conformers** = can be made identical via single-bond rotation)

   b. **configurational stereoisomers** (same atom connectivity, but not identical)

   c. **constitutional isomers** (same formula, different atom connectivity)

   d. **mirror images** (look like reflections of one another in the mirror)

4. (Check your work) There are two correct answers to the previous question.

Model 2: Chiral Centers (Stereogenic Atoms)

It turns out there are two ways to arrange four different groups around a tetrahedral atom (usually C), and that these two arrangements will be mirror images of one another. Such a carbon is often marked with an *, and called a **stereogenic carbon** because its presence generates two configurational stereoisomers. Stereogenic atoms are also called **chiral centers** (*cheir* is Greek for "handed").

stereogenic carbon (chiral center) has 4 different groups | 1-chloro-1-fluroroethane | lactic acid | 2-chlorobutane | naproxen (Aleve™)

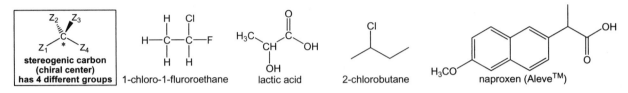

Construct Your Understanding Questions (to do in class)

5. There is one **chiral center** in each molecule in Model 2. Mark this chiral center with an *.

6. Figure 8.1 shows both possible configurational stereoisomers of 2-chlorobutane.

   a. Make a model of 2-chlorobutane so that it looks like the stereoisomer **on the left side of Figure 8.1** (Check your model by holding it up to the drawing.).

   b. Now make your model look like the drawing on the right side of Figure 8.1. Can you do this without breaking bonds?

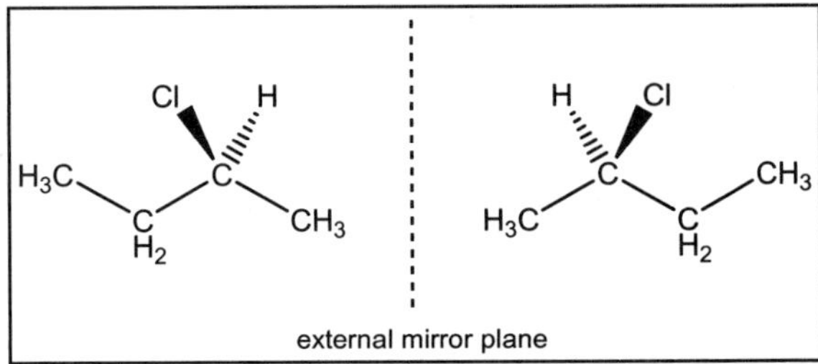

Figure 8.1: Stereoisomers of 2-chlorobutane

Memorization Task 8.1: Drawing the mirror image of molecules with one chiral center

You can transform your model of the stereoisomer on the left into its mirror image (the stereoisomer on the right) by **switching any two groups attached to the chiral center**.

This trick can be used to generate a mirror image of any molecule with one chiral center. One way to show this on paper is to change a wedge bond into a dash bond or vice versa.

7. How can you tell that the H on the chiral carbon on the left in Memorization Task 8.1 is going into the paper?

Memorization Task 8.2: Terms for Describing Chiral Molecules

Term	Definition	Term describes…	Origin
chiral	= not identical to its mirror image	a property *(like "round" or "green")*	*cheir* is Greek for "handed"
enantiomer	= the mirror image of a chiral object or molecule	a relationship *(like "son" or "sister")*	*enantio* is Greek for "opposite"
racemic mixture	= a 1:1 mixture of a pair of enantiomers	a special sample *(like the passengers on Noah's Ark)*	From *racemic acid* found in wine (*acin* is Latin for grape)

8. Below each molecule, sketch its mirror image. Circle each molecule that is chiral.

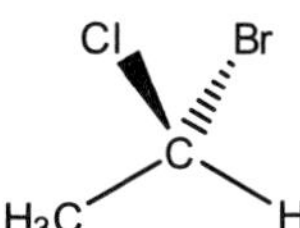

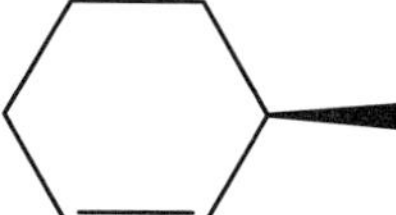

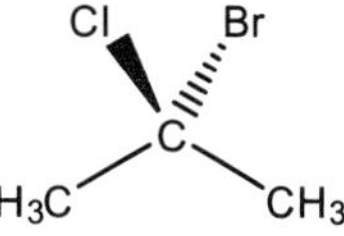

9. (Check your work) Explain why the far right example above does not have an enantiomer (and is not an enantiomer). [If you made accurate drawings, the first two pairs *are* enantiomers of each other.]

10. Explain why there is likely a **racemic mixture** (or very close to one) of shoes in this classroom.

Model 3: Internal Plane of Symmetry (Internal Mirror Plane)

It is always true that an object or molecule with an **internal mirror plane** is not chiral.

Not chiral = **achiral** = identical to its mirror image.

An object or molecule with no internal mirror plane is **chiral** = different from its mirror image.

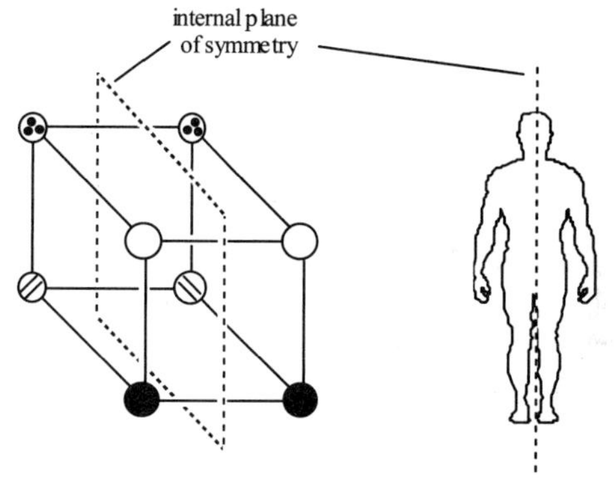

Construct Your Understanding Questions

11. Object C has an obvious mirror plane, marked with a dotted line, and a not-so-obvious one. (Assume the circles are spheres of equal size.)

    a. Where is the second mirror plane on Object C?

    b. Is Object D chiral? Explain your reasoning.

    c. Draw a modification on one of Object D's spheres so as to make it chiral.

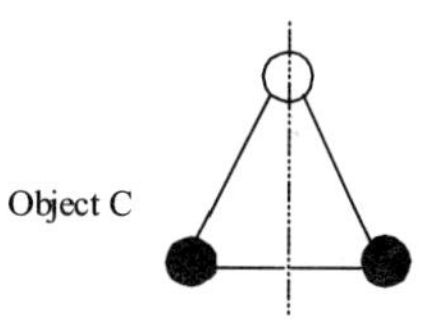

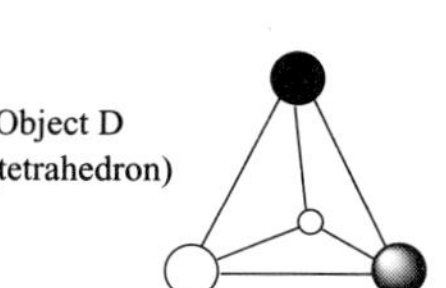

12. Does the molecule at right have an internal mirror plane? If so, where?

13. Based on the definition in Model 3, is the molecule at right chiral?

14. (Check your work) Based on the definition in Memorization Task 8.2, is the molecule at right chiral? … Is the carbon at right a **chiral center**?

15. Apply the internal mirror plane test to the molecules at the top of the page. In which one do you expect to find a mirror plane? (If you cannot find this mirror plane, read the first bullet in Common Points of Confusion.)

Model 4: Chiral Center on a Ring

Not Chiral

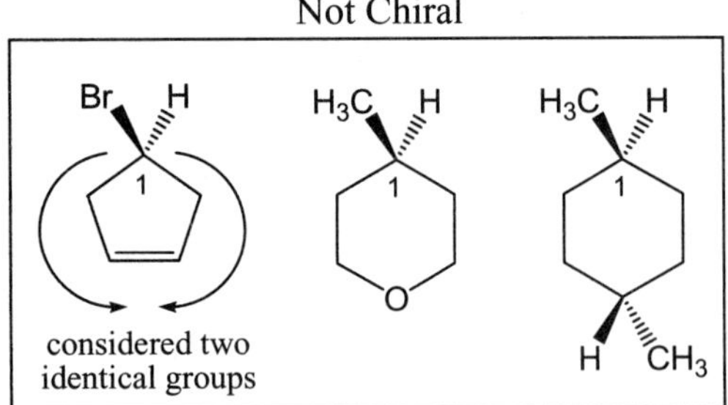

Chiral

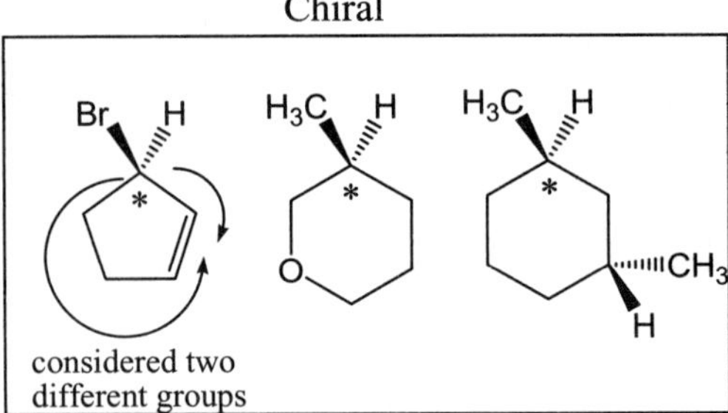

Extend Your Understanding Questions (to do in or out of class)

16. Confirm that each molecule in the "Not Chiral" box has an **internal mirror plane** and that each molecule in the "Chiral" box does not have an internal mirror plane. (If you cannot find the internal mirror planes in the Not Chiral box, read the first bullet in Common Points of Confusion.)

17. Construct an explanation for why each carbon indicated with a "1" is considered to have **two identical groups**, while each carbon with an * is a chiral center with **four different groups**.

Model 5: Molecules with Two Chiral Centers

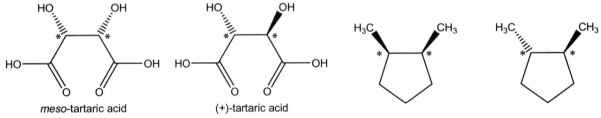

meso-tartaric acid

(+)-tartaric acid

Extend Your Understanding Questions (to do in or out of class)

18. Confirm that each * in Model 5 marks a chiral center with four different groups attached.

19. Two of the molecules in Model 5 are **not chiral even though they have chiral centers**! Identify the two achiral molecules and mark each one's internal mirror plane.

Memorization Task 8.3: Meso Compounds

meso compound = molecule with chiral centers that is **not chiral**

a meso compound always has two or more chiral centers and an **internal mirror plane**

20. Label each *meso* compound in Model 5, and explain your reasoning.

Model 6: Chiral Molecules with No Chiral Centers

Very rarely, we will encounter a **chiral molecule with no chiral centers**.

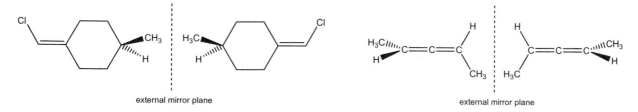

external mirror plane

external mirror plane

Extend Your Understanding Questions (to do in or out of class)

21. Confirm that there are no chiral centers in any of the molecules in Model 6.

22. Is either pair of molecules in Model 6 an identical pair? Explain your reasoning.

23. Do any of the molecules in Model 6 have an internal mirror plane? If so, describe where it is.

24. Which molecules in Model 6 are chiral? Explain your reasoning.

Confirm Your Understanding Questions (to do at home)

25. Draw skeletal structures of both configurational stereoisomers of lactic acid (shown in Model 2).

26. Shade or mark the spheres on the object at right so as to generate a chiral object.

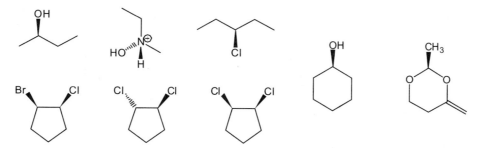

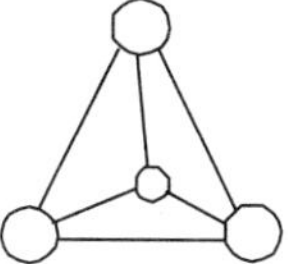

27. Consider the following structures, some of which are chiral.

   a. Label each chiral center with an *. (There are nine chiral centers.)
   b. Circle the **three** structures at the top of the page that are NOT chiral, and for each circled structure, indicate the internal plane of symmetry (mirror plane).
   c. One of the structures you circled is a *meso* compound. Label it *meso*.

28. Consider the structures below.

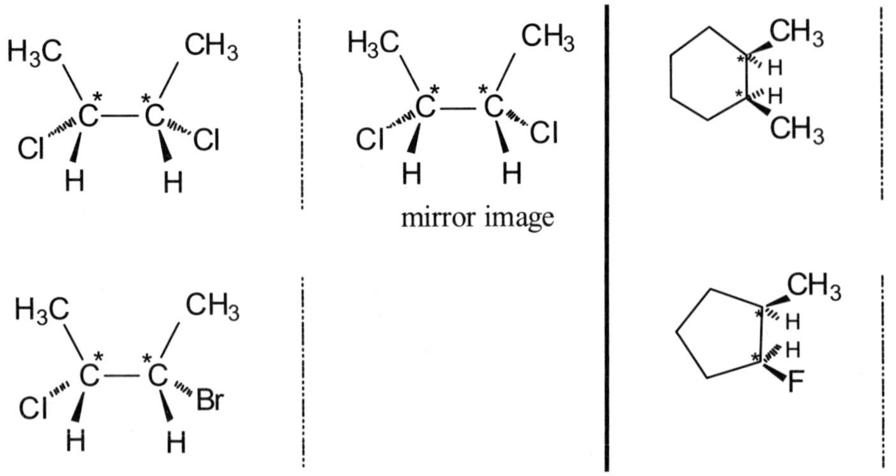

mirror image

a. Draw the mirror image of each molecule (the first one is done for you.)

b. Circle any structure that is not chiral, and use a dotted line to show the <u>internal</u> symmetry plane. Label any **meso compounds**.

Ask your instructor if you are responsible for the following information about chiral molecules.

Historical Notes on the Discovery of Chiral Molecules

In the early 1800's, French physicist Jean-Baptiste Biot discovered that shining a beam of polarized light through a solution of certain pure substances caused a rotation in the polarization plane of the light.

Such molecules were therefore called optically active.

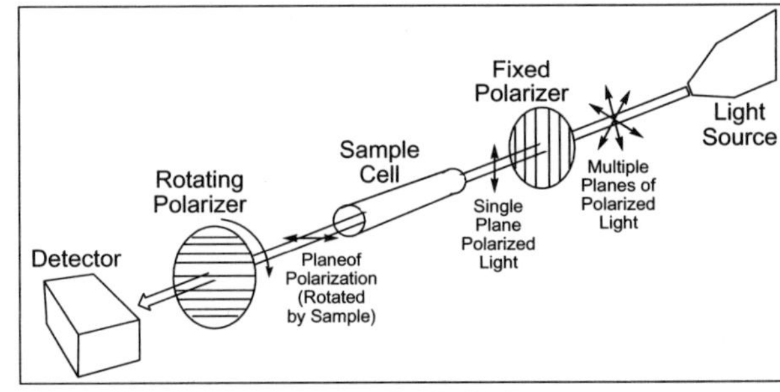

Schematic of a Polarimeter

Polarimetry studies show that nearly all biological molecules are optically active. In fact, many molecules have one biological function while their enantiomer has a totally different function.

It took nearly the whole 19th century for scientists to reach the following conclusions: A solution of...

- a given **<u>chiral</u>** molecule will rotate the plane of polarized light some unpredictable angle (α)
- its **<u>enantiomer</u>** will rotate the plane of polarization an equal angle in the **<u>opposite direction</u>** ($- \alpha$)
- an **<u>achiral</u>** molecule will NOT rotate the plane of polarization.
- a **<u>racemic</u>** mixture will NOT rotate the plane of polarization.

There is no reliable way to look at a structure and predict the size or direction of the rotation of the plane of polarization unless you have already measured alpha (α) for its enantiomer.

The Old System for Naming Chiral Molecules: (+) and (–)

Based on these early studies, enantiomers were first named based on whether the plane of polarization was rotated clockwise (+) or counterclockwise (–).

There is now a new system (which we will learn in the next activity) in which molecules are categorized as either **right-handed** or **left-handed** based on their structure. There is no correlation between the right/left system (called **R/S** for the Latin words for ***rectus*** and ***sinister***) and the old (+)/(–) system.

Physical Properties of Chiral Molecules

In a laboratory without a polarimeter, it is very difficult to tell enantiomers apart because they have the same basic physical properties such as melting point, boiling point, solubility, density, etc.; yet your nose can distinguish some enantiomers. For example, (+) and (–)-**carvone** are the flavors caraway and spearmint. (+)-Carvone fits a specific receptor in your nose the way only a right hand fits properly in a right glove. Many drugs, including ibuprofen, have biological activity while their enantiomer is inert or even toxic.

Confirm Your Understanding Questions (to do at home)

29. (+)-**tartaric acid** rotates **plane-polarized light** 12° clockwise ($\alpha_D = +12$). What is α_D for (–)-tartaric acid?

30. In 1848 French chemist Louis **Pasteur** noticed that a solution of a chemical derived from wine grapes, tartrate (the conjugate base of tartaric acid, which is also called racemic acid from the Latin word for grapes, *racemus*), did NOT rotate polarized light. By looking through a microscope Pasteur found that crystals made from this solution had two different shapes—and that the shapes were mirror images of each other. He painstakingly separated the two crystal types (shown at right) into piles and made a solution from each pile. The resulting two solutions exhibited optical activity of equal size, but opposite sign.

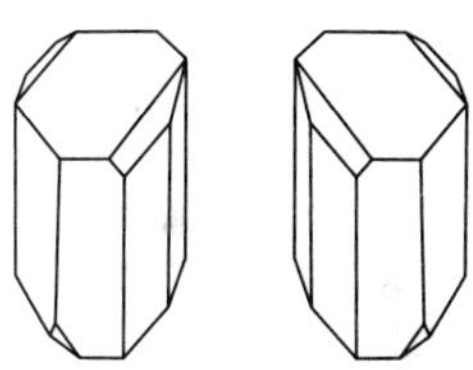

Tartrate Crystals

   a. Construct an explanation for why Pasteur's original tartrate solution was NOT optically active.

   b. Explain the origin and definition of the term **racemic mixture**?

   c. What term describes the relationship between the molecules that Pasteur painstakingly separated on the basis of crystal shape using tweezers and a microscope?

Read the assigned sections in your text, and do the assigned problems.

The Big Picture

Ordinary laboratory chemistry is not chiral. That is, when new a chiral center is generated via a synthetic reaction in a beaker, except in very special circumstances, the result is a racemic mixture of both enantiomers. It is very difficult and expensive for synthetic organic chemists to control the stereochemistry of a reaction and selectively generate one configurational stereoisomer over another. Nevertheless HUGE resources are devoted to this goal because, if the target of the synthesis is related to a biological system (e.g., a drug) the product will not work correctly if it does not have exactly the right stereochemistry at each chiral center.

The purpose of this activity is to introduce you to the complex chiral world around and inside of us. Almost all biological molecules, including interactions between drugs and living systems, are chiral interactions. That is, usually only one configurational stereoisomer of a drug, hormone, protein, or other biomolecule will have the desired effect. In some cases one enantiomer of a drug has therapeutic effects while its mirror image is toxic.

A fascinating and personally relevant example of this is the drug, thalidomide. One configurational stereoisomer is a highly effective anti-nausea drug and was taken in the 1950's by many pregnant women experiencing debilitating morning sickness. Unfortunately, its enantiomer causes terrible birth defects. Tens of thousands of children whose mothers took a racemic mixture of the drug during pregnancy were born with deformed or missing body parts. The public and scientific image of thalidomide was so damaged by this tragedy that the drug was banned worldwide, and further research was halted. As it turned out, this decision had a profound impact on my father's life, and the lives of many others who could have benefited from the safe therapeutic properties of the drug.

The birth defect tragedy did lead to strengthening of regulations and testing procedures for all drugs, especially in Europe where the drug was used on wide scale. In the United States, the FDA (Food and Drug Administration) had refused to license thalidomide, and this limited, but did not prevent tragedy here; the weak protections of the time allowed the maker of the drug to distribute millions of sample pills to doctors even though there was no FDA approval, resulting in many US babies born with defects.

Recent research, initially in violation of the ban, led to the discovery that thalidomide is a potent inhibitor of blood vessel growth. Years of research and testing overcame the stigma of the 1950s and thalidomide was finally approved by the FDA for use in the treatment of leprosy (1998), and multiple myeloma (2006). The drug has revolutionized treatment of both diseases. Patient education and frequent pregnancy tests for women taking the drug minimize the risk of further birth defects, but some health officials still argue that the ban should remain in effect due to the impossibility of perfect protection against pregnant women taking the drug. You may wonder, why not just produce the safe/therapeutic enantiomer. Besides the large cost associated with doing this, it turns out that humans have an enzyme that converts some of the therapeutic enantiomer into its dangerous mirror image.

My own father, Eric Straumanis, died of multiple myeloma in 1996, at the age of 57. About that same time, the first clinical trials of thalidomide as a treatment for the disease were taking place. Before thalidomide, the median survival of multiple myeloma patients was under two years, with a 100% mortality rate. Results from the thalidomide trials were so convincing that the "trials" were expanded to include all multiple myeloma patients, and so effective that many are still calling it a cure. Of course, I cannot help but think about the decades when thalidomide research was idled in response to the birth defects of the 1950s, and wonder what my father's fate would have been if this research had been accelerated by just one year.

Silverman, W. "The Schizophrenic Career of a "Monster Drug" (2002) Pediatrics 110 (2): 404–406.
World Health Organization website, http://www.who.int/lep/research/thalidomide/en/index.html, accessed 09/23/2010
"Thalidomide: controversial treatment for multiple myeloma" Health News, March 10, 2006.

Common Points of Confusion

- The last Construct Your Understanding Question (Q15) asks you to confirm that the molecule at right is not chiral using the internal mirror plane test. Many students do not, at first, recognize that this molecule has an internal mirror plan (shown with a dotted line). This is because the standard drawing (the first structure at right) does NOT accurately depict the molecule.

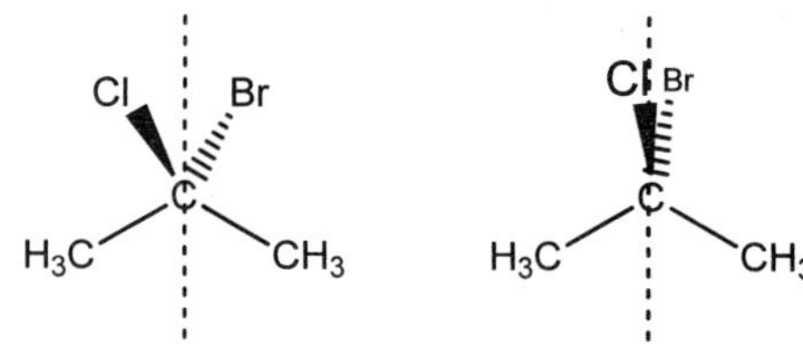

  The second structure (above, far right) is a little better because it emphasizes that when the three carbons are all in the plane of the paper the Cl and Br are lined up with each other. The internal mirror plane (shown with a dotted line) cuts right through the middle of the Cl (shown larger since it is closer) and the Br (shown smaller since it is farther away). Note that an even more accurate drawing would have the Cl exactly *on top of* the Br. Both drawings above show the molecule with some twist so you can see both halogens.

- There are two types of mirror plane tests used in this activity: the internal mirror plane test, and the external mirror plane test. Students sometimes confuse the two.
  - The internal mirror plane test is generally more useful. It involves looking for an internal mirror plane in a molecule to determine if it is chiral. <u>A molecule with an internal mirror plane is not chiral</u>.
  - The external mirror plane test can also be also be used to determine if a molecule is chiral. To do this, draw (or make a model of) the mirror image of a molecule then try to rotate or flip the new molecule so as to superimpose it on the original structure. If the mirror image is identical to the original then the molecule is not chiral.

Notes

ChemActivity 9: Absolute Configuration (R/S)

BUILD MODELS: <u>TWO PER GROUP</u> of trans 1,2-dimethylcyclopentane

(How do you name the two different kinds of *trans*-1,2-dimethylcyclopentane?)

Model 1: Sample of Molecules with Molecular Formula C_7H_{14}

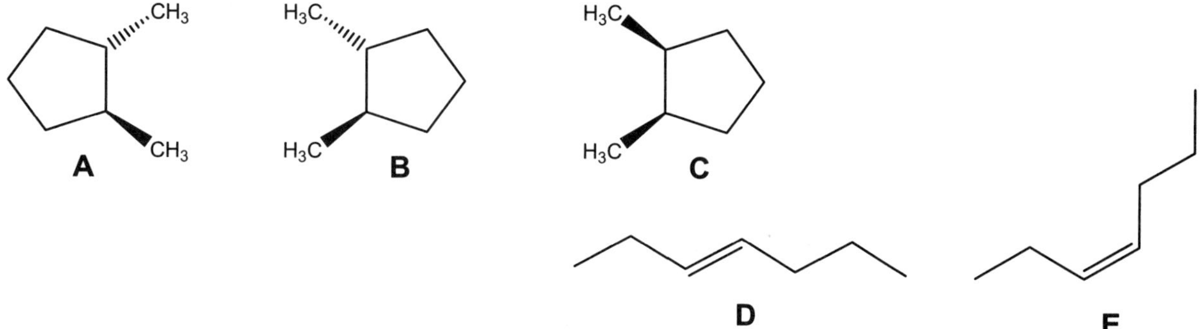

Construct Your Understanding Questions (to do in class)

1. Label A-E in Model 1 with the terms *cis, trans,* and/or *meso,* as appropriate. <u>Two terms may apply</u>. (Review) *meso* = molecule with chiral centers <u>and</u> an internal mirror plane such that it is not chiral.
2. **<u>Make separate models of A and B.</u>** Check to make sure the wedge and dash bonds on the paper match the methyl groups on your models by holding each completed model up to the paper.
3. Are A and B the same? **<u>Confirm your answer by testing if your models are superimposable.</u>** (That is, can they be placed on top of one another so that all their atoms match up?)

Memorization Task 9.1: Diastereomers = two objects that are configurational stereoisomers **<u>but not enantiomers</u>**	a relationship *(like "son" or "sister")*

4. What term best describes the relationship between each of the following? Choose from:

same/conformers **enantiomers** **diastereomers** **constitutional isomers** **different formula**

a. B & C c. D & E e. A & E

b. A & C d. A & D f. A & B

5. (Check your work) If you decided that A and B in Model 1 are the same/conformers, go back and build models of A and B and use the drawings to carefully check that your models are correct.

6. **T** <u>or</u> **F**: All chiral objects have exactly one enantiomer. If false, cite a molecule from among A-E that has more than one enantiomer.

7. **T** <u>or</u> **F**: All diastereomers are chiral. If false, cite a molecule from among A-E.

Memorization Task 9.2: Right-Handed and Left-Handed Chiral Centers

A and B (on the previous page) are different (they are enantiomers), so they must have different names.

By analogy to the pair of enantiomers you carry with you (your hands), every chiral center is designated as right handed or left handed. (We will soon learn how to determine which is which.)

The Latin words for left and right are used:

- **right**-handed chiral centers are called "**R**" for *rectus,* a Latin word for right, and
- **left**-handed chiral centers are called "**S**" for *sinister,* a Latin word for left.

Model 2: Physical Properties of Enantiomers and Diastereomers

Our right and left hands can be used to model R and S chiral centers, and, as we will see in Model 3, they can even be used to *determine* if a chiral center is R or S.

Demonstration: Making Hand Models

Turn to the person next to you and shake hands with them (right hand to right hand). Each of your **right** hands represents an **R chiral center**, (each **left** hand will be an **S chiral center**), therefore we will call shaking hands the normal way "**R, R**".

Construct Your Understanding Questions (to do in class)

8. Now shake with your left hands (left hand to left hand).
   a. In terms of R and S, how should we name this new "molecule" made with two left hands?
   b. What term describes the relationship between your R,R-handshake and your S,S-handshake?
   c. (Check your work) Make both the R,R "molecule" and the S,S "molecule" simultaneously, and identify the <u>external</u> mirror plane between the R,R-handshake and the S,S-handshake.
   d. Take note that your handshakes (R,R and S,S) *feel* the same. That is, the S,S-handshake is the same as the R,R-handshake in every way except that it is made from left hands instead of right hands. This will help you remember part of Memorization Task 9.3, that two <u>enantiomers have the same physical properties</u> (e.g. melting point, boiling point, etc.) as one another.

9. **Let go of your partner**. Now, one of you offer a **<u>left hand</u>** to be shaken by the other's **<u>right hand</u>**.
   a. What should we call this new "model" in terms of R and S?
   b. What term describes the relationship between this R,S-handshake and the R,R-handshake you made a moment ago?
   c. The R,S-handshake feels very different (like holding hands instead of shaking hands) from the R,R-handshake. How might this help you remember part of Memorization Task 9.3?

Memorization Task 9.3: Physical Properties of Enantiomers and Diastereomers

enantiomers (e.g. R and S; or R,R and S,S) have the <u>same</u> physical properties (e.g. mp, bp, etc.)

diastereomers (e.g., R,S and S,S) have <u>different</u> physical properties (mp, bp, etc.)

Model 3: Absolute Configuration (Naming Chiral Centers R or S)

The following section describes one way to designate a chiral center R or S (its **absolute configuration**).

There are a number of different systems for assigning R and S. Make sure you learn one and can use it consistently. The system described here, unlike some others, does not require you to redraw a molecule to determine R or S.

Step 1: Rank the four attached groups 1 (largest) to 4 (smallest)

Use the **Cahn-Ingold-Prelog rules** outlined in ChemActivity 5. Recall that priority is assigned based on atomic number with a "card game" system for breaking ties (*see example below)*.

Construct Your Understanding Questions

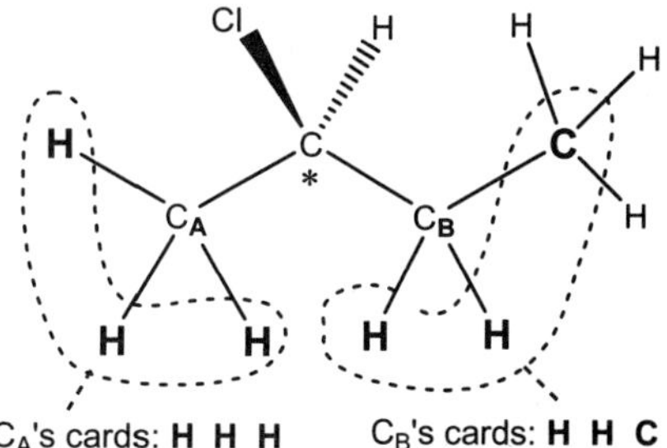

10. Which carbon (C_A or C_B) has a higher rank? Explain your reasoning.

11. Label the four atoms attached to the chiral carbon (*) one to four to show their ranking.

Step 2: Point your RIGHT thumb (hitchhiker style) toward the "4" group (usually an H)

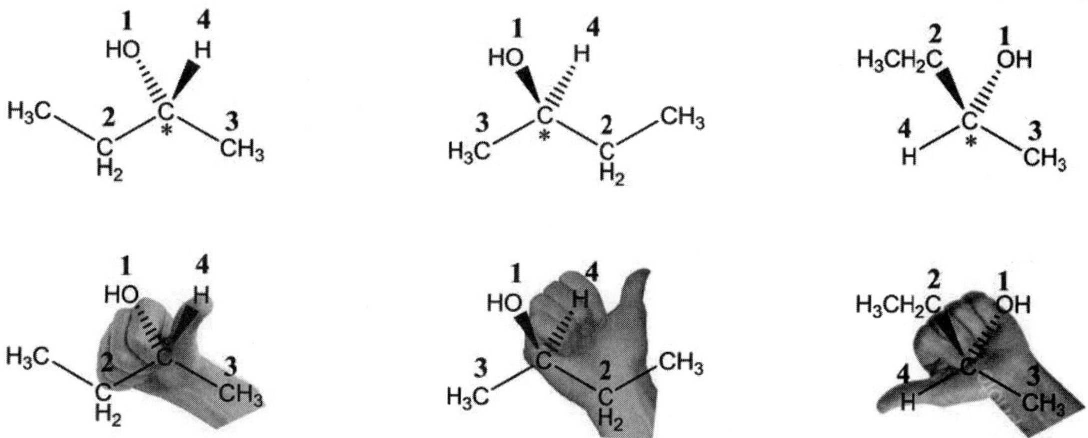

Step 3: Curl your wrist and imagine the tip of your finger touching group 1 then 2 then 3

If the curl of your **RIGHT** hand matches the progression from 1→2→3 then **chiral center is R**.

If curl does not match, **chiral center is** S. Confirm by touching 1→2→3 with your **LEFT** hand.

Warning! Don't let your thumb drift. Keep it pointing in the direction of group 4 (e.g., H) at all times.

Construct Your Understanding Questions (to do in class)

12. Make a model of (R)-2-chlorobutane. Use the hand method to confirm the R assignment. Now change your model so it is S, and use the hand method to confirm the absolute configuration as S.

13. Determine the absolute configuration of each chiral center on this page, and label each R or S.

14. Determine the absolute configuration of each chiral center in the following structures.

Recall that D (^2H) gets priority over H.

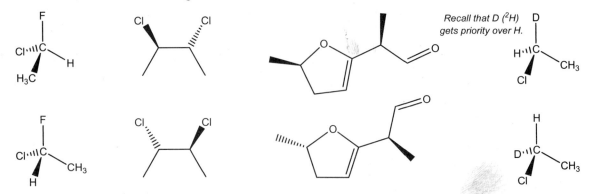

15. Check your work: The structures in the first row are the enantiomers of the structures in the second row. What happens to the absolute configuration of a chiral center if you switch two groups?

Model 4: Fischer Projections

In a **Fischer projection**, assume:

horizontal bonds = wedge bonds

vertical bonds = dash bonds

The carbon backbone must be drawn vertically. For standard sugars CHO should be at the top and CH_2OH should be at the bottom.

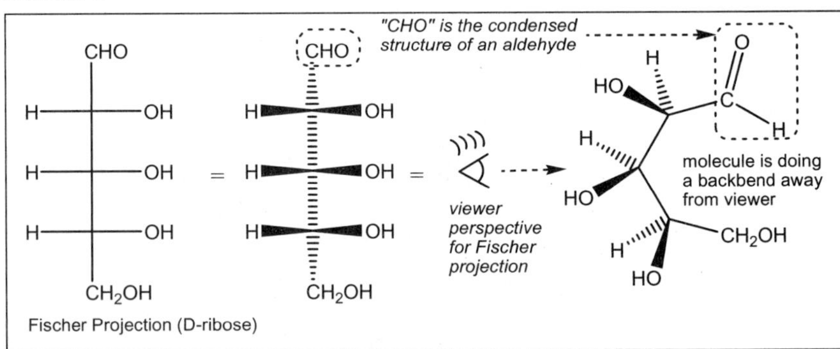

Fischer Projection (D-ribose)

The D/L naming system for sugars is not equivalent to the (+)/(–) or R/S naming systems. A sugar is D if the bottom chiral center on its Fischer projection has an OH to the right, and it is L if that last chiral OH is to the left.

Extend Your Understanding Questions (to do in or out of class)

16. Mark each chiral center on the Fischer projection of D-ribose (above) with an R or S.

17. Mark each chiral center on the Fischer projections at right with an R or S.

18. How many configurational stereoisomers exist of D-erythrose (including D-erythrose)?

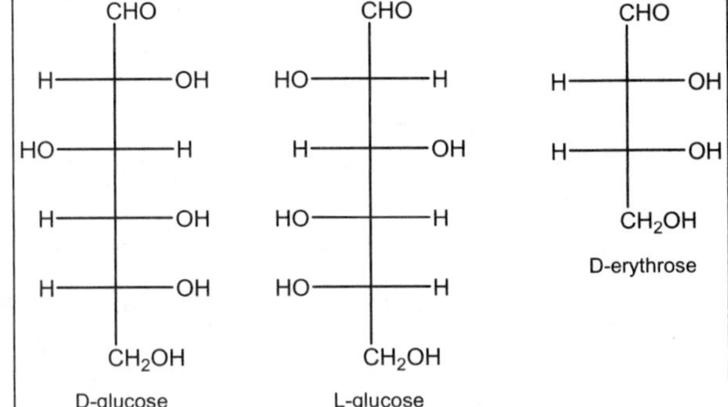

Memorization Task 9.4: The 2^n Rule

For many molecules (*watch out for exceptions!*)
[No. of configurational stereoisomers] = 2^n
where n = [No. of chiral centers] + [No. *E/Z* π bonds]

19. How many configurational stereoisomers exist of D-glucose?

20. How many configurational stereoisomers exist for (R,*E*)-4-chloro-2-pentene→ and what are their names?

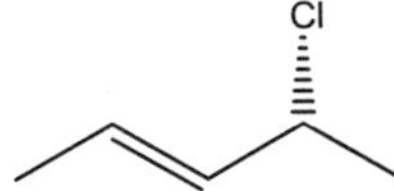

21. Another quick hand game: Give a partner a R,R-"high five," and freeze at the point where your palms are touching. (Assume your hands are identical. That is, ignore differences between your right hand and your partner's right hand, and between your left hand and your partner's left hand.)

  a. Does this R,R "model" have an internal mirror plane? If so, where?

  b. Now make a R,S-high five. Does *this* "model" have an internal mirror plane? Where?

  c. Which of these two molecules (R,R-high five, or R,S-high five) is a *meso* compound?

  d. Now simultaneously make S,R and R,S by giving your partner a "high ten," and confirm that S,R is the same as R,S (The two models can be super-imposed on one another—try it.).

22. A *meso* compound such as the one below can be assigned either R,S or S,R depending on whether you assign the chiral centers from left to right or right to left. So which name is correct? The answer is that both names are logically equivalent and describe the same compound. In fact, the following molecule has four acceptable names, all of them refer only to this one molecule). What are the four acceptable names for this molecule?

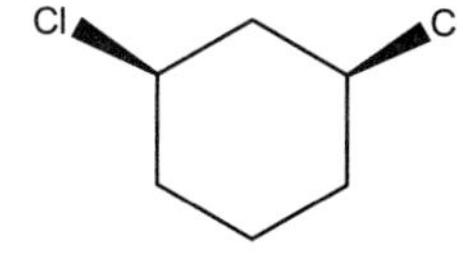

23. How many different configurational stereoisomers exist (including the one shown) for each of the following molecules? (For homework you will be asked to draw and name each one.)

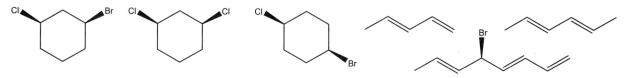

24. There are four possible configurational stereoisomers of 2,3-dichloro**pen**tane. Draw **and name** all four.

  a. There are only *three* possible configurational stereoisomers of 2,3-dichloro**but**ane. Draw and name all three.

  b. Explain why there are three configurational stereiosomers in the dichlorobutane set and four in the dichloropentane set.

25. The name *cis*-1,3-dichlorocyclohexane is explicit (refers to exactly one molecule), but the name *trans*-1,3-dichlorocyclohexane is not explicit. Explain.

Confirm Your Understanding Questions (to do at home)

26. Draw an example of a pair of configurational stereoisomers that are NOT enantiomers. What is the name for the relationship between such structures?

27. Draw three different configurational stereoisomers of 1,3-dimethylcyclopentane, and label one *meso*.

28. Each Fisher projection in Model 4 has a group whose condensed structure is "CHO." What does this group look like if you draw out all the bonds?

29. **Fisher projections** of D-galactose and the reduced form of D-galactose are shown below.

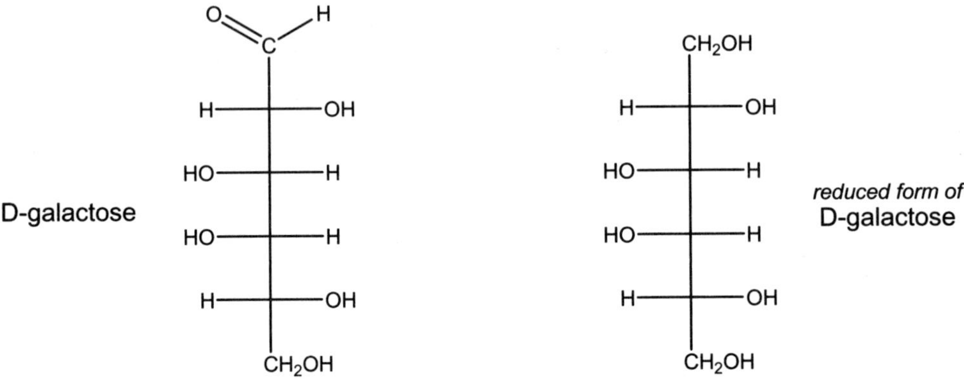

a. Speculate: why is it called the "reduced" form of D-galactose?

b. How many chiral centers are there in each molecule?

c. How many possible configurational stereoisomers of D-galactose (including D-galactose) are possible?

d. The reduced form of D-galactose has fewer possible configurational stereoisomers. Construct an explanation for why.

e. Specify the stereochemistry of all possible D-galactose configurational stereoisomers (e.g., D-galactose is R,S,S,R).

f. There are only 10 unique configurational stereoisomers of the reduced form of D-galactose. Specify the stereochemistry of each of these. (Careful! This question is complicated.)

30. Sometimes (as in the example at right) the first round of the card game results in a tie. When this happens you must play another round using the next atoms in the chain.

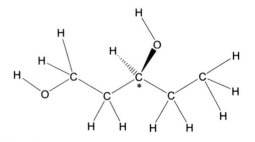

a. Confirm that both C_2 and C_4 hold the same cards, CHH, and therefore tie.

b. Play the card game with the next atoms in the chain to break the tie between C_2 and C_4.

c. Label the four atoms attached to the chiral carbon one to four to show their ranking, and assign R or S.

31. Sometimes a molecule with one or more chiral centers is drawn without giving enough informa-tion to determine if the chiral center(s) are R or S. Mark each chiral center below with an *.

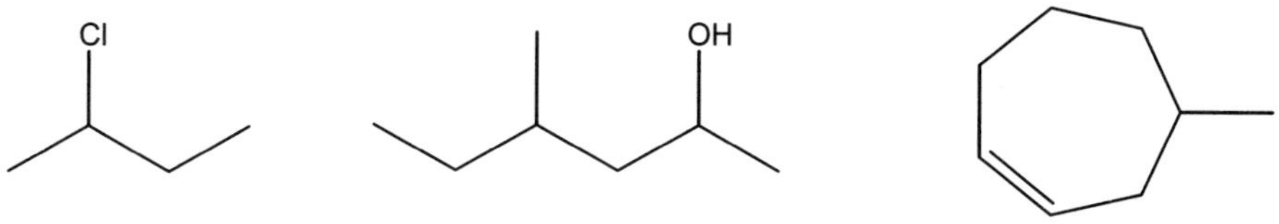

Cl OH

By convention, if enough information is not given to assign R or S, we are to assume that the structure represents an even mixture of all stereochemical possibilities. For each structure above, draw all configurational stereoisomers implied by the drawing.

32. Label every chiral center on the following molecules with an R or S and state whether the molecule as a whole is chiral.

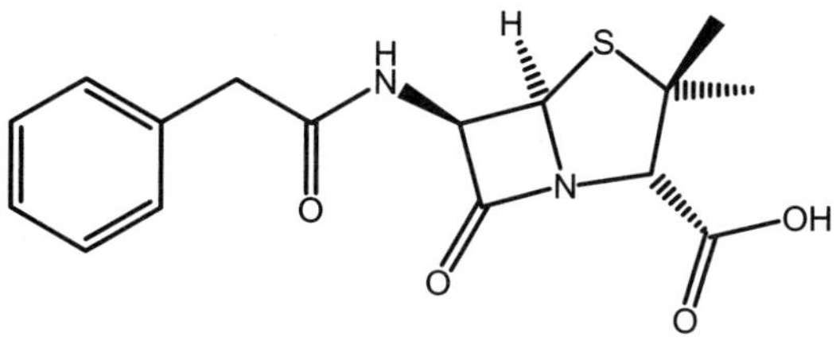

H N H S N O O OH O OH O O

Penicillin G

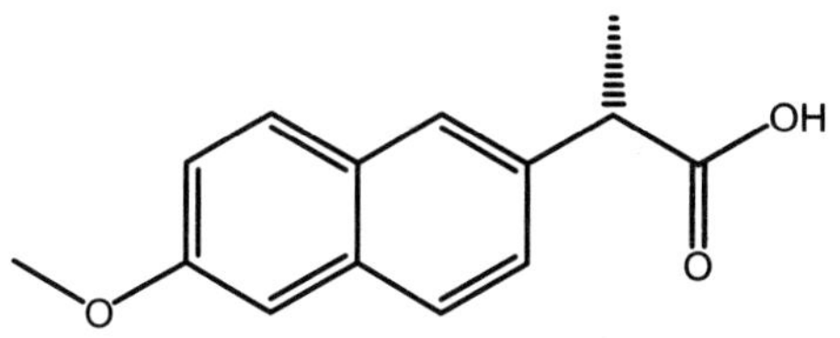

Aleve (Naproxen)

33. How many different configurational stereoisomers of each molecule in the previous question are there (including the one shown)?

34. Draw a flow chart connecting the following terms: isomers, stereoisomers, constitutional isomers, configurational stereoisomers, conformers (conformational stereoisomers), diastereomers, enantiomers, *E/Z* isomers, R/S isomers, *cis/trans* isomers. Label the connection arrows between the terms with questions (e.g. "Do the two structures have the same atom connectivity?") and answers (e.g. "Yes" or "No") to make your flow chart into a decision tree.

Read the assigned sections in your text, and do the assigned problems.

The Big Picture

The immediate purpose of this activity is to refine your skill at seeing molecules in three dimensions. Accurately assigning R or S to a drawing *requires* that you see the drawing in three dimensions. Seeing drawings in 3D and assigning R and S is a critical skill that takes a bit of practice. It is very frustrating for some students at first, but stick with it! Use the following three tools in conjunction until you can assign R/S consistently: 1) a drawing, 2) a model, 3) a friend or TA who can confidently assign R/S.

In terms of chemistry, you may have already learned reactions that affect the stereochemistry of a molecule, and soon you will be learning more.

Common Points of Confusion

- Another reminder is to watch out for the term "**stereoisomer**" used by itself. In conversation, and even in some textbooks this term is used when "**configurational stereoisomer**" is appropriate. (Recall that configurational stereoisomers are the *E/Z, cis/trans*, **R/S, D/L,** +/- enantiomers, diastereomers, etc. that we have been working with in this ChemActivity.) The strictly correct use of the term "stereoisomer" includes all stereoisomers. That is, both configurational stereoisomers and conformational stereoisomers (conformers).

- The definitions in this activity cause some confusion. Enantiomer is just another word for mirror image, but ONLY when you are talking about a chiral molecule. Students have more trouble with **diastereomer**. It may help to think of diastereomer as the "catch all" configurational stereoisomer term. If two molecules have the same connectivity, but do not fall into any other category (conformer/same or enantiomer), then they are diastereomers.

- Students tend to get caught up in *chiral* configurational stereoisomers, and forget that a pair of *cis/trans* or *E/Z* configurational stereoisomers are diastereomers of each other **even if they are not chiral!**

- All students struggle with assigning R/S. Students who are persistent in working with both a model and pencil/paper eventually can assign R and S with a high degree of accuracy. There are three main types of difficulties: **1)** Errors assigning priorities [*Remedy*: Review priority rules in ChemActivity 5.], **2)** difficulty seeing the chiral center in 3D [*Remedy*: Work with a model and a drawing at the same time.], and **3)** Mixing up different numbering systems or methods of assigning R and S [*Remedy*: Invest in one method. They all work, but learning more than one causes confusion.].

- When counting isomers it is tempting to use the **2^n rule** without thinking. You must be mindful of symmetrical molecules with the possibility for *meso* compounds among the configurational stereoisomers.

Notes

ChemActivity 10: One-Step Substitution (S_N2)

(What are the characteristics of a favorable S_N2 reaction?)

Model 1: Examples of One-Step Nucleophilic Substitution Reactions

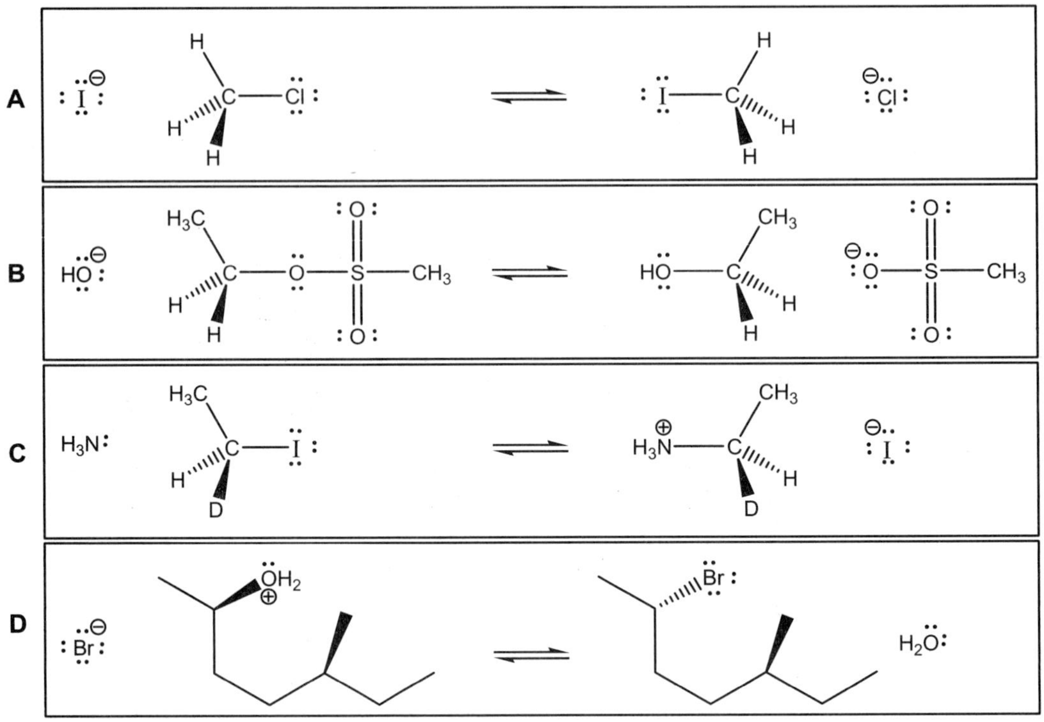

A	$:\ddot{I}:^{\ominus}$ + H_3C–Cl (CH₂-wedge drawing: H, H, H on C bonded to :Cl:)	⇌	:I–C(H)(H)(H) + $:\ddot{Cl}:^{\ominus}$
B	$H\ddot{O}:^{\ominus}$ + CH_3CH_2–O–SO_2–CH_3	⇌	HO–CH(H)(H)CH_3 + $^{\ominus}$:O–SO_2–CH_3
C	$H_3N:$ + CH_3CH(D)–I	⇌	$H_3\overset{\oplus}{N}$–CH(D)CH_3 + $:\ddot{I}:^{\ominus}$
D	$:\ddot{Br}:^{\ominus}$ + alkyl–$\overset{\oplus}{O}H_2$	⇌	alkyl–$\ddot{Br}:$ + $H_2\ddot{O}:$

Construct Your Understanding Questions (to do in class)

1. In each reaction in Model 1, an "incoming group" displaces (*substitutes for*) a **leaving group**.
   a. (E)Circle each "incoming group" among the reactants on the left side of Model 1.
   b. (E)Put a box around each leaving group among the reactants on the left side of Model 1.
   c. (E)Can leaving groups and incoming groups consist of more than one atom?
   d. (E)**True or False**: an incoming group must be negatively charged.
   e. **True or False**: a neutral leaving group will be negatively charged after it leaves, and a positively charged leaving group will be neutral after it leaves.
2. Add a δ+ to the most electrophilic (electron loving /wanting/deficient) carbon among each pair of reactants in Model 1. Label this molecule containing the **electrophilic carbon** the "**electrophile**." (Hint: consider bond polarity and polarizability.)
3. Use curved arrows to illustrate a one-step mechanism that will accomplish each substitution reaction in Model 1. (*Hint*: You will need to draw two curved arrows for each reaction.)
4. Label the reactant acting as a **nucleophile**. (Check your work) Is your answer is consistent with the fact the name **nucleophile** means "nucleus loving" or "+ charge loving?" Explain.

Model 2: Transition State

The **transition state (t.s.)** of a reaction is the highest potential energy species between the reactant and the product. The diagram below shows two possible transition states for Reaction A in Model 1.

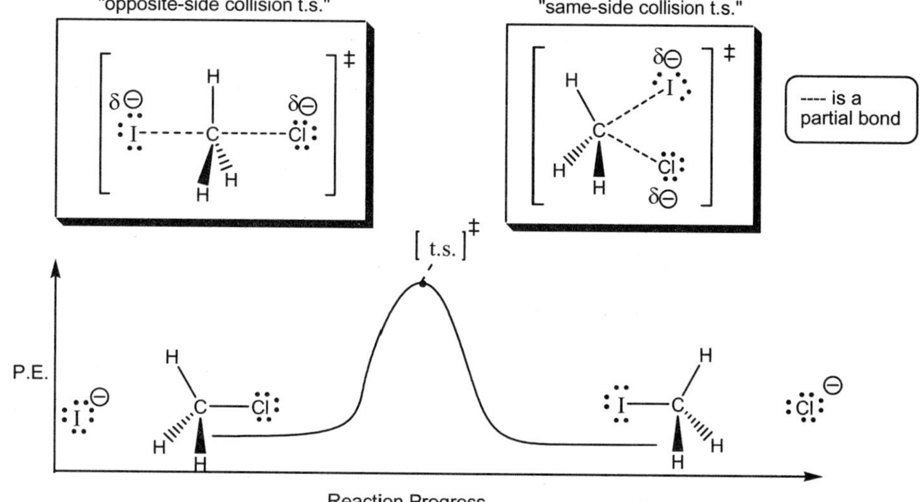

Construct Your Understanding Questions (to do in class)

5. Assign an R or S to each chiral center (reactants and products) among Reactions A-D in Model 1.
6. Explain why the data in Model 1 (specifically, the R/S assignments you made for Reactions C and D) support the opposite-side-collision transition state hypothesis.
7. The reactions in Model 1 are commonly called **S_N2 reactions**. Does the "2" in S_N2 refer to the number of steps in the reaction? Explain your reasoning.
8. Each reaction in Model 1 takes place in a single step so this one step must be the rate determining step (also called the slow step). How many separate molecules are involved in the slow step of an S_N2 reaction (see examples in Model 1)?

Memorization Task 10.1: One-step Nucleophilic Substitution Reaction = "S_N2 Reaction"

- **S** stands for **S**ubstitution
- **N** stands for **N**ucleophilic
- **2** stands for bimolecular—**2** molecules (nucleophile & electrophile) involved in the rate-determining step (in this case, the only step)

9. Soon we will learn about another type of nucleophilic substitution reaction called an S_N1. By analogy to S_N2, what can you infer about an S_N1 reaction based on its name? (*See* Mem. Task 10.1)

Model 3: Review of Acid-Base Reactions and pK_a

For an acid (H—Z) the pK_a value is a measure of the energy difference between the conjugate acid (H—Z) and the conjugate base ($Z^{\ominus}$) in water.

This means pK_a can be used as an estimate of the…

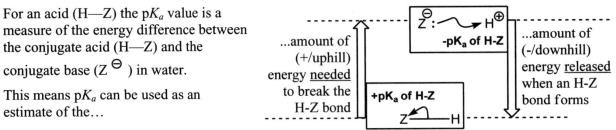

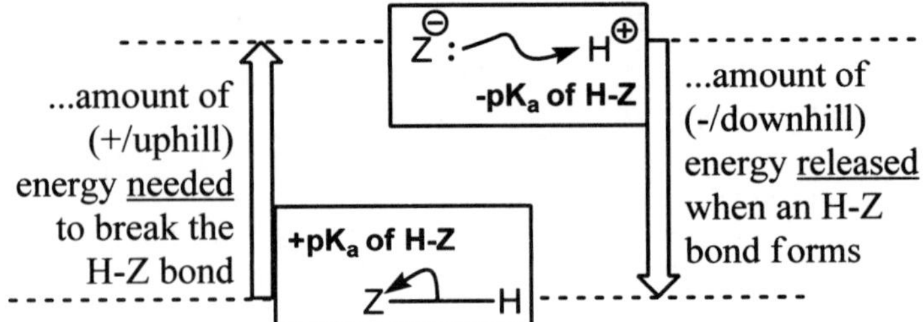

Memorization Task 10.2: Energy Differences between Conjugate Acid-Base Pairs

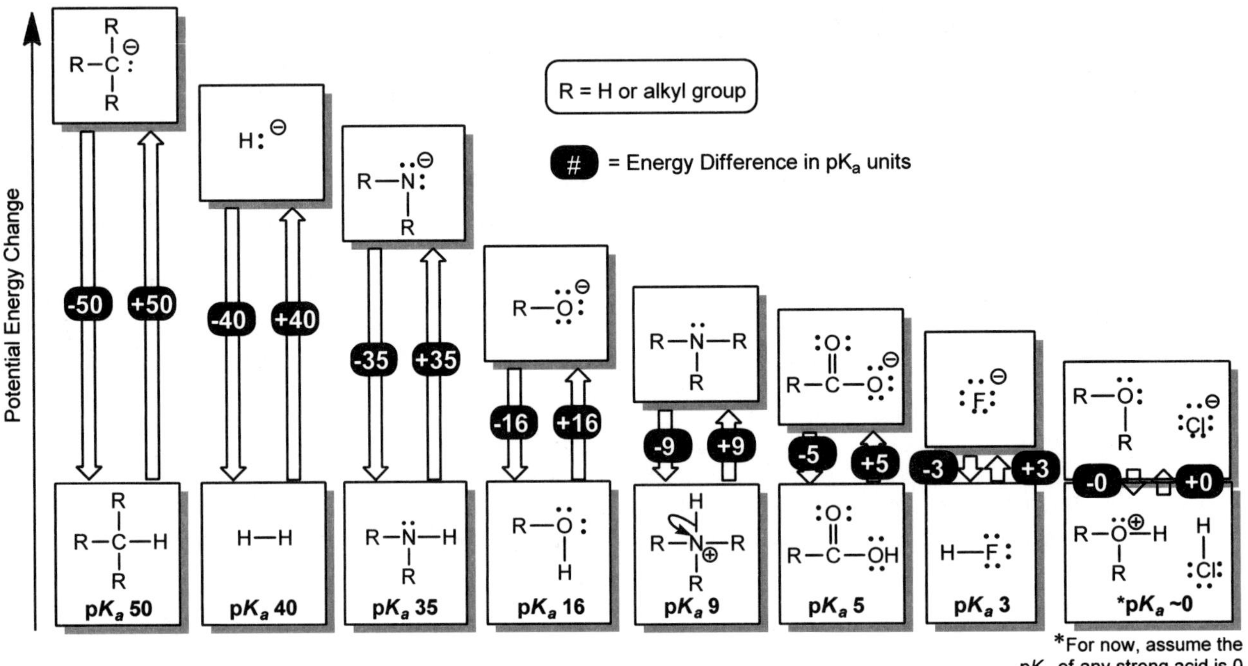

Construct Your Understanding Questions (to do in class)

10. For each column in Table 10.1, identify the **conjugate acid** and the **conjugate base**.

11. How much energy is released when the lone pair on $H_2N^{\ominus}$ makes a bond to an $H^{\oplus}$ to become H_3N? (Give the value and sign of the energy change in pK_a units.)

12. A student answers "–9" to the previous question. What error did this student make?

13. Find the *one* curved arrow on Table 10.1. What is the energy change associated with this arrow?

14. According to Model 3, how much energy are you to assume it takes to break an HCl, HBr, or HI bond? Explain how this is consistent with the fact that HCl, HBr, and HI are strong acids?

15. For the following reaction, explain how the value of $\Delta H_{rxn} = -13$ pK_a units was calculated.

-16 +3

$H_3C—\ddot{O}:^{\ominus}$ + $H—\ddot{F}:$ → (ΔH_{rxn} = -13 pK_a units) $H_3C—\ddot{O}—H$ + $:\ddot{F}:^{\ominus}$

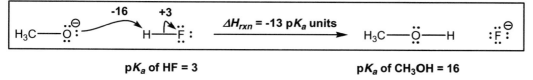

pK_a of HF = 3 pK_a of CH_3OH = 16

Model 4: Using pK_a of H—Z to Make Predicitons About C—Z

pK_a values give the energy required for polar breakage of an H—Z bond in water. However, similarities between C and H allow us to use the pK_a of the acid H—Z to make predictions about a C—Z bond.

For example, the pK_a's of HI (*assume* ≈ *0*) and H_4N^+ can be used to estimate the relative amounts of…

- **(+) energy required to break a C—I bond** in the following S_N2 reaction
- **(–) energy released when a H_3N forms a new bond to C** in the following S_N2 reaction

$H_3N:$ + H_3C, H, D–C–I: ⇌ $H_3\overset{\oplus}{N}$–C(CH_3, H, D) + $:\ddot{I}:^{\ominus}$

Construct Your Understanding Questions (to do in class)

16. Based on the "C is like H" assumption in Model 4, label each curved arrow in the S_N2 reaction above, with a sign (+ or –) and an appropriate pK_a value to give an estimate of the energy change (in pK_a units) associated with each curved arrow.

17. (Check your work) Is your answer to the previous question consistent with the following estimate of the total change in heat for the reaction in Model 4? $\Delta H_{rxn} \approx -9$ pK_a units (assume pK_a of HI = 0)

18. Based on the "C is like H" assumption in Model 4, which is a better leaving group in an S_N2 reaction: **hydroxide** or **water**? Circle one and explain your reasoning.

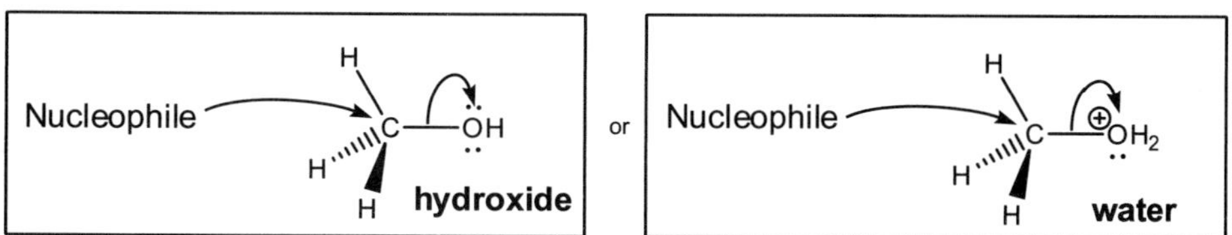

19. (Check your work) The pK_a value associated with hydroxide leaving is 16—**add this along with the correct sign (+ or –)** to the appropriate curved arrow above. What pK_a value is associated with breaking the bond to allow water to leave? Label the appropriate curved arrow with this value.

20. Based on what you know so far, what type of groups (give characteristics and examples) are expected to be good leaving groups in an S_N2 reaction? Explain your reasoning.

Extend Your Understanding Questions (to do in or out of class)

21. (Check your work) Is your answer to the previous question consistent with Table 10.1, below?

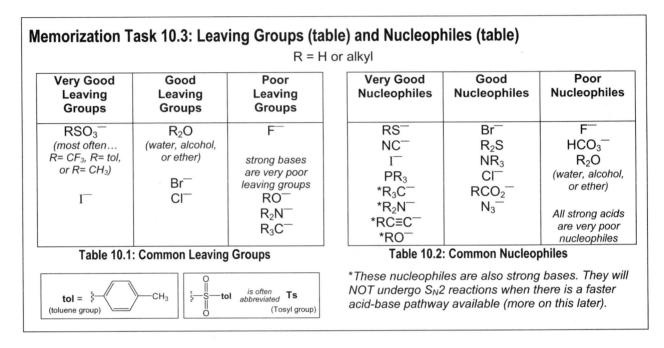

Memorization Task 10.3: Leaving Groups (table) and Nucleophiles (table)

R = H or alkyl

Very Good Leaving Groups	Good Leaving Groups	Poor Leaving Groups
RSO_3^- *(most often… R= CF_3, R= tol, or R= CH_3)* I^-	R_2O *(water, alcohol, or ether)* Br^- Cl^-	F^- *strong bases are very poor leaving groups* RO^- R_2N^- R_3C^-

Table 10.1: Common Leaving Groups

Very Good Nucleophiles	Good Nucleophiles	Poor Nucleophiles
RS^- NC^- I^- PR_3 $*R_3C^-$ $*R_2N^-$ $*RC{\equiv}C^-$ $*RO^-$	Br^- R_2S NR_3 Cl^- RCO_2^- N_3^-	F^- HCO_3^- R_2O *(water, alcohol, or ether)* *All strong acids are very poor nucleophiles*

Table 10.2: Common Nucleophiles

tol = (toluene group) —C_6H_4—CH_3

—$S(=O)_2$—tol *is often abbreviated* Ts (Tosyl group)

These nucleophiles are also strong bases. They will NOT undergo S_N2 reactions when there is a faster acid-base pathway available (more on this later).

22. Answer the following questions about a very good leaving group, LG $^{\ominus}$

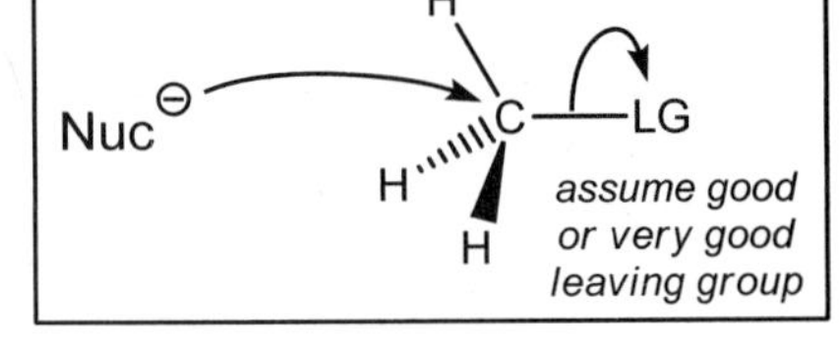

   a. It is **hard** or **easy** [circle one] to break the C—LG bond.

   b. LG $^{\ominus}$ is the conjugate base of a **weak** or **strong** [circle one] acid.

   c. LG $^{\ominus}$ is a very **strong** or **weak** [circle one] base.

   d. The pK_a of the conjugate acid (H—LG) is **high** or **low** [circle one].

23. <u>Without</u> looking at the list at the top of this page, answer the following questions about a very good nuclophile (Nuc $^{\ominus}$).

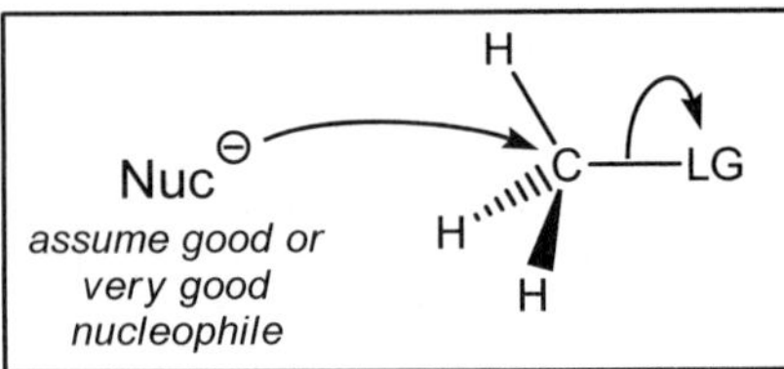

   a. The S_N2 reaction will be downhill if the new Nuc—C bond is **strong** or **weak** [circle one].

   b. A **large** or **small** [circle one] amount of energy will be released when the Nuc—C bond is formed.

   c. Nuc $^{\ominus}$ will be a **strong** or **weak** [circle one] base.

   d. The pK_a of the conjugate acid (Nuc—H) will be **high** or **low** [circle one].

24. (Check your work) For each part of Question 22 the second answer is correct. Logic would therefore dictate that the *first* answer in each part of Question 23 should be correct. That is, the trends for nucleophiles should be the opposite of the trends for leaving groups.

   a. However, it turns out that there are two types of good nucleophiles. One type (the **strong base nucleophile**) is profiled by circling the first answer in each part of Question 23. Go back to Question 23 and make a note in the margin that says "Characteristics of a strong base nucleophile."

   b. The other type (the **soft base nucleophile**) is described in Memorization Task 10.4. Go back to Table 10.2 and categorize each very good nucleophile as either a **strong base nucleophile** or a **soft base nucleophile**. [Note: NC^- does not fit in either category.]

Memorization Task 10.4: Soft bases are unexpectedly good nucleophiles

Soft bases, in particular thiols (SR_2) thiolates (RS^-), phosphines (PR_3), and the larger halide ions (I^-, Br^- and Cl^-) have big "fluffy" polarizable orbitals that can stretch out and begin to make a bond to the electrophilic carbon from far away. This leads to low P.E. transition states and faster S_N2 reactions. In general, nucleophilicity increases as you go down the periodic table.

   c. What ion has the unusual (and seemingly contradictory) distinction of being BOTH a very good nucleophile and a very good leaving group?

   d. Cite a pair of nucleophiles from Table 10.2 that illustrate the **periodic trend** described in Memorization Task 10.4

Synthetic Transformation 10.1: Halogenation of an Alcohol using HX (X = Cl, Br, or I)

$$R-CH_2-OH \xrightarrow{H-X} X^{\ominus} \quad R-CH_2-\overset{\oplus}{O}H_2 \longrightarrow R-CH_2-X \quad H_2O$$

Synthetic Transformations 10.2a-c: Examples of Synthetic Uses of S_N2 Reactions

Any good leaving group from Table 10.1 can be replaced with a good nucleophile from Table 10.2.

a $$R-CH_2-LG \xrightarrow{\text{any nucleophile in Table I.2}} R-CH_2-Nuc$$
LG = Cl, Br, I or OTs

b $$R-CH_2-LG \xrightarrow{\text{NaOH or KOH}} R-CH_2-OH$$
LG = Cl, Br, I or OTs

c $$R-C{\equiv}C-H \xrightarrow[\text{strong base)}]{NaNH_2} \xrightarrow[\text{LG = Cl, Br, I or OTs}]{R-CH_2-LG} R-C{\equiv}C-CH_2-R$$

Confirm Your Understanding Questions (to do at home)

25. Determine the absolute configuration of each chiral molecule below.

a. Explain why the reaction above would NOT produce the organic product shown in a one-step nucleophilic substitution reaction.

b. Draw the *correct* product of an S_N2 reaction for the reactants above.

26. The product shown above, (R)-2-butanol, can be made via reaction of (S)-2-iodobutane and NaOH.

a. Use curved arrows to show this reaction.

b. Draw an energy diagram for this reaction. Include on your energy diagram wedge-and-dash drawings of the reactants, transition state, and products. In your transition-state drawing, use a dotted line to indicate a partial bond and δ– to indicate partial negative charge on an atom.

c. Write the rate expression for this reaction.

27. One-step nucleophilic substitution reactions at a chiral carbon are characterized by the inversion of the absolute configuration of this carbon. Your model set is not very useful for simulating this **chiral inversion**. Explain how the model set falls short and how an umbrella in the wind can sort of simulate this inversion.

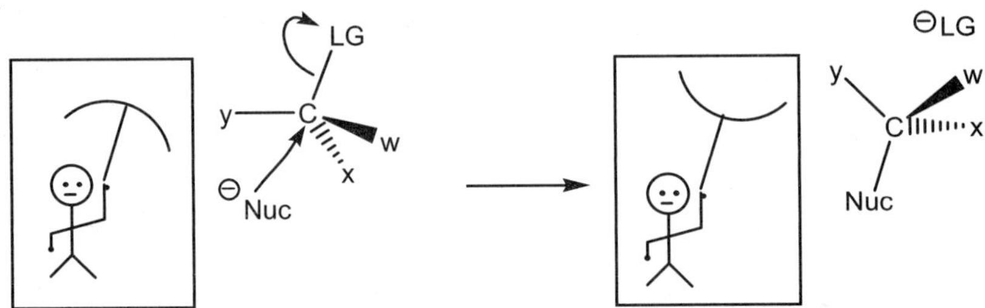

28. Show the mechanism and neutral products of an S_N2 reaction between water and methyl iodide.

29. Use curved arrows to show the mechanisms of the following two acid-base reactions.

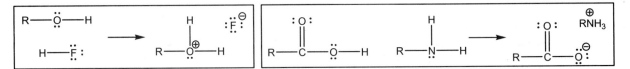

a. Above each curved arrow write a positive or negative number that gives the energy change associated with that arrow. Use the pK_a values in Table 10.1 and assume that the pK_a of a strong acid is approximately zero—this should make sense since it takes almost no energy (~0 pK_a units) to remove an H from a strong acid such as H_3O^+.

b. Calculate ΔH_{rxn} in pK_a units, and write this value above each reaction arrow.

30. Add curved arrows to show the mechanism of both **mechanistic steps** in Synth. Transf. 10.1.

a. What is the pK_a of the conjugate acid of hydroxide (HO^-)?

b. Explain why the following reaction does NOT work.

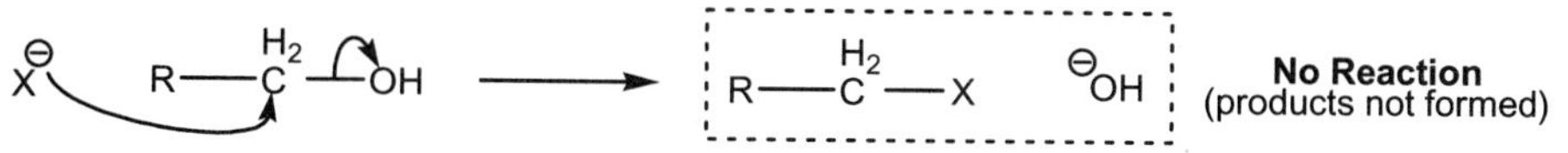

$X^{\ominus}$ R—CH_2—OH ⟶ R—CH_2—X $^{\ominus}$OH **No Reaction** (products not formed)

c. An acid can transform the OH of an alcohol (R—OH) from a terrible leaving group into an excellent one. What is this excellent leaving group, and what is the pK_a of its conjugate acid?

31. Add curved arrows to each of the following S_N2 reactions. Label each arrow with an energy change, and calculate an estimated ΔH_{rxn} in pK_a units (based on H—Z pK_a values).

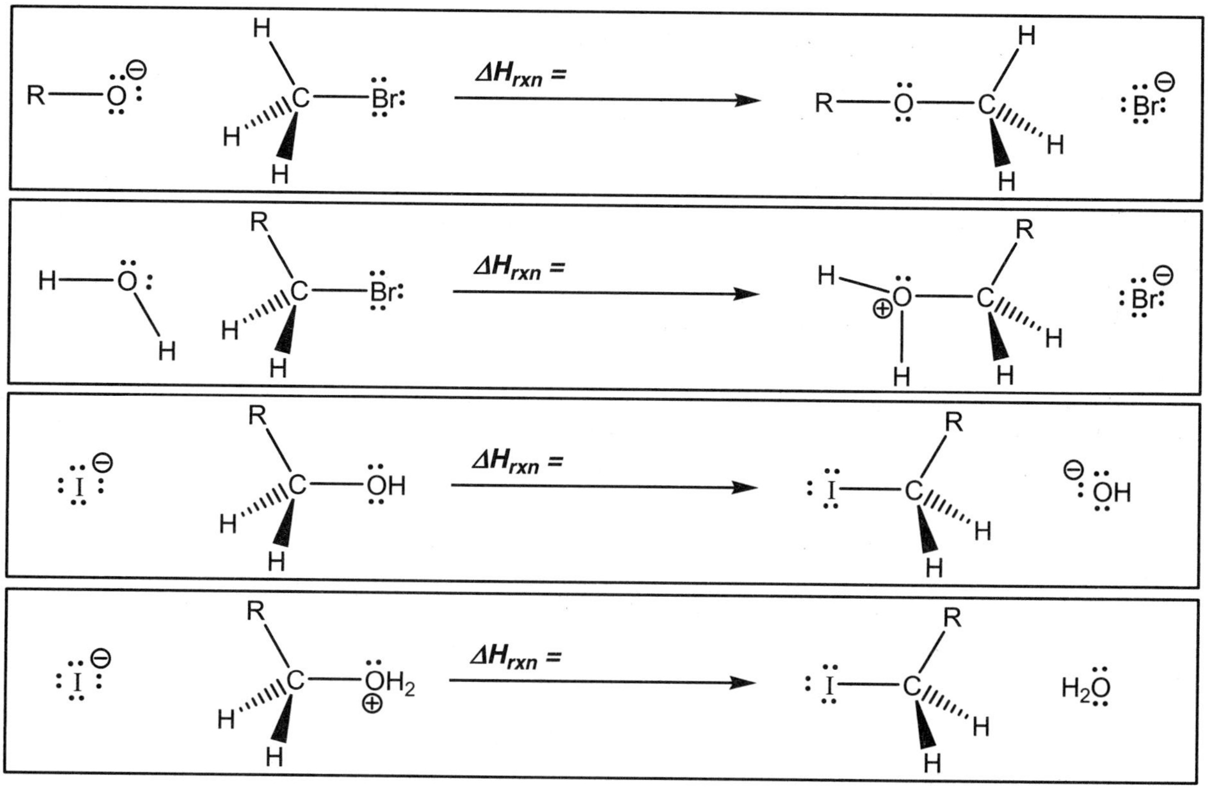

R—$O^{\ominus}$ + CH_3—Br	ΔH_{rxn} = ⟶	R—O—CH_3 + $Br^{\ominus}$
H—O—H + RCH_2—Br	ΔH_{rxn} = ⟶	H—$O^{\oplus}$(H)—CH_2R + $Br^{\ominus}$
$I^{\ominus}$ + RCH_2—OH	ΔH_{rxn} = ⟶	I—CH_2R + $^{\ominus}$OH
$I^{\ominus}$ + RCH_2—$OH_2^{\oplus}$	ΔH_{rxn} = ⟶	I—CH_2R + H_2O

32. Write "No Reaction" next to the one reaction in the previous question that is very unlikely to produce the products shown.

33. For each pair, circle the better leaving group (assume breakage of the bond highlighted in **bold**).

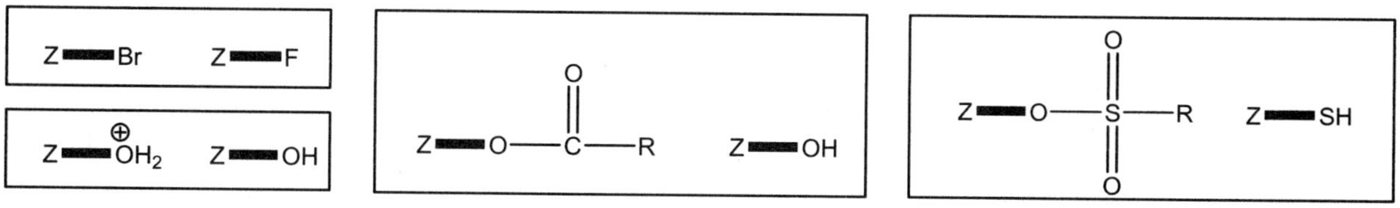

Z—Br	Z—F
Z—$OH_2^{\oplus}$	Z—OH
Z—O—C(=O)—R	Z—OH
Z—O—S(=O)(=O)—R	Z—SH

34. For each pair, circle the better nucleophile and make either a pK_a argument (citing relevant pK_a data) or a soft base argument (citing a periodic trend).

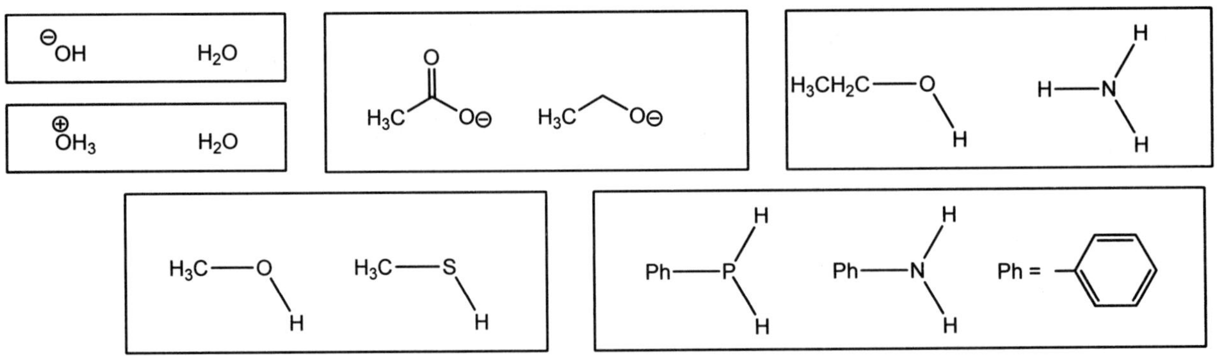

35. Draw the full structure of each reactant and reagent; then draw the products of each reaction.

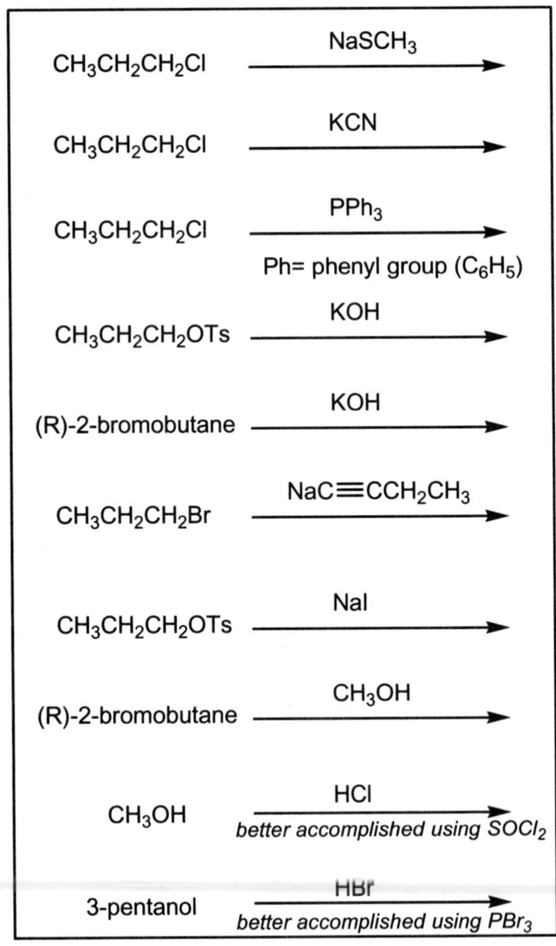

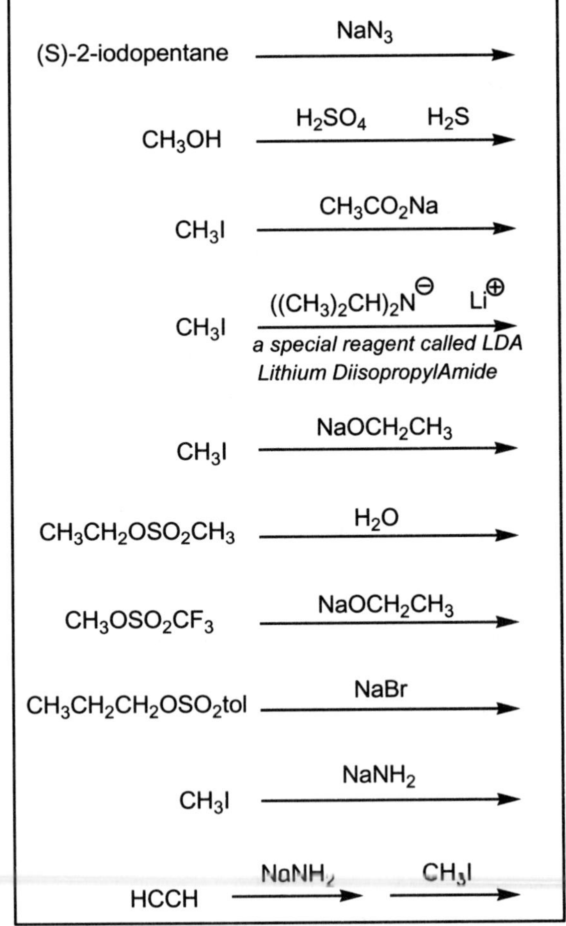

34. For each pair, circle the better nucleophile and make either a pK_a argument (citing relevant pK_a data) or a soft base argument (citing a periodic trend).

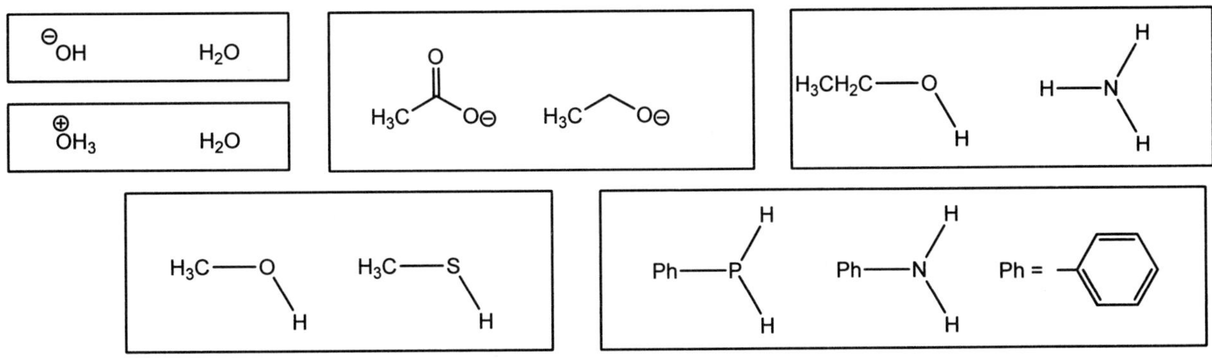

35. Draw the full structure of each reactant and reagent; then draw the products of each reaction.

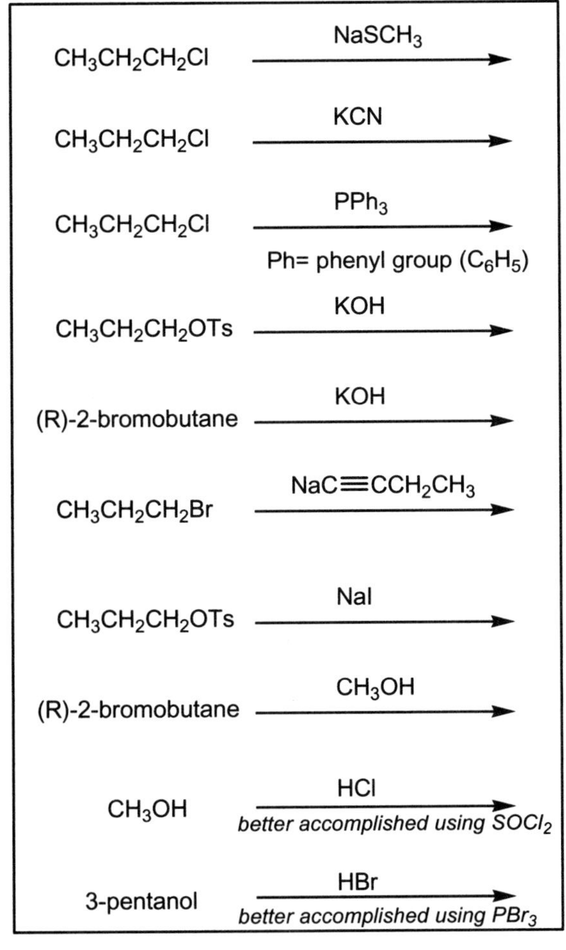

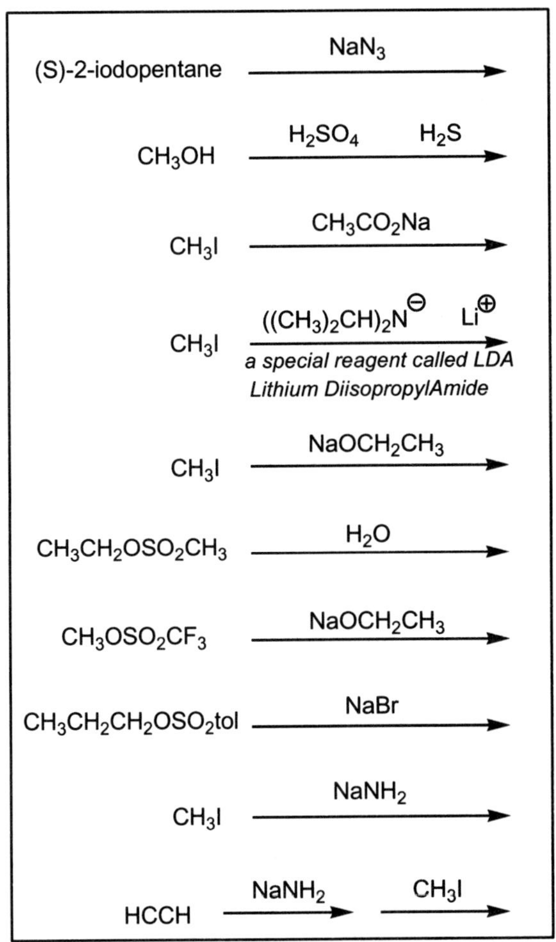

30. Add curved arrows to show the mechanism of both **mechanistic steps** in Synth. Transf. 10.1.

    a. What is the pK_a of the conjugate acid of hydroxide (HO^-)?

    b. Explain why the following reaction does NOT work.

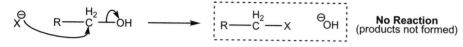

$X^{\ominus}$ + R—CH_2—OH ⟶ [R—CH_2—X + $^{\ominus}$OH] **No Reaction** (products not formed)

    c. An acid can transform the OH of an alcohol (R—OH) from a terrible leaving group into an excellent one. What is this excellent leaving group, and what is the pK_a of its conjugate acid?

31. Add curved arrows to each of the following S_N2 reactions. Label each arrow with an energy change, and calculate an estimated ΔH_{rxn} in pK_a units (based on H—Z pK_a values).

R—$O^{\ominus}$ + H_3C—Br ⟶ (ΔH_{rxn} =) R—O—CH_3 + $Br^{\ominus}$

H—O—H + RCH_2—Br ⟶ (ΔH_{rxn} =) $H_2O^{\oplus}$—CH_2R + $Br^{\ominus}$

$I^{\ominus}$ + RCH_2—OH ⟶ (ΔH_{rxn} =) I—CH_2R + $^{\ominus}$OH

$I^{\ominus}$ + RCH_2—$OH_2^{\oplus}$ ⟶ (ΔH_{rxn} =) I—CH_2R + H_2O

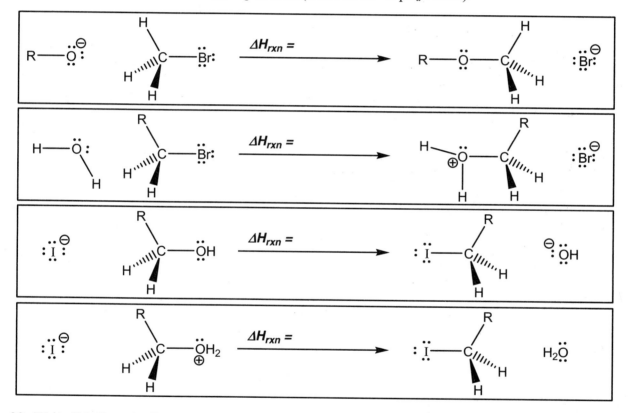

32. Write "No Reaction" next to the one reaction in the previous question that is very unlikely to produce the products shown.

33. For each pair, circle the better leaving group (assume breakage of the bond highlighted in **bold**).

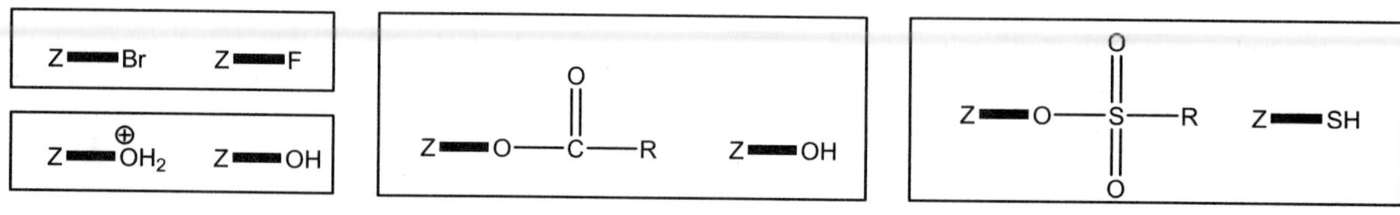

36. Consider a reaction between NaOH and 2-iodopropane and a reaction between NaOH and 2-chloropropane.

   a. Draw the reactants.

   b. Add curved arrows to show a one-step nucleophilic substitution reaction.

   c. Draw the most likely products of each reaction.

   d. Predict, based on the data in Table 10.1, which reaction will be faster.

   e. Draw an energy diagram showing both reactions on the same set of axes. Draw the 2-iodopropane reaction profile with a dotted line and the 2-chloropropane reaction profile with a solid line, and <u>assume that the potential energies of the reactants and the potential energies of the products are the same in both reactions</u> (i.e., the only difference is in the energies of the two different transition states).

37. Leaving groups (Table 10.1) exactly follow predictions based on pK_a data. However, only some nucleophiles follow pK_a, trends. Cite at least four species in Table 10.2 that are expected to be poor nucleophiles based on pK_a data, but instead are good or very good nucleophiles because they are soft bases.

38. Fill each blank with one of the following words (some may be used twice, some not at all). **weak, strong, low, high, up, down, large, small, good, poor, soft,** or **hard**.

Leaving groups follow pK_a trends, so, since a good leaving group is generally a _______ base, the conjugate acid of a good leaving group will have a _________ pK_a.). The opposite is true for one type of nucloephiles (_________ base = good nucleophile), but there is another type of nucleophile called a ______ base nucleophile. These species, notably the larger halogens, are __________ bases but unexpectedly __________ nucleophiles. In each column of the periodic table, nucleophilicity increases as you go _________. This means that HS^- and I^- are very ________ nucleophiles. A partial explanation for this trend is that atoms in rows three, four and five of the periodic table have ________ fluffy orbitals. Such nucleophiles are called __________ nucleophiles or bases. (Counterexample: A nucleophile like F^- has ________ orbitals and is therefore called a ________ nucleophile or base.) _________ nucleophiles are malleable and tend to make lower-energy transition states and thus fast substitution-reaction rates.

39. Show a synthesis of each target starting from a primary or methyl alkyl halide and any reagents containing three or fewer carbons.

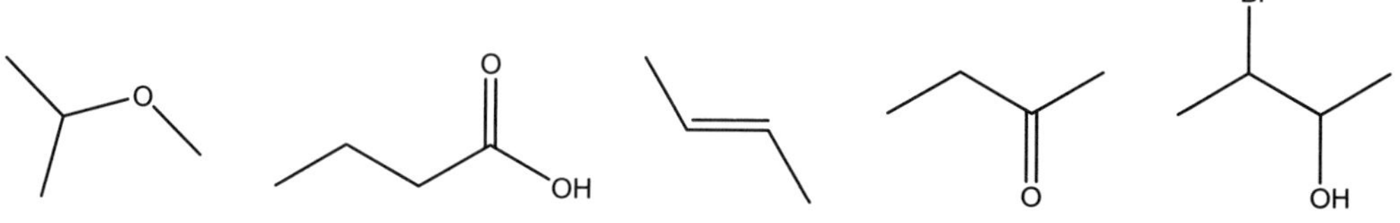

O OH Br O OH

Read the assigned pages in the text, and do the assigned problems.

The Big Picture

S_N2 reactions are one member of a family of substitution and elimination reactions (S_N2, S_N1, E2, E1) that are often in competition with one another. One of the key challenges of this course (and the basis for many exam questions) is to figure out which type of reaction will occur given a particular set of reactants. If you are doing these activities in order, the next four activities will be devoted to exploring competition between these four reaction types.

This will get quite complex, so make sure you are very comfortable with the concepts in this activity, including knowing the mechanism of an S_N2 reaction. Now is a good time to ramp up to daily study of organic chemistry. Even if do not have class every day, it is still a very good idea to study a little bit of organic chemistry every day. This is far less stressful, and far more effective than letting the work pile up and cramming at the last minute.

Common Points of Confusion

- Chiral inversion in an S_N2 reaction ONLY occurs at the electrophilic carbon because it is caused by the nucleophile making a new bond and the leaving group leaving. This will be clear if you go through the mechanism of the reaction. Unfortunately, students sometimes cut corners and simply memorize that S_N2 causes **chiral inversion**. When faced with a reaction such as the one below, such students incorrectly assume that the result of the reaction is to invert ALL chiral centers.

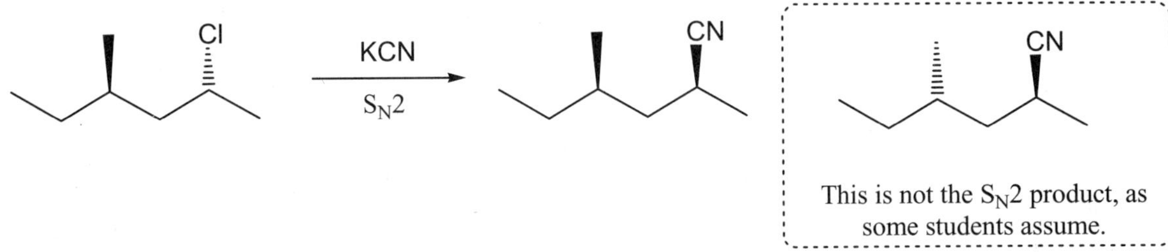

- Some students want further explanation about why we can use pK_a values in place of energy differences. In short, a pK_a value is minus the log of an equilibrium constant, and ΔG^o is minus the ln of an equilibrium constant, so the two are proportional.
    - Another way of thinking about this is that an equilibrium constant between two states (e.g. reactants and products) is a function of the energy difference between these states.
    - Here is a simplified proof based on the key energy formulas found in most general chemistry textbooks:

$\Delta \mathbf{G^o} = -RT \ln K_{eq}$ and $\Delta G^o = \Delta H^o - T(\Delta S^o)$

Assumption: $\mathbf{\Delta S} \approx 0$ *for most organic reactions, therefore* $\Delta G^o \approx \Delta H^o$

$\mathbf{pK_a = -log\ K_a}$, and under standard conditions $K_a \propto K_{eq}$ ($\propto$ = "proportional to")

$\boldsymbol{\Delta G^o = -RT\ \ln K_{eq}}$, since $(\log x) \propto (\ln x)$ it follows that $\mathbf{pK_a \propto \Delta G^o}$

ΔG^o is a measure of energy difference, therefore **pK_a is a measure of energy difference.**

Notes

ChemActivity 11: Two-Step Substitution (S_N1)

(How can we explain nucleophilic substitution reactions at tertiary carbons?)

Model 1: S_N2 Reactions that are Downhill but Slow

The four reactions below share the same nucleophile and leaving group, and each is steeply downhill based on pK_a data (ΔH = -16 pK_a units). Yet they have very different **S_N2 reaction rates.**

I RO⁻ + CH_3Br → very fast → $ROCH_3$ + Br⁻

II RO⁻ + CH_3CH_2Br → fast → $ROCH_2CH_3$ + Br⁻

III RO⁻ + $(CH_3)_2CHBr$ → slow → $ROCH(CH_3)_2$ + Br⁻

IV RO⁻ + $(CH_3)_3CBr$ → no reaction → $ROC(CH_3)_3$ + Br⁻

(so slow that none of this product is observed)

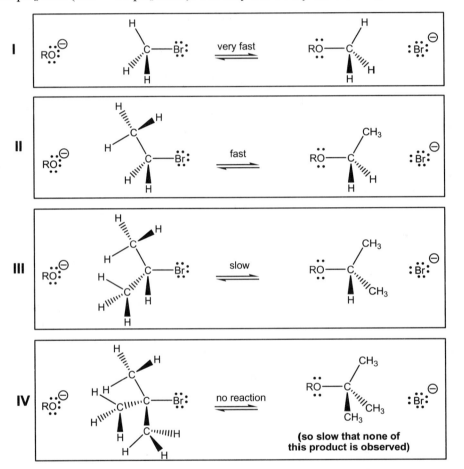

Construct Your Understanding Questions (to do in class)

1. Add curved arrows to each reaction in Model 1 showing an S_N2 reaction.
2. Label each electrophilic C (the carbon with Br attached) in Model 1 as methyl (0°), primary (1°), secondary (2°) or tertiary (3°).
3. Construct an explanation for each of the following statements:
   a. **Reaction IV** is slower than **Reaction III; Reaction III** is slower then **Reaction II**; etc.

   b. For any of the reactions above, the rate slows down as R gets bigger (i.e. as the nucleophile gets bigger). For example, Reaction II is slower when R = CH_3 vs. when R = H.

Model 2: Two-Step Nucleophilic Substitution (S_N1) Reactions

It turns out that substitutions *can* occur at tertiary electrophilic carbons, just not by an S_N2 mechanism. Each pair of reactants below undergoes a two-step substitution mechanism called S_N1.

V

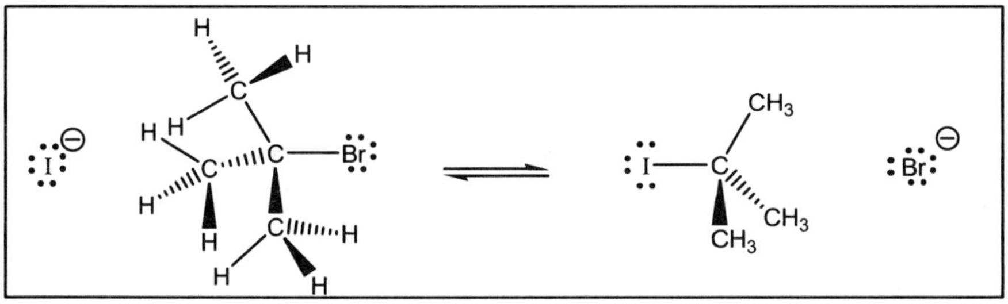

VI

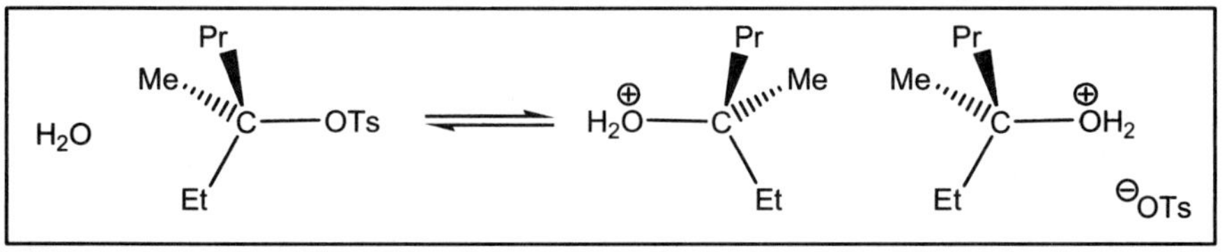

VII

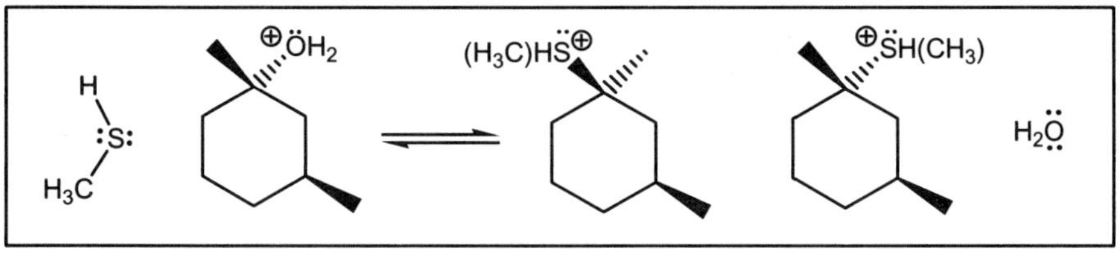

Construct Your Understanding Questions (to do in class)

4. In the first step of an S_N1 reaction, the leaving group leaves, giving a carbocation intermediate. Based on this information, use curved arrows to illustrate the mechanism of **Reaction V** in Model 2. **Be sure to draw the carbocation intermediate.**

5. The reactants in Model 2 cannot undergo a one-step nucleophilic substitution (S_N2) because there is no room for the nucleophile to make a bond to the electrophilic carbon. Explain how the first step of an S_N1 reaction "makes room for" the nucleophile.

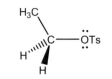

6. Sketch the carbocations that would result *if* the following electrophiles were to undergo S_N1 reactions (*though they do not undergo S_N1*).

   a. Label the carbocations you drew above and the carbocation in your mechanism on the previous page as methyl (0°), primary (1°), secondary (2°) or tertiary (3°) as appropriate.

   b. (Review) Rank the carbocations that you labeled from most favorable (most likely to form) to least favorable (least likely to form).

   c. Construct an explanation for why the pairs of reactants in *this* question DO NOT undergo S_N1 reactions while those in **Reactions V – VII** (on the previous page) *do* undergo S_N1.

7. Mark the point on each S_N1 reaction pathway <u>below</u> that represents the carbocation intermediate. Mark one of these points, "**1°/too high PE to form**" and the other point "**3°/very likely to form.**"

Model 3: S_N1 vs. S_N2 at a 1° vs. 3° Electrophilic Carbon

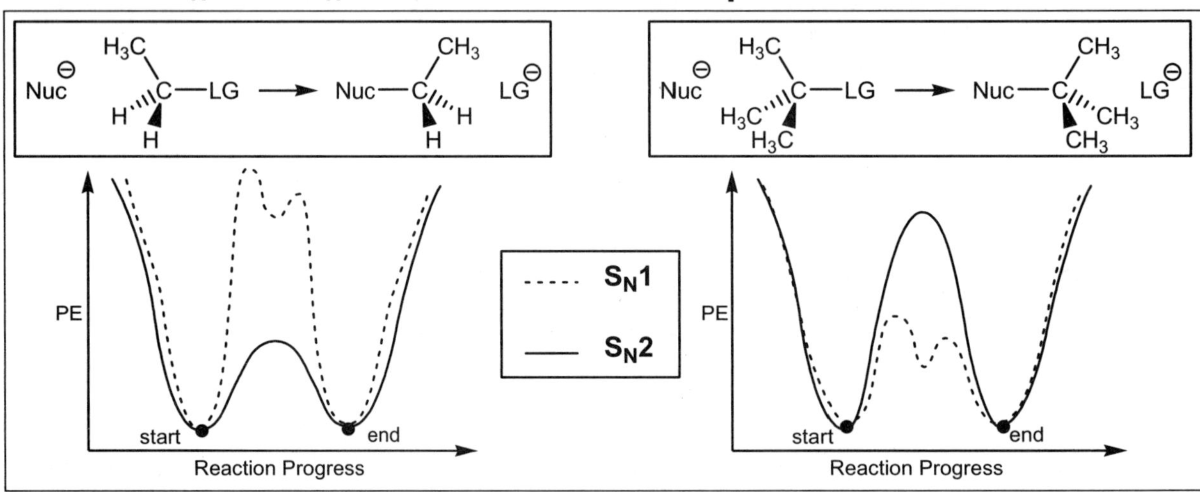

Construct Your Understanding Questions (to do in class)

8. According to Model 3, which step (first or second) is rate-determining in an S_N1 reaction?

9. Use this information to write a possible rate expression for an S_N1 reaction.

10. Will changing the concentration of the nucleophile change the rate of an S_N1 reaction? Explain your reasoning.

Extend Your Understanding Questions (to do in or out of class)

11. (Check your work) Is your answer to the previous question consistent with the following rate data?

Nucleophile e.g., I^-	Electrophile—LG e.g., $(CH_3)_3C$—Br	Relative Reaction Rate
1 M	1 M	1
2 M	1 M	1
1 M	2 M	2
2 M	2 M	2

Table 11.1: Effect of concentration on rate for S_N1 reaction

12. (Review) Explain why a two-step nucleophilic substitution reaction is called an $S_N\underline{1}$ reaction, and a one-step nucleophilic substitution is called an $S_N\underline{2}$ reaction What do the "1" and "2" stand for (if not the number of steps)?

13. Consider the S_N1 reaction below:

    a. Add a curved arrow to illustrate how the **S** product could be generated in step 2 of this reaction, and label this arrow "**S-forming**."

    b. Draw a second curved arrow showing how the **R** product is formed and label it "**R-forming**."

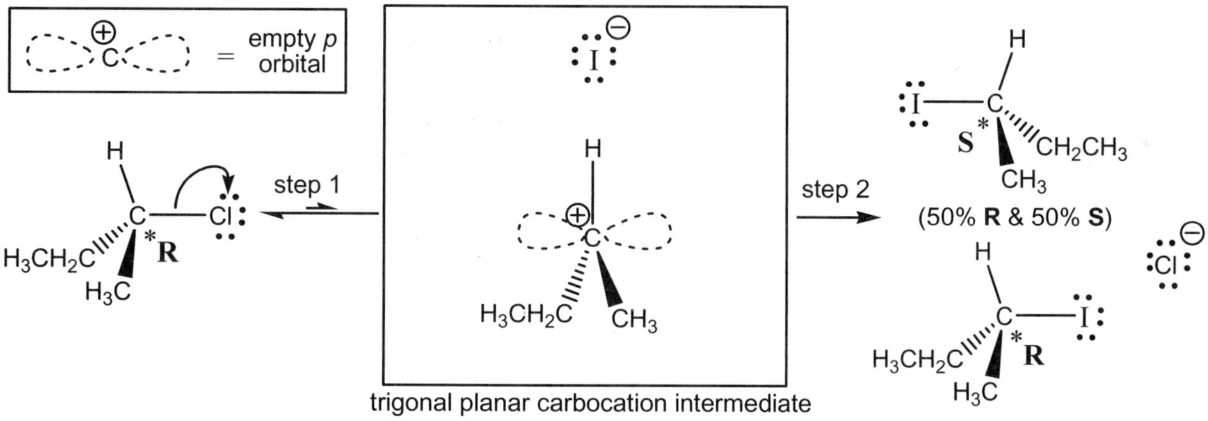

14. Explain why an S_N1 reaction at a chiral electrophilic carbon does NOT lead to chiral inversion the way an S_N2 reaction at a chiral center is known to do.

15. What is the special name of a 1:1 mixture of enantiomers like the product mixture produced in the reaction above (and in **Reaction VI** of Model 2)?

Model 4: S_N1, S_N2 or a Mixture of the Two?

- One pair of reactants in the box below goes to products exclusively by an S_N2 mechanism.
- One pair goes exclusively by an S_N1 mechanism.
- One pair goes by a mixture of S_N1 and S_N2 mechanisms.

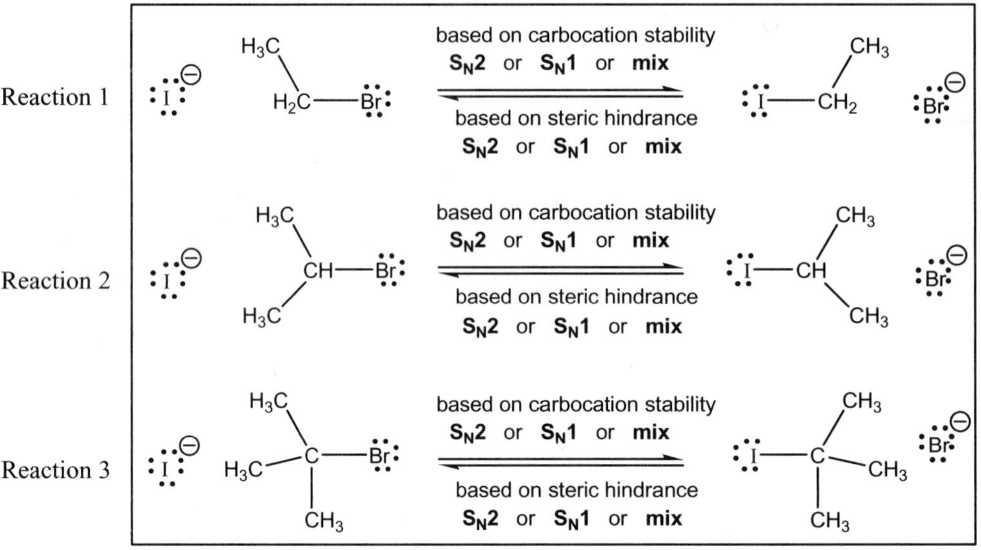

Extend Your Understanding Questions (to do in or out of class)

16. Which alkyl bromide CANNOT form a favorable carbocation? Cross out "S_N1" and "mix" ABOVE the reaction arrow for this reaction to indicate that S_N1 cannot occur.

17. Which alkyl bromide will form the lowest potential energy (most favorable) carbocation when Br leaves? Circle "S_N1" above the appropriate reaction arrow to show that S_N1 is favorable.

18. Which alkyl bromide is least crowded and so MOST likely to proceed via a one-step (S_N2) mechanism? Circle "S_N2" below the appropriate reaction arrow.

19. Which alkyl bromide is too crowded to proceed via a one-step (S_N2) mechanism? Cross out "S_N2" and "mix" below the appropriate reaction arrow.

20. Fill in each of the blanks in the following summary sentences with either "**S_N1**" or "**S_N2.**"

    a. Reaction 1 in Model 4 must proceed via an _____ mechanism. It cannot undergo _____ because it would have to form a very unfavorable carbocation. Conveniently, it turns out to be perfect for _____ because the electrophile is so uncrowded.

    b. Reaction 3 in Model 4 must proceed via an _____ mechanism. It cannot undergo _____ because the transition state would be too crowded. Conveniently it turns out to be perfect for _____ because the electrophile forms a very favorable carbocation when Br leaves.

21. When it comes to deciding whether a reaction will be S_N1 or S_N2, are **electronic factors** (carbocation stability) and **steric factors** generally in agreement or in competition? Explain.

Model 5: Solvent Effects

(Check your work) Steric and electronic factors are in agreement. Normally, both push a reaction toward one or the other substitution mechanism (S_N1 or S_N2).

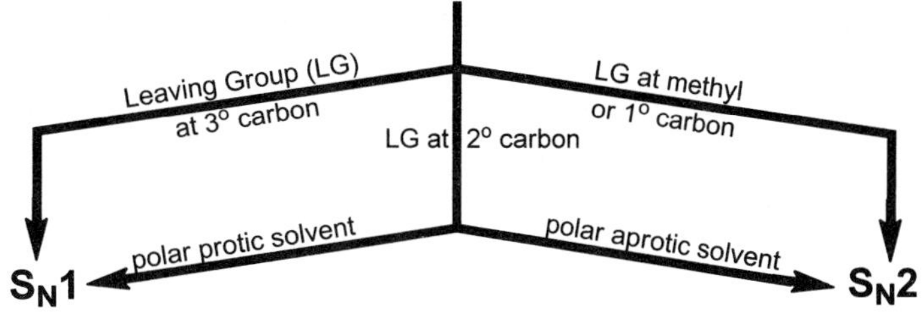

However, there are cases (such as Reaction 2 in Model 4) where a mixture of S_N1 or S_N2 is likely. In such cases, the solvent can push the reaction toward one or the other mechanism.

Extend Your Understanding Questions (to do in or out of class)

22. [E]According to Model 5, which type of solvent, **polar protic** or **polar aprotic** [circle one] is best for an S_N1 reaction?

23. A good way to think about a polar protic solvent is as a super-polar solvent. Recall that based on the general solvent rule of "like dissolves like" a super-polar solvent will be good at dissolving (and stabilizing) ions. Construct an explanation for why a polar protic solvent is best for an S_N1 reaction.

24. Is your answer to the previous question consistent with the energy diagram below, right? Explain.

25. On the energy diagram below, left, use block arrows to show the activation energy of the S_N2 reaction in aprotic solvent vs. protic solvent. In which type of solvent is an S_N2 reaction faster?

26. It turns out that a polar protic solvent both speeds an S_N1 reaction (by stabilizing the carbocation intermediate), and impedes (slows) an S_N2 reaction. Construct an explanation for why the polar protic solvent lowers the potential energy of the nucleophile in the S_N2 reaction depicted below, left. [Note that by stabilizing/deactivating the nucleophile, the polar protic solvent increases the activation energy and thereby slows the reaction.]

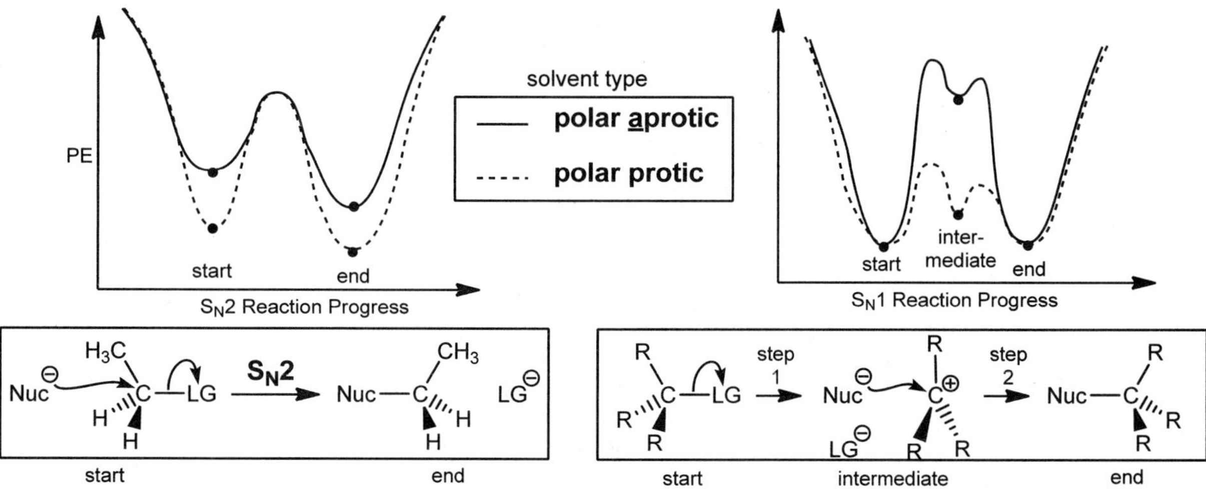

Confirm Your Understanding Questions (to do at home)

27. In general a strong base (with its high potential energy) is a good nucleophile. Construct an explanation for the following exception. Both nucleophiles shown (hydroxide and *tert*-butoxide) are strong bases, but only one of them is good nucleophile. Explain why.

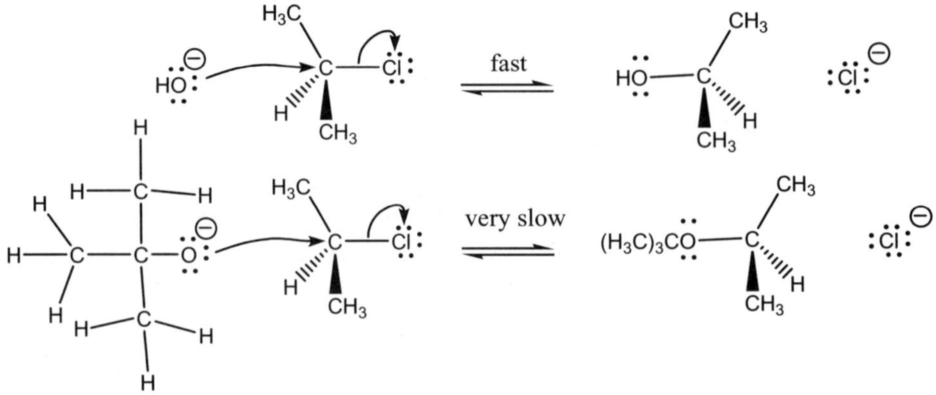

fast

very slow

28. Consider the following relative rate data.
    a. Construct an explanation for why Rxn A is slower than Rxn B.
    b. Construct an explanation for why Rxn A is slower than Rxn C.
    c. Construct an explanation for why Rxn A is slower than Rxn D.

A.  Rel. Rate = 1

B. 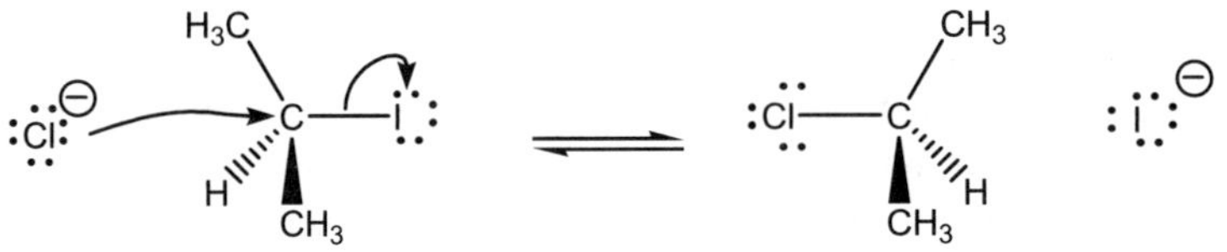 Rel. Rate = 1000

C.  Rel. Rate = 10

D. 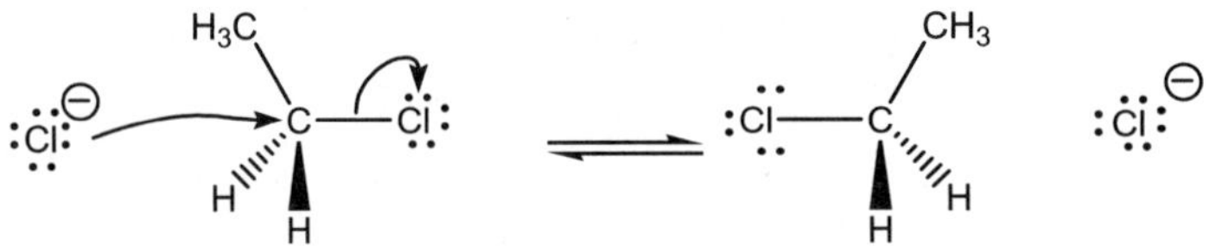 Rel. Rate = 100

29. Use curved arrows to show an S_N1 and S_N2 pathway for Reaction V in Model 2.
    a. Draw an energy diagram with both reaction profiles.
    b. Write a rate expression for both reaction pathways.
    c. Explain why the S_N2 pathway is so much slower.

30. Each energy diagram below shows two reaction profiles. The dotted line shows the original reaction, and the solid line shows the reaction after a change has been made. Match each of the following changes with the correct energy diagram below.

   (1) Change to a nucleophile that is a stronger base.

   (2) Change from a secondary electrophilic carbon to a primary electrophilic carbon.

   (3) Change to a leaving group that is a weaker base.

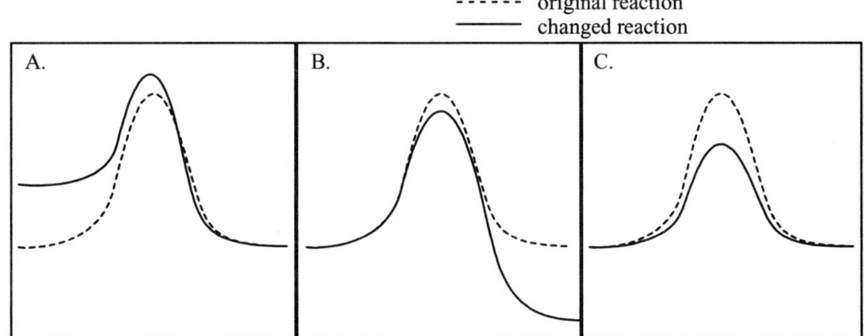

31. The products formed in Reaction VII in Model 2 are not a racemic mixture. What term best describes the relationship between the two chiral products shown, and why is this not a racemic mixture?

32. Show a synthesis of each target starting from a tertiary alkyl halide and any reagents containing three or fewer carbons.

SH O CH3 N(CH3)3 ⊕

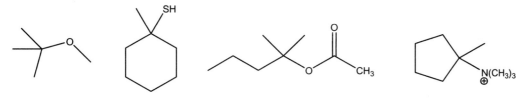

33. Ordinary primary alkyl halides cannot undergo S_N1. However, in a polar protic solvent the alkyl halides below *can* undergo S_N1. What is special about these primary alkyl halides allows them to undergo S_N1? Hint: draw the carbocation intermediate including important resonance structures.

Br Cl

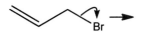

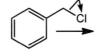

34. Like many allylic and benzylic alkyl halides, the examples above can undergo either S_N1 or S_N2 (depending on the solvent). However, this is not always the case. Explain why the following allylic and benzylic alkyl halides CANNOT undergo an S_N2 regardless of solvent.

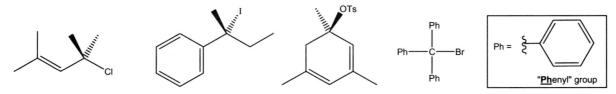

Cl I OTs Ph Ph—C—Br Ph

Ph = "Phenyl" group

35. Explain why a polar protic solvent speeds the rate of an S_N1 mechanism.

36. Explain why a polar protic solvent <u>slows</u> the rate of an S_N2 mechanism.

37. For each reaction, draw the most likely product, and indicate which of the following is most likely: [S_N1], [S_N2], [acid-base (H^+ transfer) with no substitution reaction], [acid-base followed by S_N1], [acid-base followed by S_N2], or [S_N1 with carbocation rearrangement likely].

Br $\xrightarrow{NaSCH_3}$	(S)-3-deutero-1-iodopentane $\xrightarrow{NaN_3}$
$CH_3CH_2CH_2Cl \xrightarrow{KCN}$	$CH_3OH \xrightarrow{H_2SO_4 \quad H_2S}$
$-OSO_2CH_3 \xrightarrow{H_2S}$	$CH_3I \xrightarrow{CH_3CO_2Na}$
$H_3CO-S(=O)_2-C_6H_4-CH_3 \xrightarrow{KOH}$	$CH_3OH \xrightarrow{((CH_3)_2CH)_2N^{\ominus} \quad Li^{\oplus}}$
$CH_3I \xrightarrow{KOH}$	$-OSO_2CH_3 \xrightarrow{HOCH_3}$
O, OH $\xrightarrow{NaC\equiv CCH_3}$	$CH_3CH_2OSO_2CH_3 \xrightarrow{H_2O}$
$CH_3CH_2OTs \xrightarrow{NaC\equiv CCH_3}$	$F_3C-C(=O)-OH \xrightarrow{NaOCH_2CH_3}$
$-OH_2^{\oplus} \xrightarrow{KOH}$	$-OTs \xrightarrow{NaBr}$
$-OH \xrightarrow{HCl}$	$CH_3I \xrightarrow{NaN(CH_3)_2}$
(S)-1-deutero-1-iodopentane $\xrightarrow{KCN}$	$(H_3C)_3CO-S(=O)_2-C_6H_4-CH_3 \xrightarrow{HOCH_3}$

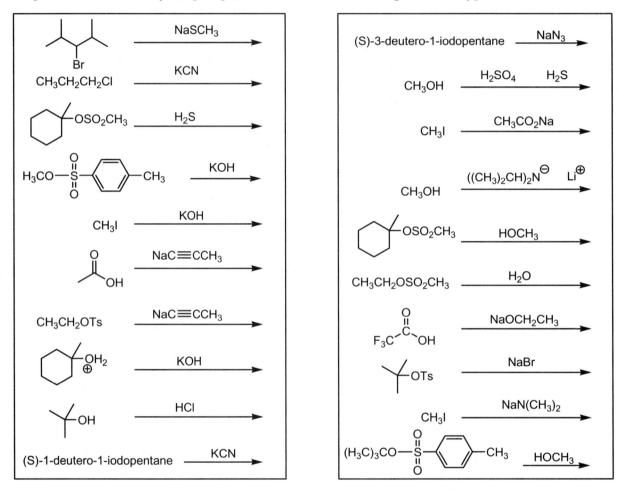

38. Give an example of a pair of reagents that are more likely to undergo S_N1 than S_N2, and vice versa.

Read the assigned pages in the text, and do the assigned problems.

The Big Picture

As you will see in the next activity, substitution is not the only reaction that can occur when a nucleophile is mixed with a molecule containing a good leaving group, especially when the nucleophile is a strong base. A totally different reaction pathway (called an elimination) is often more likely. In the next three activities we will learn about two types of eliminations (called E1 and E2) that have many similarities to the two substitution reactions that we have encountered. Be sure you are very comfortable with both the S_N1 and S_N2 mechanisms, and be able to explain why a given pair of reactants would go by one mechanism versus the other.

Common Points of Confusion

- $S_N\underline{2}$ reactions are <u>one</u>-step while $S_N\underline{1}$ reactions are <u>two</u>-step. No, we are *trying* to confuse you. The "1" and "2" in S_N1 and S_N2 DO NOT refer to the number of steps in the reaction. They refer to the <u>number of species involved in the slowest step of each reaction</u>.
- A simple analysis of the difference between S_N1 and S_N2 will lead, at this point, to the correct conclusion that S_N1 is favored when there is a tertiary (allylic or benzylic) leaving group, and S_N2 is favored when there is a primary (or methyl) leaving group. It is fine to memorize this distinction at this point, but be aware that for the next set of reactions (called E1 and E2) this distinction does not hold up. Specifically, E2 reactions, though they are similar to S_N2 reactions in some ways, can occur when there is a *tertiary* leaving group (or any other type of leaving group except methyl). This trips up many students and is explained in greater detail in the Common Points of Confusion section of the ChemActivity on E2.
- In the answers to Model 4 (shown below) it is important that, above the arrow for Reaction 1, S_N1 and mix are crossed out—instead of S_N2 being circled. This emphasizes that, due to the impossibility of generating a primary carbocation, S_N1 *cannot* occur. Complementing this, under the arrow, S_N2 is circled because the low steric hindrance of this primary alkyl halide *allows* S_N2. Steric hindrance does not *disallow* S_N1, so S_N1 and mix are *not* crossed out below the arrow.
- Similarly, for Reaction 3, above the arrow S_N1 is circled, and below the arrow S_N2 and mix are crossed out. Try to explain to your study partner why this is the correct answer (instead of crossing out S_N2 and mix above the arrow and circling S_N1 below the arrow).

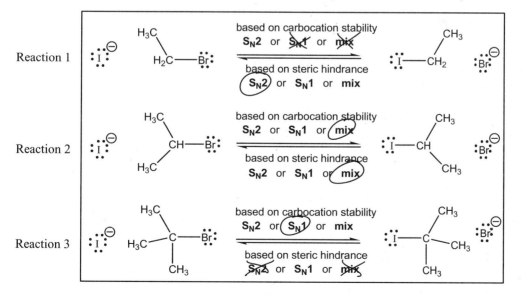

	Reactants	Conditions	Products
Reaction 1	I^- + H_3C–H_2C–Br	based on carbocation stability: S_N2 or ~~S_N1~~ or ~~mix~~ / based on steric hindrance: (S_N2) or S_N1 or mix	I–CH_2–CH_3 + Br^-
Reaction 2	I^- + $(H_3C)_2CH$–Br	based on carbocation stability: S_N2 or S_N1 or (mix) / based on steric hindrance: S_N2 or S_N1 or (mix)	I–$CH(CH_3)_2$ + Br^-
Reaction 3	I^- + $(H_3C)_3C$–Br	based on carbocation stability: S_N2 or (S_N1) or mix / based on steric hindrance: ~~S_N2~~ or S_N1 or ~~mix~~	I–$C(CH_3)_3$ + Br^-

Notes

ChemActivity 12: Two-Step Elimination (E1)

(By what mechanism does a molecule that can form a favorable carbocation become an alkene?)

Model 1: Hyperconjugation

Adjacent alkyl groups stabilize a carbocation. One argument is that alkyl groups donate electron density to the carbocation as described below, left. A methyl carbocation does not experience this effect, called **hyperconjugation** (by analogy to conjugation, which is observed between two *p* orbitals).

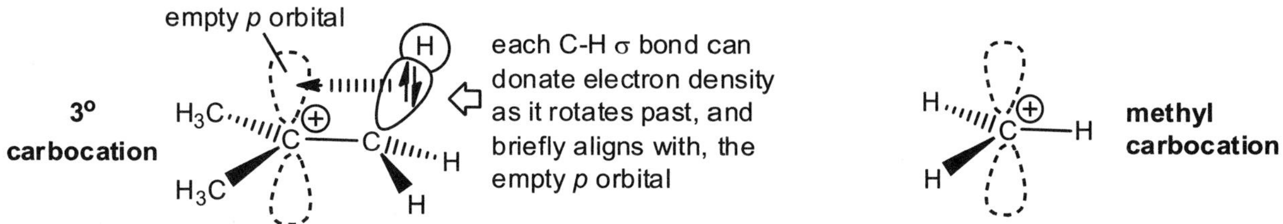

Hyperconjugation spreads out the + charge and helps complete the octet of the carbocation carbon.

Construct Your Understanding Questions (to do in class)

1. Explain why the methyl carbocation in Model 1 is much less stable than the 3° carbocation shown. *[Questions 1-9 of ChemActivity 6 provide carbocation terminology you are expected to know.]*

2. Construct an explanation for why a C–H σ bond on the 3° carbocation in Model 1 is momentarily <u>weakened</u> while it is aligned with the neighboring empty *p* orbital.

3. It turns out that this effect is powerful enough to lower the pK_a of most carbocations to zero.

   a. (Review) A pK_a of zero means it is **easy** <u>or</u> **difficult** (circle one) to pull an H off this molecule.

   b. All nine H's on the carbocation in Model 1 are equivalent, so any one could be associated with the pK_a of zero. This is not the case for the carbocation at right → <u>Only four of these H's are easy to pull off</u>. Circle these four H's, and explain your reasoning.

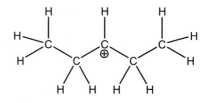

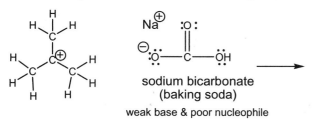

4. A tertiary carbocation is such a strong acid it will react with even a very weak base.

   sodium bicarbonate (baking soda)
   weak base & poor nucleophile

   a. Use curved arrows to show the mechanism and the products of this acid-base reaction.

b. It is possible to draw the organic product of the reaction in (a) with all zero formal charges. Have you drawn this most important representation of the C_4H_8 product? If not, draw it now.

Model 2: Two-Step Elimination

Certain groups act as leaving groups (LG).

When carbocation formation is favorable, the carbon-LG bond can break as shown below…

Very Good Leaving Groups	Good Leaving Groups	Poor/Very Poor Leaving Groups
RSO_3^- *(most often… R= CF_3, R= toluene [also called OTs], or R= CH_3)* I^-	R_2O *(water, alcohol, or ether)* Br^- Cl^-	F^- *All strong bases are very poor leaving groups…* e.g., RO^- R_2N^- R_3C^-, …etc.

…to form a carbocation that can undergo loss of an H. (Often the solvent is a strong enough base to pull off the H.) This elimination of a LG and H to form a double bond is a **two-step elimination.**

step 1 step 2

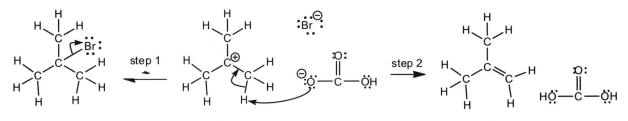

Construct Your Understanding Questions (to do in class)

5. [(E)]Circle the **leaving group** and **hydrogen** that are "eliminated" in the reaction above.

6. Why it is less favorable for a leaving group to leave when it is attached to a lower degree carbon? Hint: See Model 1. (Note: Ordinary primary and methyl carbocations are too unfavorable to form.)

methyl carbon — methyl carbocation does not form

primary carbon — primary carbocation does not form

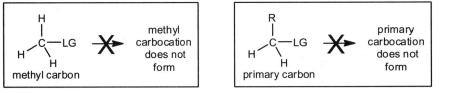

Memorization Task 12.1: Alpha (α) and Beta (β) Hydrogens

- The **alpha-carbon** (C_α) is the carbon attached to the leaving group
- The **beta-carbon** (C_β) is the next carbon down each chain
- an H_α is any H on a C_α; an H_β is any H on a C_β, etc.

7. Draw the carbocation that would result if the leaving group on the molecule above ↑ were to leave. Carry the α, β, γ, and δ labels to this new structure, and circle any H that *could* be eliminated from this carbocation. What Greek letter is associated with all of the H's you circled?

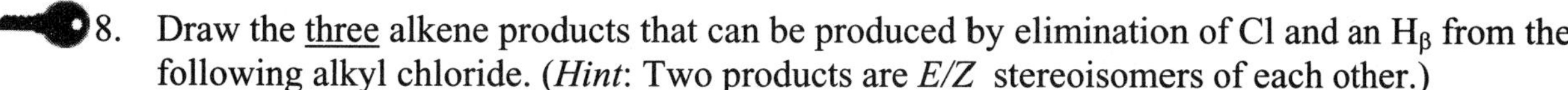

8. Draw the <u>three</u> alkene products that can be produced by elimination of Cl and an H_β from the following alkyl chloride. (*Hint*: Two products are *E/Z* stereoisomers of each other.)

Cl

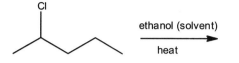

ethanol (solvent)

heat

9. Explain why the following molecule CANNOT undergo elimination even though it readily forms the very favorable, resonance stabilized (benzylic) carbocation shown.

benzylic carbocation

Base:

NO ELIIMINATION REACTION

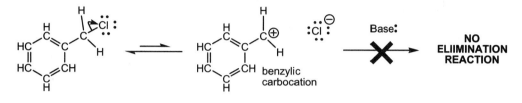

Model 3: Mono-, Di-, Tri-, and Tetra-substituted Alkenes

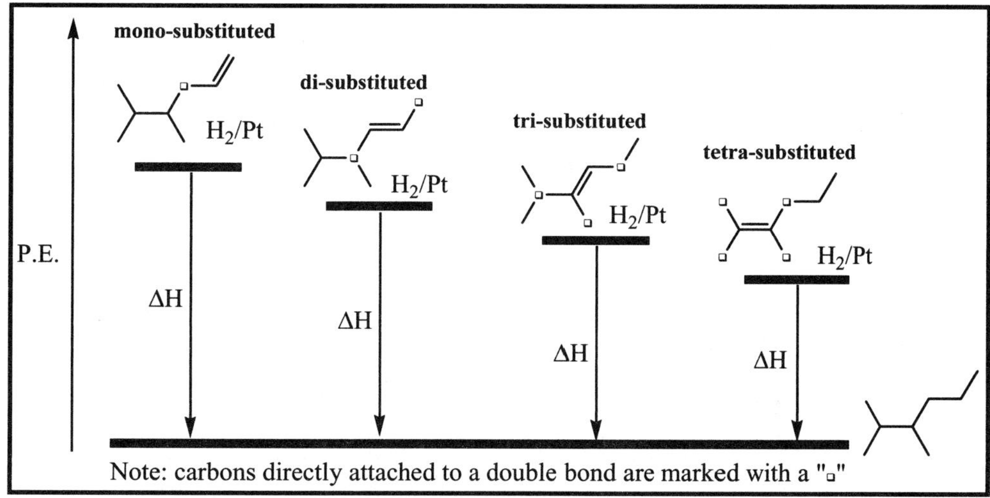

Construct Your Understanding Questions (to do in class)

10. (E)How many carbons are directly attached to the two carbons of the C=C double bond in a mono-substituted alkene? ...a di-substituted alkene? ...a tri-substituted alkene? ...a tetra-substituted alkene?

11. The reaction at the top of this page produces the following three products, but one of them is produced in very small amounts. Based on the information in Model 3, write "minor product" below one of the following alkene products, and explain your reasoning.

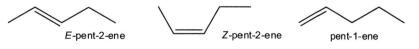

E-pent-2-ene   *Z*-pent-2-ene   pent-1-ene

Memorization Task 12.2: Zaitsev's Rule

In a two-step elimination reaction the most substituted (lower PE) alkene product(s) will dominate the product mixture.

12. (Check your work) All three structures below are produced by the two-step elimination reaction on the previous page, but <u>the two in the box are observed in much larger quantities</u>. Is this consistent with **Zaitsev's rule?** (And with your answer to the previous question?) **Explain your reasoning.**

(Note: E dominates over Z because of steric factors; the alkyl groups are more spread out in the transition state leading to E.)

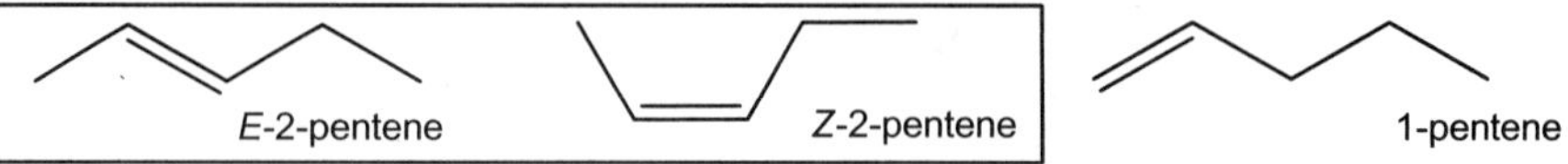

Extend Your Understanding Questions (to do in or out of class)

13. A common type of two-step elimination involves heating an alcohol in acid. (Recall that an acid will protonate an —OH group, turning it into a good leaving group [water].) Draw <u>all five</u> possible alkene products of this reaction, circle the two you expect to dominate the product mixture, and cross out the one you expect will be least abundant. *Check your work: See the first Confirm Your Understanding Question AFTER you have completed your answer to this question.*

OH

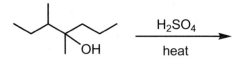

H_2SO_4 → heat

14. Consider the following rate data for a two-step elimination (like the one in Model 2).

[R—Br]	[HCO_3^-] (bicarbonate)	Relative Rate
1 M	1 M	1
2 M	1 M	2
1 M	2 M	1

Table 12.1: Rate dependence on concentration for Two-Step Elimination

a. Is the reaction rate dependent on the concentration of the starting material (R—Br)? Is the reaction rate dependent on the concentration of base (bicarbonate)?

b. Write a rate expression for this two-step elimination.

c. A **two-step elimination** is usually called an "**E1**" reaction. The "E" stands for elimination, but clearly the "1" does not stand for the number of steps in the reaction. Speculate: What does the "1" tell you about this reaction? *(Check your work: See Common Points of Confusion section.)*

d. Draw an energy diagram for this E1 reaction. (*Hint*: Which is the rate-limiting step?)

Synthetic Transformation 12.1: Alkyl Halide (R—X) or Tosylate (R—OTs) to Alkene (via E1)

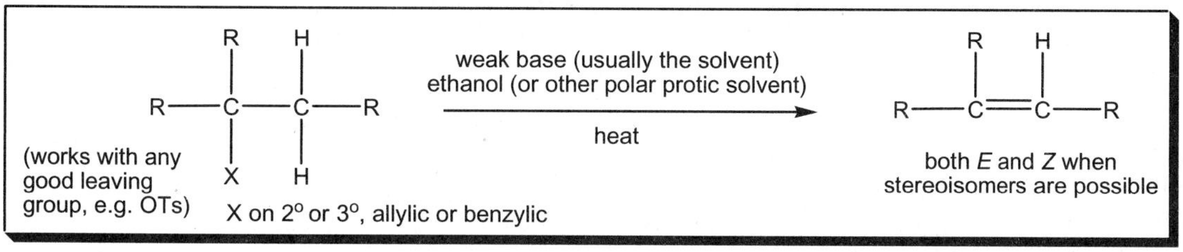

R—C(R)(X)—C(H)(H)—R → R—C(R)=C(H)—R

weak base (usually the solvent)
ethanol (or other polar protic solvent)
heat

(works with any good leaving group, e.g. OTs)

X on 2° or 3°, allylic or benzylic

both *E* and *Z* when stereoisomers are possible

Synthetic Transformation 12.2: 2° or 3°, allylic or benzylic Alcohol to Alkene (via E1)

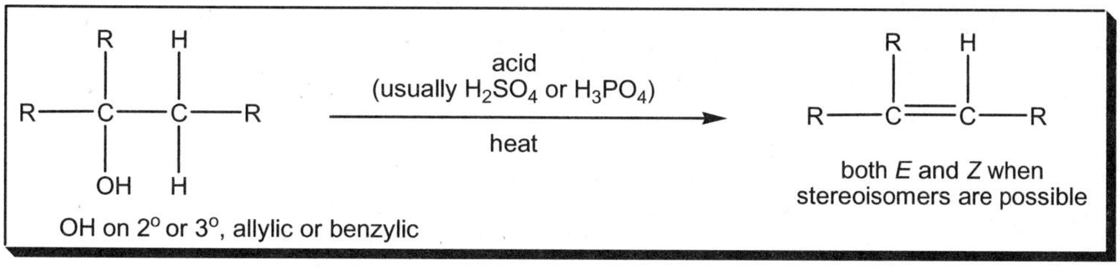

R—C(R)(OH)—C(H)(H)—R → R—C(R)=C(H)—R

acid
(usually H_2SO_4 or H_3PO_4)
heat

OH on 2° or 3°, allylic or benzylic

both *E* and *Z* when stereoisomers are possible

Confirm Your Understanding Questions (to do at home)

15. (Check your work) Are the following product ratios consistent with your answer to Question 13?

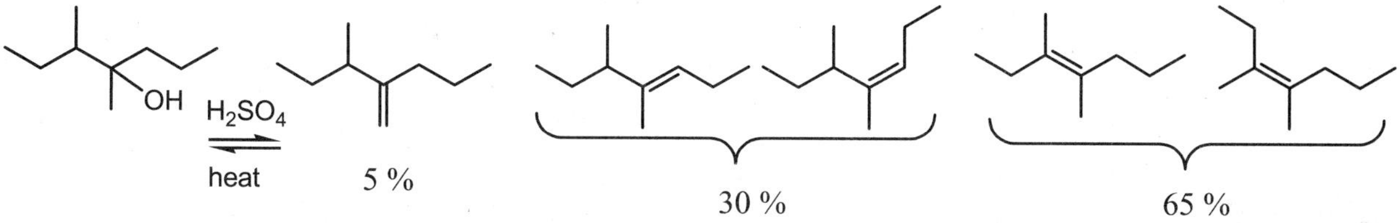

OH H_2SO_4 ⇌ heat

5 %

30 %

65 %

a. Label each alkene product with a word describing the level of substitution of the double bond.

b. Label each double bond among the products as *E, Z,* or neither.

c. Use curved arrows to show the mechanism of formation of the disubstituted product and a representative tri- and tetra-substituted product.

16. Show the mechanism of Synthetic Transformation 12.1.

a. Draw an energy diagram showing this reaction.

b. Write a rate expression for this reaction.

17. Show the mechanism of Synthetic Transformation 12.2, and draw an energy diagram (assume the acid-base reaction is much faster than either step in the elimination).

18. A good nucleophile (bromide) *is* present in the reaction in Model 2. Use curved arrows to show Br^- making a bond to the carbocation, and draw the resulting product. Explain why this competing reaction pathway (S_N1) is not a problem with respect to the desired elimination reaction.

19. Which point (of A-E) is the transition state of step 2 of an E1 reaction? Draw this transition state using dotted lines to represent partial bonds and δ's to represent partial charges.

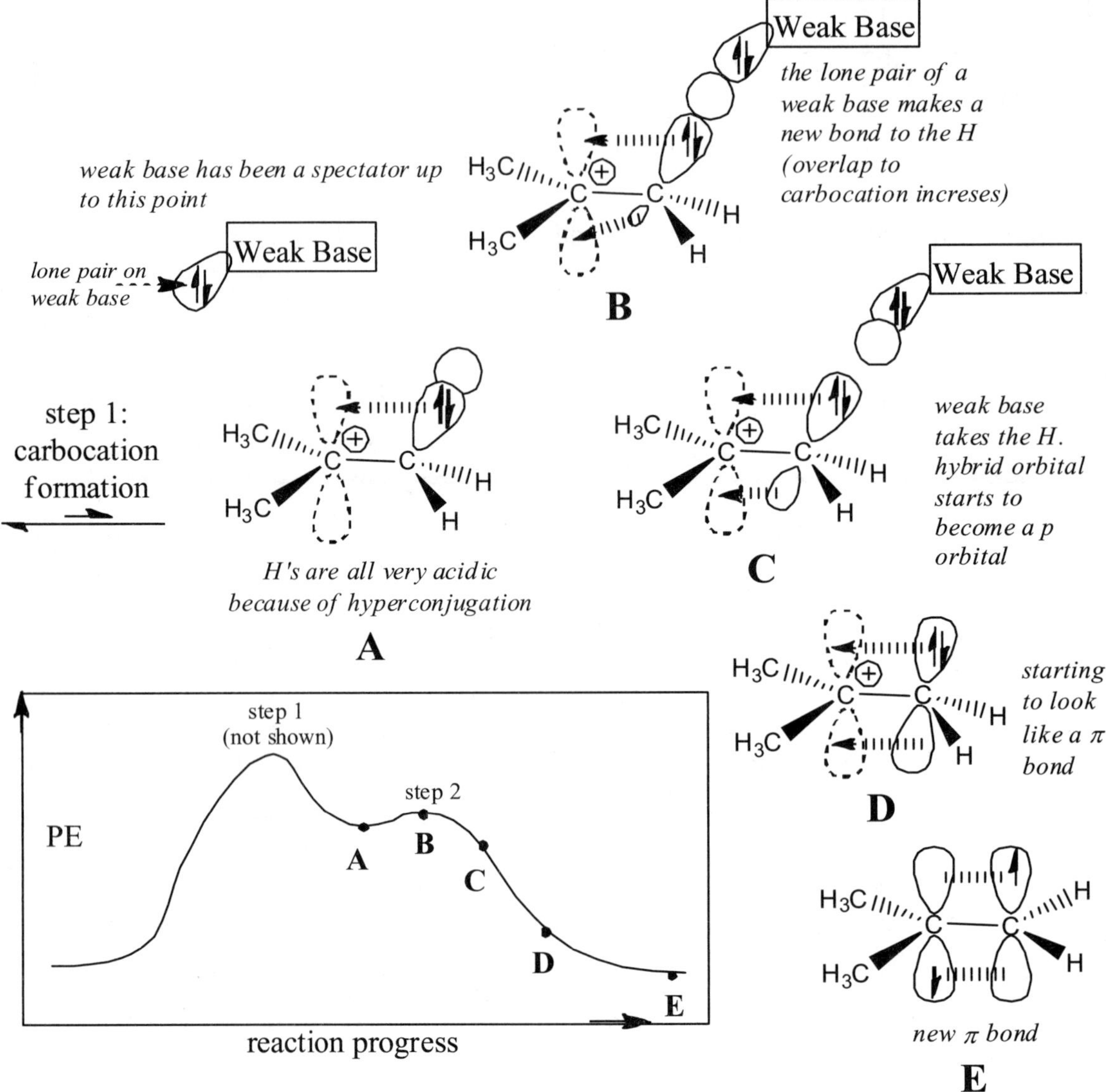

20. One carbon involved in the reaction above changes hybridization state during this step in the reaction. Label this carbon with its hybridization state in A and in E.

21. An **un-substituted** alkene has a molecular formula of C_2H_4. Draw the structure of this molecule and explain why it is called **unsubstituted**. (This is the only "unsubstituted" alkene.)

22. According to the information in Model 3, what is the relationship between the level of substitution of a double bond and the potential energy of the alkene? (Note that most textbooks do not give an explanation for this correlation. Nevertheless, it is important to remember this fact.)

23. The result of mixing *t*-butyl alcohol and sulfuric acid is shown below. Use curved arrows to show the entire mechanism of this reaction, and draw the two intermediate products.

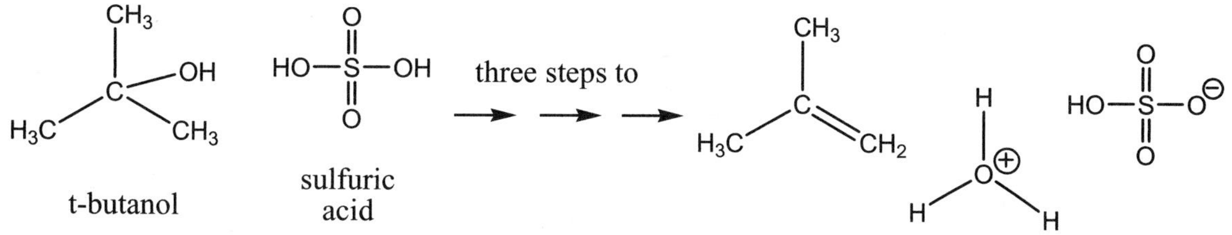

CH_3, OH, C, H_3C, CH_3 — t-butanol; HO—S(=O)(=O)—OH — sulfuric acid; three steps to → → → CH_3, C, H_3C, CH_2; H, O⊕, H, H; HO—S(=O)(=O)—O⊖

24. A lab technician accidentally spills some sodium iodide into the reaction mixture above. As a result a substitution product is observed along with the elimination product. Use curved arrows to show the mechanism of this side reaction, and draw the substitution product. (Is this substitution likely to be an S_N1 or an S_N2 reaction?)

25. Draw the products that result if *t*-butyl alcohol in the reaction in Question 23 is replaced with…

a. 3-methyl-3-pentanol (<u>three</u> different alkene products).

b. 3-methyl-3-hexanol (<u>five</u> different alkene products)

c. Label the major product and the minor product in a).

26. When 3-methyl-2-pentanol is mixed with acid and heated, the same <u>three</u> products are produced as in a reaction with 3-methyl-3-pentanol, plus there is an additional product, 3-methyl-1-pentene.

a. Show a mechanism that accounts for the product 3-methyl-1-pentene.

b. Show a mechanism that accounts for the major product, *E*-3-methyl-2-pentene.

c. Show a mechanism that accounts for the product 2-ethyl-1-butene.

27. Show a mechanism to account for the major product of each reaction below.

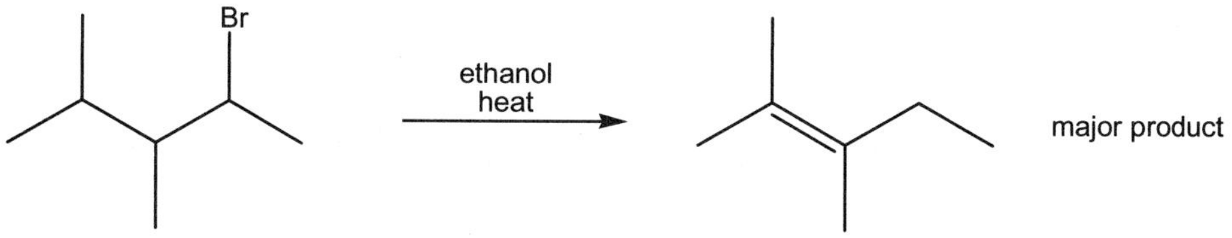

Br — ethanol, heat → major product

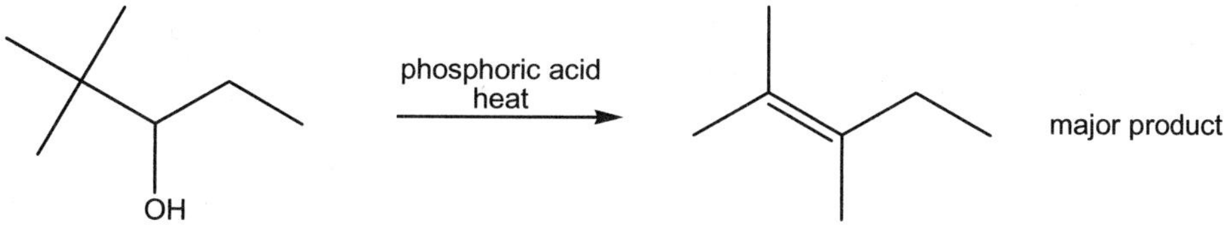

OH — phosphoric acid, heat → major product

Read the assigned pages in the text, and do the assigned problems.

The Big Picture

E1 (two-step elimination) is one of four key reactions that are commonly in competition with one another (S_N1, S_N2, E1, E2). Depending on the topic sequence in your course, you may have already learned about substitution reactions (S_N1 and S_N2). The next topic is another elimination reaction (E2).

Up to this point, many students have been assuming that reactions are all or nothing. In fact, all products are in equilibrium with their reactants (which means no reaction goes 100.00...% to completion), and that most reactions are in competition with other possible reaction pathways.

In an introductory course, these "side reactions" are often ignored, but in the laboratory organic chemists spend a huge portion of their time and creative energy trying to minimize these side reactions, and maximize the yield of the desired product. The challenge (and fun) of exploring substitution and elimination is that we can speculate knowledgably about, and devise mechanisms for sever different competing reactions leading to products with very different structures.

Common Points of Confusion

- Crazy organic chemists!! Why would we call a two-step elimination reaction E1?! Just remember that the "1" does NOT refer to the number of steps, but to the number of species involved in the slow step. This same system is used to name substitution reactions, and the next reaction, E2, which is a one-step process with TWO species interacting with each other in the slow step.
- E1 reactions always obey **Zaitsev's rule** (that is, they give the most substituted alkene product as the major product). This is true regardless of the size of the weak base used. [Re-read this sentence after doing the next activity.] E1 reactions obey Zaitsev's rule because they are **reversible**. In essence, E1 products tend to "undo" themselves since the activation energy (E_{act}) going forward is similar to the activation energy going backwards. Eventually the reaction finds the lowest potential energy product. This lowest potential energy product is most likely to persist because the reverse pathway has the largest activation energy. In other words, the lowest potential energy product is least likely to "undo" itself. (see reaction diagram below)

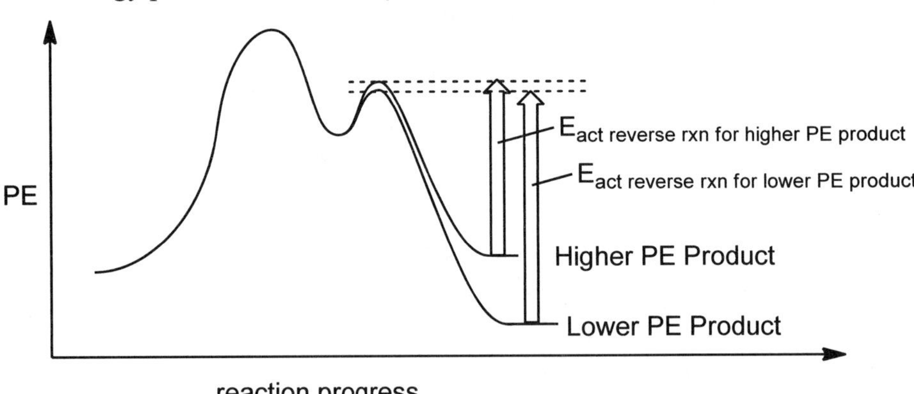

Notes

ChemActivity 13: One-Step Elimination (E2)

(How can an elimination reaction happen in just one step?)

Model 1: "Patient" and "Impatient" Bases

A weak base (such as bicarbonate) is too weak to pull an H from an electrophile until the leaving group has left. In the words of the patient/impatient analogy, <u>a weak base must wait patiently for a carbocation to form</u>. Note: Carbocation formation takes a long time relative to most other types of reactions.

weak base is patient

no reaction until Br leaves and the carbocation forms

takes H_β once carbocation forms

E1 step 2

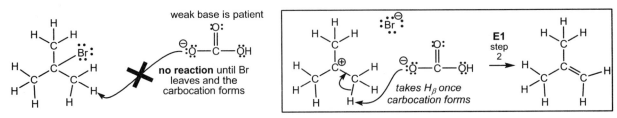

Figure 13.1: Weak base must wait patiently for a carbocation to form

In contrast a strong base can remove an H_β off an electrophile without waiting for the leaving group to leave. In the words of the analogy, <u>a strong base is impatient and does not wait for a carbocation to form</u>. By taking this H_β, the **strong base** initiates a **one-step elimination (E2)** reaction.

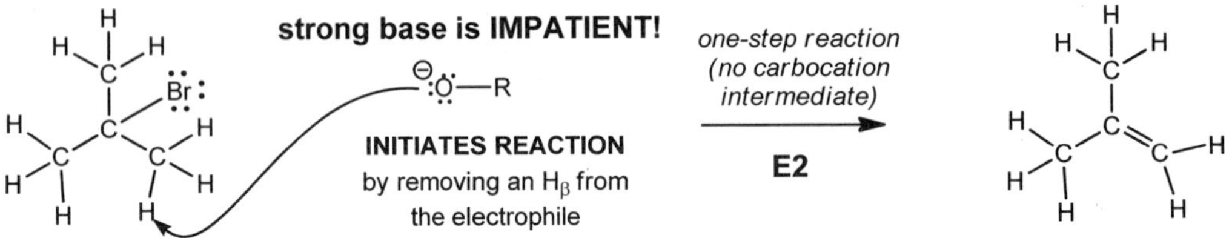

Figure 13.2: Strong base initiates a reaction by taking a beta hydrogen

Strong Base = species with a negative charge localized on a C, N, or O (see examples in Model 2)

Construct Your Understanding Questions (to do in class)

1. Add two more curved arrows to Figure 13.2 to complete the one-step elimination with a strong base.

2. Consider the following rate data for the reaction in Figure 13.2.

[R—X]	RO⁻ (base)	Relative. Rate
1 M	1 M	1
2 M	1 M	2
1 M	2 M	2
2 M	2 M	4

Table 13.1: Rate data for one-step elimination

a. Is the rate dependent on the concentration of R—X? … base (RO⁻)?

b. Write a rate expression for this reaction.

c. A **one-step elimination** is usually called an "**E2**" reaction. Explain both the "E" and the "2," and why this is an appropriate name for this reaction.

Model 2: E2 Reactions—Strong Bases (definition of small vs. large)

Different product ratios are sometimes achieved when using a large (vs. a small) base in an E2 reaction. There is no clear definition of large/small bases, but the following definitions work reasonably well.

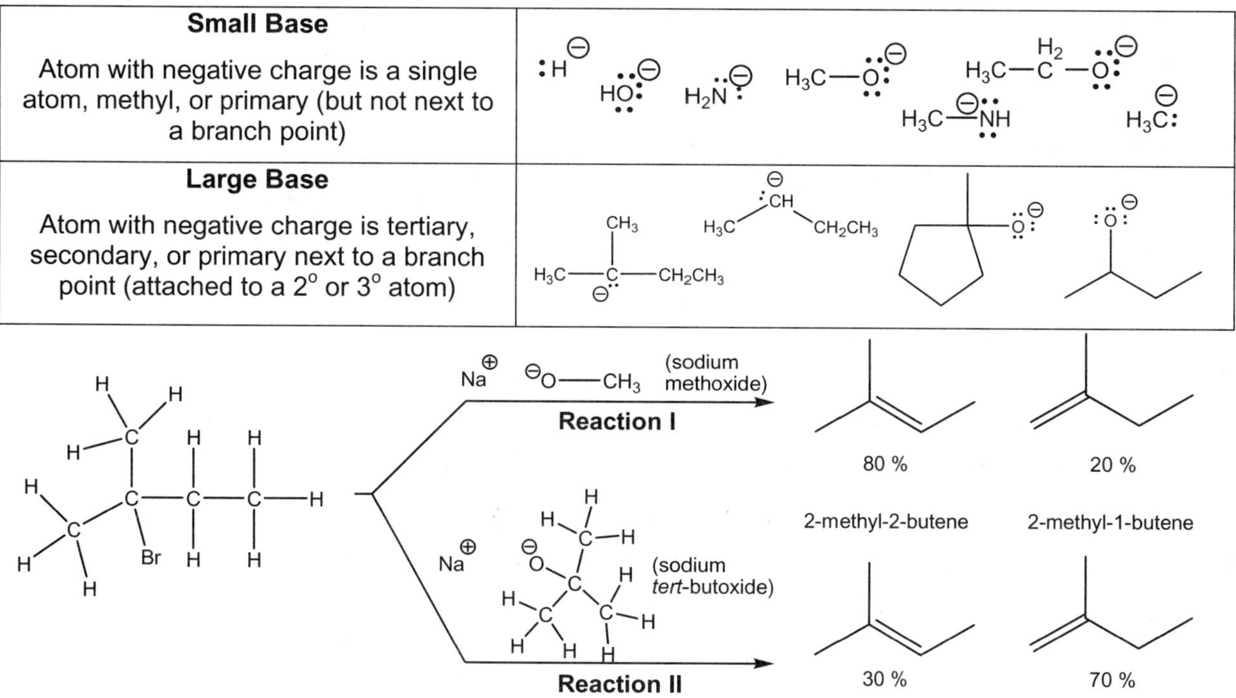

Figure 13.3: Example of how base size can change product ratios for E2 reactions

Construct Your Understanding Questions (to do in class)

3. Label the base in Reactions I and Reaction II as being small or large.
4. Which is lower in potential energy: **2-methyl-2-butene** or **2-methyl-1-butene** (circle one)?
5. For each (Reaction I and Reaction II), determine if it follows Zaitsev's rule (*see* activity 12).
6. Put a box around each H_β on the starting alkyl bromide that could be removed to produce the tri-substituted product (2-methyl-2-butene).
7. Circle each H_β on the starting alkyl bromide that could be removed to produce the di-substituted product (2-methyl-1-butene).
8. A small base (such as methoxide) is small enough to get at any H_β on the starting material.
   a. Use curved arrows to show the mechanism of Reaction I leading to the major product shown.
   b. Construct an explanation for why methoxide preferentially removes a boxed H_β.

   c. Use curved arrows to show the mechanism of Reaction II leading to the major product shown.
   d. Construct an explanation for why it is easier for *tert*-butoxide to remove a circled H_β than a boxed H_β (even though the resulting product is higher in potential energy).

Memorization Task 13.1: Zaitsev Product vs. Hofmann Product

Zaitsev product = most-substituted alkene product of an E1 or E2 reaction (*follows* **Zaitsev's rule**)

Hofmann product = least-substituted alkene product of E2 reaction (*violates* **Zaitsev's rule**)

The Hofmann product is named for 19th-century German chemist August Wilhelm von Hofmann, who developed an elimination using a $N(CH_3)_3$ leaving group that also violates the rule named for his Russian contemporary, Alexander Zaitsev.

9. Label the products in Figure 13.3 as being a Zaitsev or Hofmann product.

10. What base that would give an even higher percentage of the Zaitsev product in Reaction I in Figure 13.3 (as compared to methoxide).

Memorization Task 13.2: E2 reactions with a small base obey Zaitsev's Rule.

A small base can get to any H_β, so it selectively removes the H_β leading to the lowest PE alkene. Due to steric repulsion, a large base is forced to pull off the least-hindered H_β, leading to the Hofmann product.

11. Draw the major E2 product when each bromide is mixed with a) methoxide or b) *tert*-butoxide.

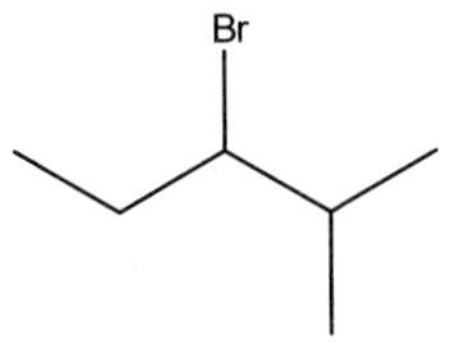

Br

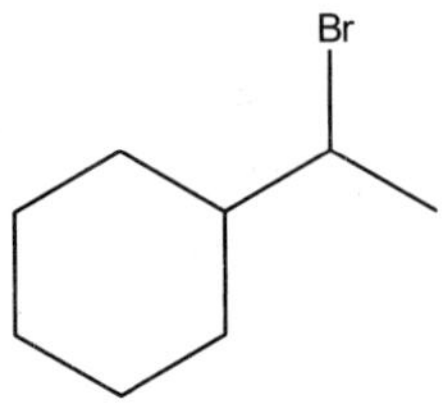

Br

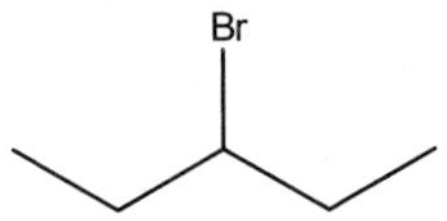

Br

12. Explain why the last reactant above gives the same major product with either a large or small base.

Extend Your Understanding Questions (to do in or out of class)

13. Use curved arrows to show the 2nd step in the following E1 reaction, and draw the major product.

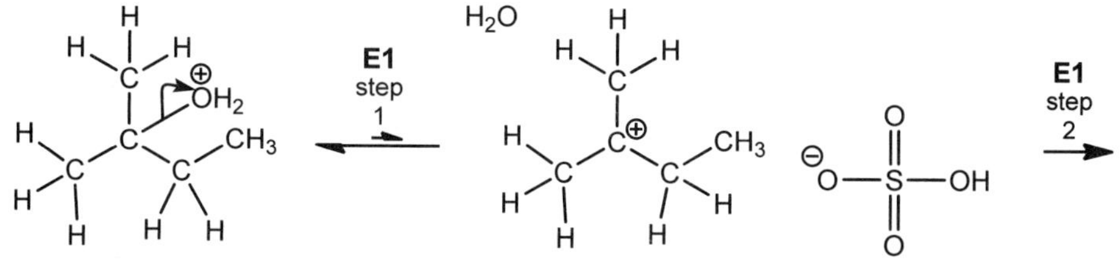

Review of Memorization Task 12.2: Zaitsev's Rule (All E1 reactions obey Zaitsev's Rule)

Regardless of size, a weak base will always favor taking the H_β that leads to most substituted alkene, due to the reversibility of E1 reactions (*see* energy diagram in the Common Points of Confusion section). Because many weak bases (e.g., the conjugate base of sulfuric acid) are quite large, students *incorrectly* assume the Hoffman product will dominate the product mixture.

14. (Check your work) Check that your answer to the previous question is consistent with Memorization Task 12.2 from ChemActivity 12 (and paraphrased above)?

Review of Newman Projection Terminology

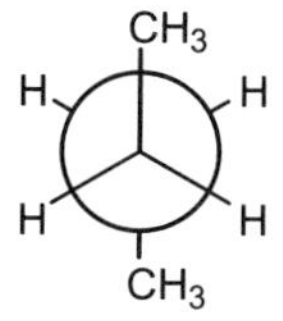

the two methyl groups are said to be **anti to one another**	the two methyl groups are said to be **gauche to one another**

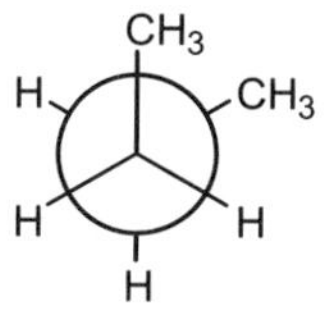

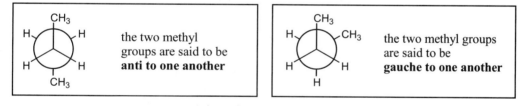

15. Make a model of (S)-2-bromobutane, and confirm that all possible staggered conformations sighting down the C_2-C_3 bond are shown below.

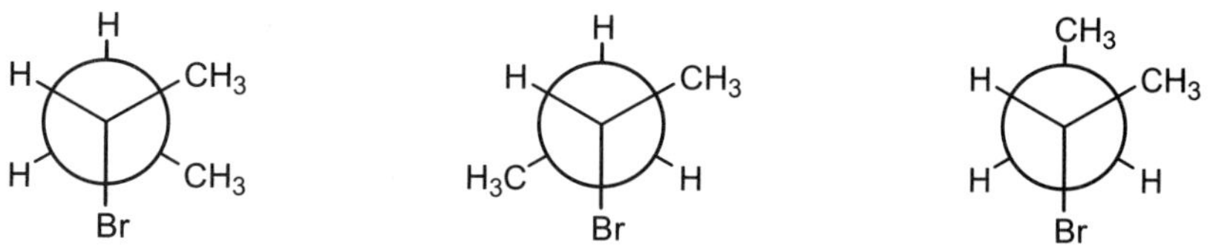

a. It turns out that an E2 reaction is only favorable when the leaving group is *anti* to the H that will be removed by the base. Based on this information, cross out the conformation above that **cannot** lead to an E2 reaction.

b. Which of the remaining two conformations (that can undergo E2) is more favorable in terms of potential energy? Label this one with the words "**lower P.E.—will spend more time in this conformation**" and explain your reasoning.

16. Below are Newman and "sawhorse" representations of the two conformations of (S)-2-bromobutane that can lead to E2 reactions, along with the products of these E2 reactions.

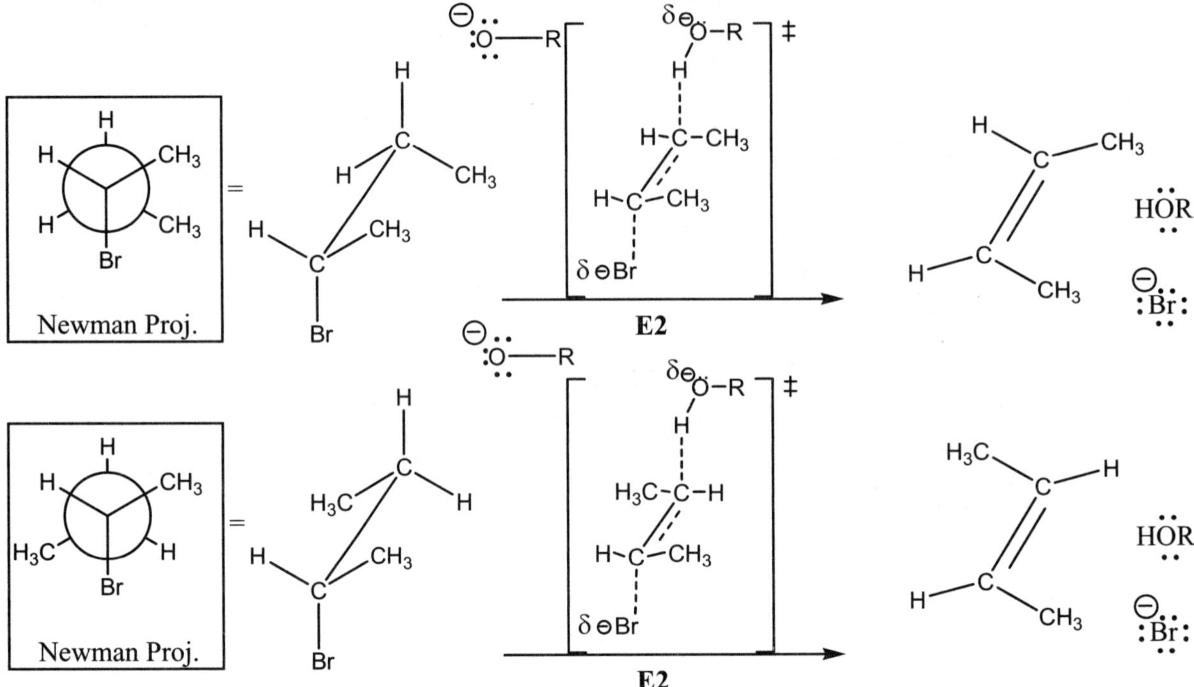

a. On the sawhorse representations of the reactants, use curved arrows to show the flow of electrons during each E2 reaction. *(S_N2 may also occur depending on temperature and R group.)*

b. (Check your work.) Are your curved arrows consistent with the electron changes depicted in the transitions state for each reaction?

c. Label one of the products *Z*-2-butene and the other *E*-2-butene.

17. Fill the blanks in the following paragraph with "*gauche,*" "*anti,*" "*E,*" or "*Z,*" as appropriate.

The reactions above are E2 reactions, so the changes happen all at once in one step. This means that...

If the base takes the H when the methyl groups are _____________ to one another, this "**traps**" the methyl groups **on the same side** of the newly forming double bond and leads to the ____ product.

In contrast...

If the base takes the H when the methyl groups are _____________ to one another, this "**traps**" the methyl groups **on opposite sides** of the newly forming double bond and leads to the ____ product.

18. Predict which will be more prevalent in the product mixture: ***E*-2-butene** or ***Z*-2-butene** (circle one), and explain your reasoning.

19. (Check your work) The answer to the previous question is closely related to Question 15b. Explain.

20. (Check your work.) Is your answer to the previous two questions consistent with the following product distribution? Explain.

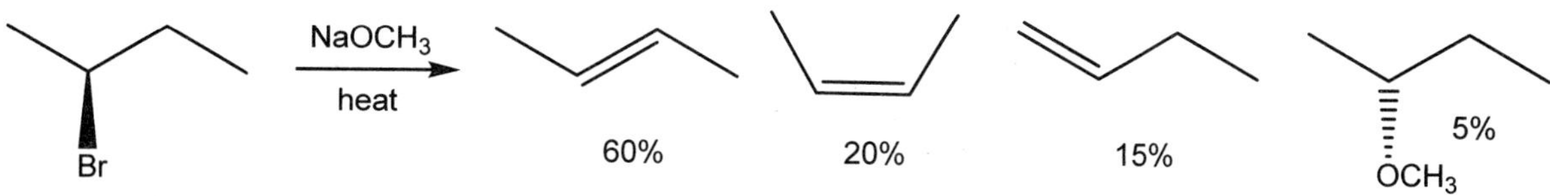

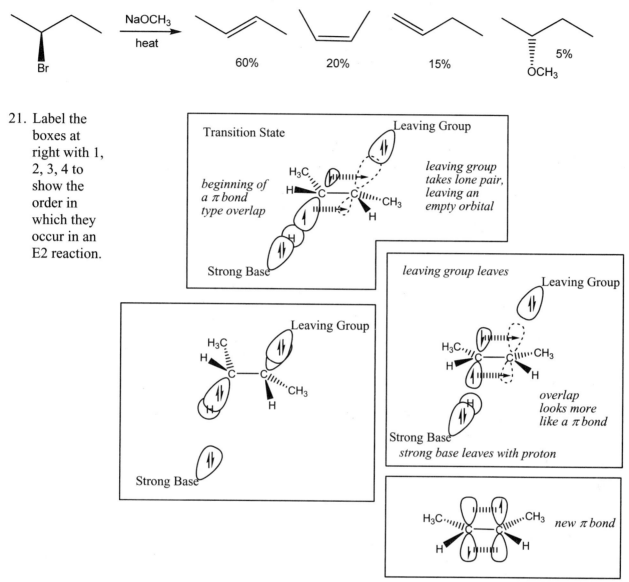

21. Label the boxes at right with 1, 2, 3, 4 to show the order in which they occur in an E2 reaction.

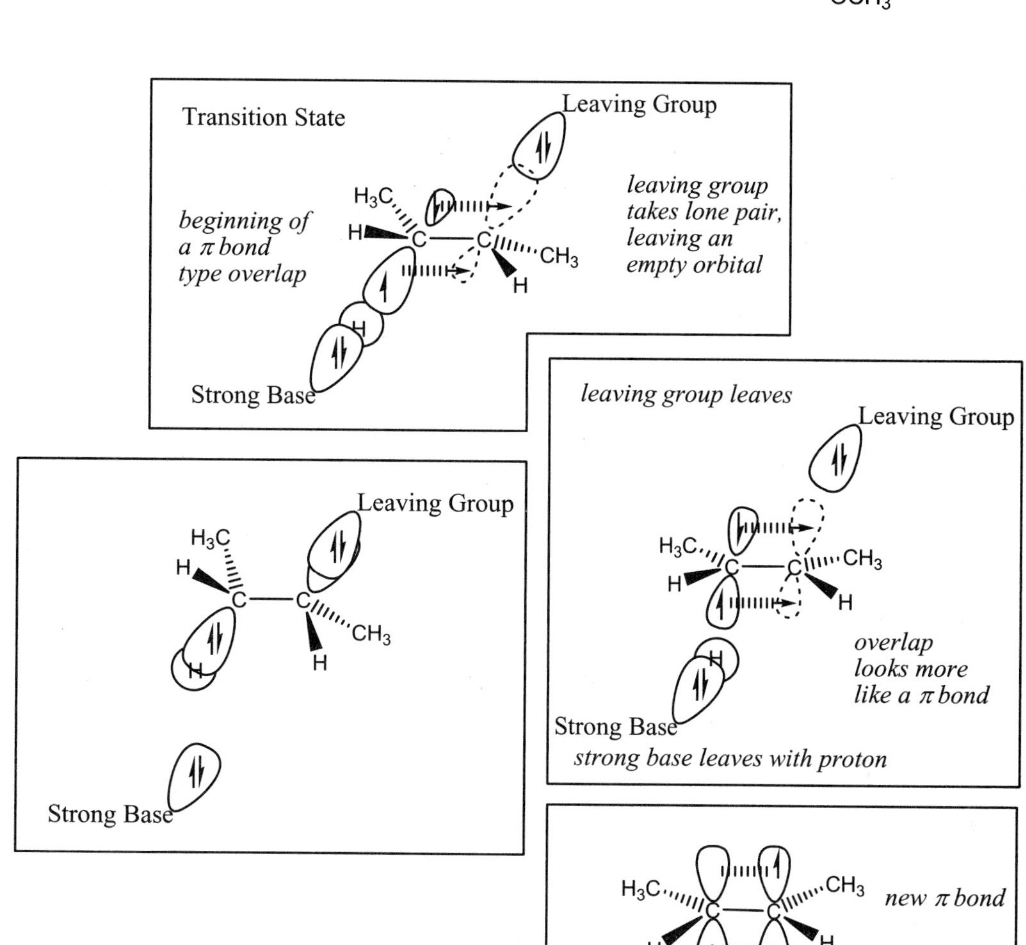

Synthetic Transformations 13.1 and 13.2: E2—Zaitsev or Hofmann

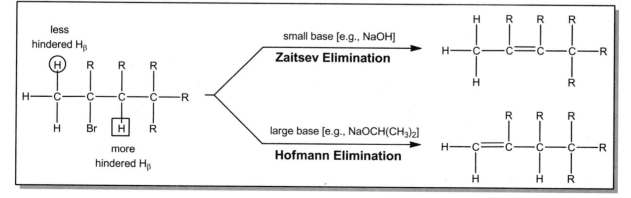

less hindered H_β

more hindered H_β

small base [e.g., NaOH]

Zaitsev Elimination

large base [e.g., $NaOCH(CH_3)_2$]

Hofmann Elimination

Confirm Your Understanding Questions (to do at home)

22. Consider two conformations of one configurational stereoisomer of 3-bromo-4-deuterohexane shown below using Newman projections. Though, in practice, D is slightly harder to remove than H (this is called an **isotope effect**), assume for this question that D has identical reactivity to H in an E2.

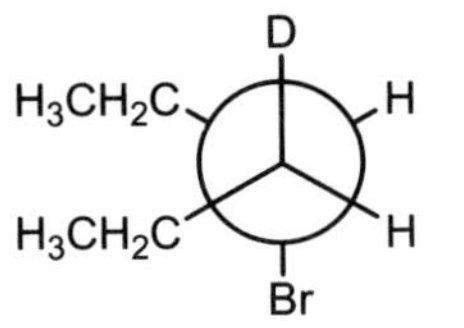

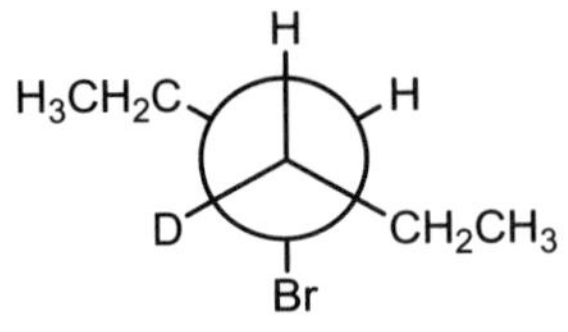

    a. Next to each, draw the product that will form if the conformation shown undergoes E2.

    b. Circle the product you expect to dominate the product mixture, and explain your reasoning.

23. Explain the following generalizations:

    a. You can't do E2 with a weak base.

    b. You can't do E1 with a strong base.

24. Use curved arrows to show the most likely mechanism and major product of this reaction.

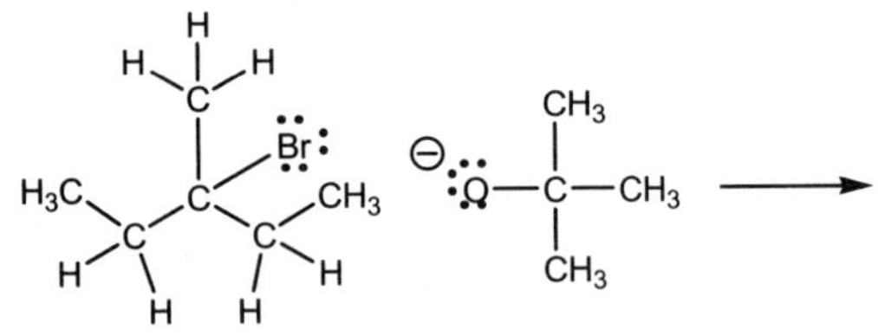

    a. Draw the transition state for the reaction above. Represent each partial bond using a dotted line, and show partial charges as $\delta-$.

    b. Is the major product of this reaction a Zaitsev or a Hofmann product?

25. In terms of orbitals, explain why an E2 reaction cannot occur via removal of an H that is gauche to the leaving group—and why the H must be anti. *Hint*: Sketch an orbital drawing of the transition state removing an H gauche to Br. Will the newly forming *p* orbitals be aligned in a way that can immediately form a double bond?

26. Assign each double bond in the following molecules as E, Z, or *neither*.

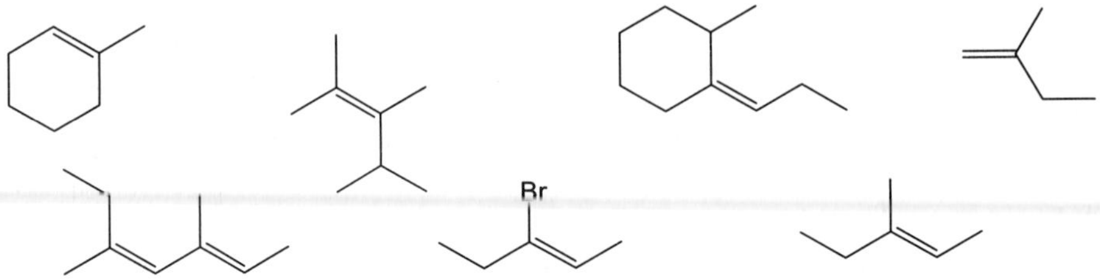

27. For each pair of reactants…
    a. Draw all possible E2 products.
    b. Circle the lowest potential energy product in each case.

i. Br :O—R ⇌

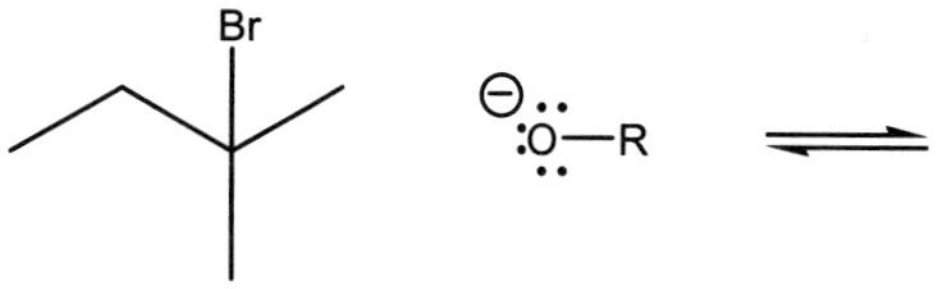

ii. Br :O—R ⇌

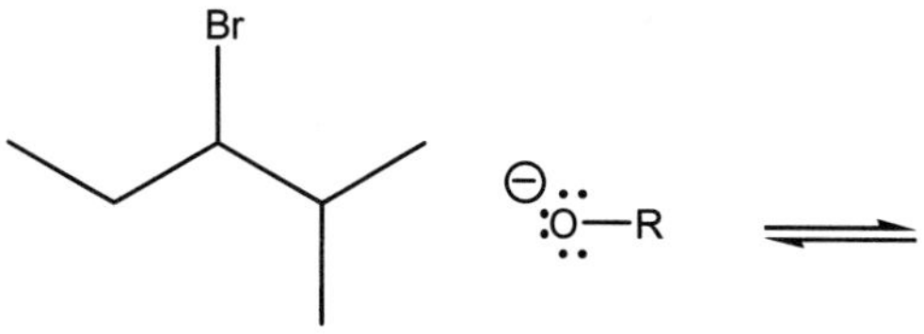

iii. Br :O—R ⇌

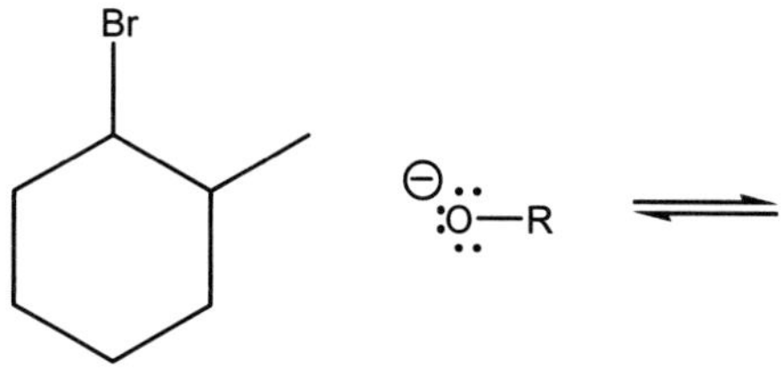

28. Complete the following Newman and sawhorse representations of (S)-3-bromohexane <u>showing a conformation that would give rise to ***trans*-3-hexene**</u>. Also draw *trans*-3-hexene, and show the mechanism of formation using curved arrows.

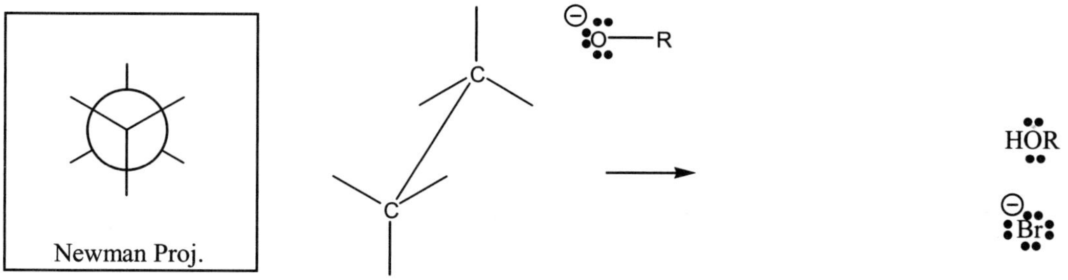

Newman Proj.

C C :O—R → HOR :Br:

"Sawhorse" representation

trans-3-hexene

29. Name the configurational stereoisomer of 3-bromo-4-deuterohexane shown in Question 22.
    a. Draw a Newman projection of the enantiomer of this molecule in its lowest potential energy conformation (what is the name of this molecule?).
    b. Draw the major product of an E2 reaction involving this enantiomer.
    c. Draw a Newman projection of the configurational stereoisomer of 3-bromo-4-deuterohexane that will give *trans*-3-hexene as the major product.
    d. What is the name of this configurational stereoisomer that gives *trans*-3-hexene?
    e. What is the relationship between this molecule and the original molecule in Question 22?

30. Consider the four possible di-substituted E2 products below.

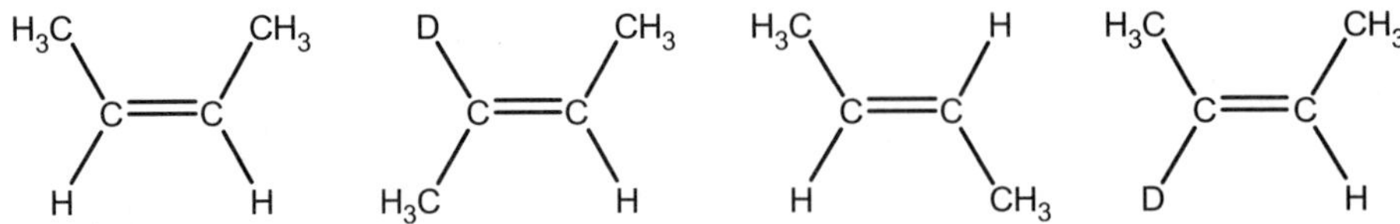

a. Circle the two that could be produced by E2 reaction of (2S,3S)-2-bromo-3-deuterobutane.

b. Mark one as the "major product" and one as the "minor product."

c. Draw a sawhorse representation of the conformation that would give rise to each of these two products.

d. Draw and name the enantiomer of (2S,3S)-2-bromo-3-deuterobutane.

e. Could E2 reaction with this configurational stereoisomer give rise to either of the two uncircled alkene products? If so, show the sawhorse representation of the conformation giving rise to each product.

f. Draw AND NAME a diastereomer of 2-bromo-3-deuterobutane.

g. Can this configurational stereoisomer give rise to the other two products? If so, show the sawhorse representation of the conformation giving rise to each product.

31. On the following drawing of (R)-2-bromobutane...

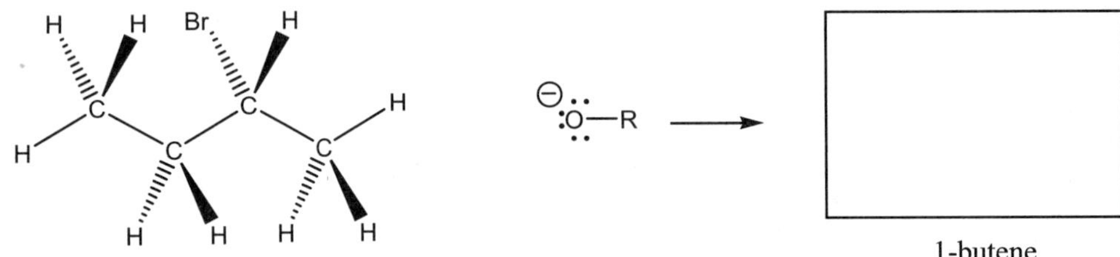

a. Circle any H that, if removed in an E2, would give *cis*-2-butene, and label it "**gives *cis*.**"

b. Circle any H that, if removed in an E2, would give *trans*-2-butene, and label it "**gives *trans*.**"

c. Draw 1-butene in the box provided.

d. Put a triangle around any H that, if removed, would give rise to 1-butene.

e. Give an example of an R group that would give 1-butene as the major product in this reaction.

f. Use curved arrows to show the mechanism of an E2 reaction leading to 1-butene.

g. A student predicts that the ratio of *trans*-2-butene to 1-butene should be 1:3 in the final product mixture. Explain how she came up with the ratio 1:3, and explain the **flaw** in her reasoning.

Read the assigned pages in the text, and do the assigned problems.

The Big Picture

If you are doing these activities in the order they appear in this book, you have now been introduced to four key reactions that are often in competition with one another: S_N1, S_N2, E1, E2.

This activity focuses on the mechanism of an E2 reaction. The key factor in producing an E2 is the use of a strong base. The combination of a good leaving group and a strong base frequently leads to E2. I is nearly impossible to have an E2 without a strong base. Furthermore, an E1 reaction will not (normally) happen in the presence of a strong base.

The next activity focuses on how to decide if a pair of reactants is most likely to undergo S_N1, S_N2, E1, or E2. This is one of the most challenging subjects for introductory students, and also happens to be a favorite topic of professors writing exam questions!

In the Extend Your Understanding Questions of this activity, you are introduced to the simple idea that a normal E2 reaction takes place when the leaving group is anti to the H being eliminated. This simple fact is not to be underestimated, and in the Confirm Your Understanding Questions here are some very challenging stereochemistry questions that stem from this fact. As with many questions involving stereochemistry, start by building a model. (It is well worth the five minutes you will spend.)

Common Points of Confusion

- The most common misconception associated with E2 is that tertiary electrophiles do not undergo E2. THIS IS NOT TRUE! **Tertiary electrophiles commonly undergo E2**. (The cause of the confusion stems from the fact that tertiary electrophiles cannot undergo $\underline{S_N2}$. Since E2 and S_N2 have other similarities, students incorrectly assume they share this trait as well.)
- Hoffman elimination is the term used to describe E2 reactions that do not follow Zaitsev's rule. In an E2 reaction with a large base, it frequently works out that the least hindered H can be removed much faster than an H at a hindered (e.g. tertiary) center. The result is that you can tailor the reaction to give the less substituted alkene product, even though a more substituted alkene is lower in potential energy.
- A common error is to assume that E1 reactions can undergo Hoffman elimination. As stated in the previous activity, E1 reactions always obey Zaitsev's rule, even when the weak base acting in the second step is very large (e.g. the conjugate base of phosphoric or sulfuric acid).

Notes

ChemActivity 14: S_N1, S_N2, E1, E2 in Competition

(Why will an S_N2 reaction often fail with a nucleophile that is also a strong base?)

Model 1: E2 in Competition with S_N2

Any small, strong base is also a good nucleophile, which opens the possibility of S_N2 competing with E2. Under ordinary conditions, the following reactant pairs give a mixture of S_N2 and E2 products.

		E2	S_N2
OTs	KOH → Reaction I		
OTs	KOH → Reaction II		

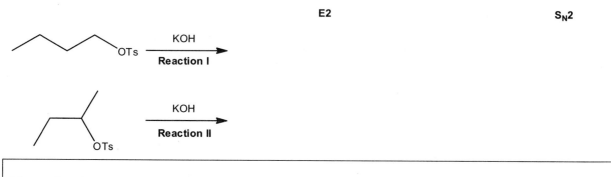

Memorization Task 14.1: Heat favors elimination (over substitution)

For reasons we will not discuss here, higher reaction temperatures favor elimination over substitution.

Construct Your Understanding Questions (to do in class)

1. Draw the <u>major E2 product</u> **and** the <u>major S_N2 product</u> for ***each*** reaction Model 1.
2. In which reaction in Model 1 is S_N2 more likely to dominate over E2 (at a given temperature)? Explain your reasoning.
3. (E)According to Mem. Task 14.1, how can a mixed-mechanism reaction be pushed toward E2?.
4. Is your answer to the previous question consistent with the fact that Reaction II in Model 1 gives an alkene as the major product at 80°C and an alcohol as the major product at 10°C?
5. Drawn at right are the E2 and S_N2 transition states (ts‡) for Reaction II in Model 1.

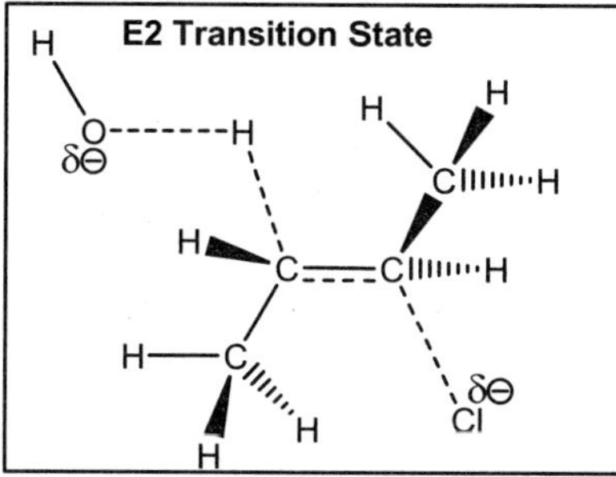

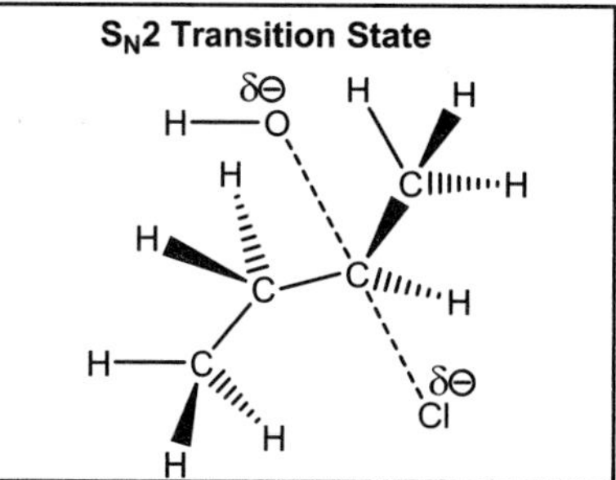

   a. In which ts‡ does HO^- experience more steric repulsion while making a new bond?

   b. Construct an explanation for why increasing the size of the base/nucleophile in a mixed E2/S_N2 reaction will increase the amount of E2 product relative to S_N2 product. (e.g., replacing the H of OH with an alkyl group leads to more E2, and less S_N2).

Memorization Task 14.2: S_N1, S_N2, E1, E2 (Decision Tree)

Don't try to memorize this by brute force. *Start by answering the questions on the next page with your group. You many then want to try to draw the trees from memory without looking at the original.*

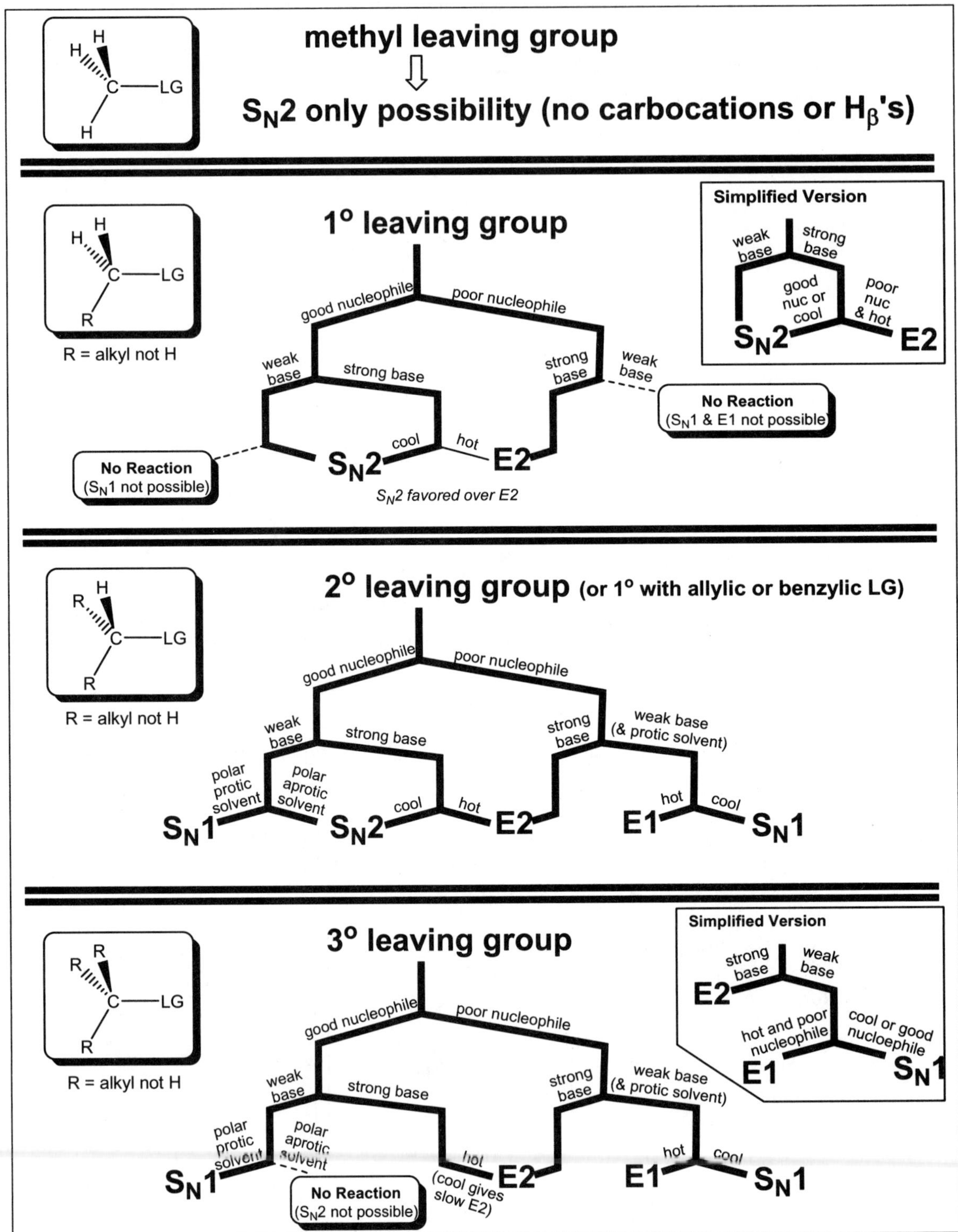

Use this diagram to answer the questions on the following page.

Construct Your Understanding Questions (to do in class)

6. Construct an explanation for each of the following generalizations:
   a. S_N1 and E1 are not possible with 1° leaving groups (unless allylic or benzylic).

   b. S_N2 is not possible with 3° leaving groups.

   c. Strong base favors E2 when a *beta* hydrogen is present (except when the leaving group is primary).

   d. Polar-protic solvents favor S_N1 and E1.

   e. Methyl alkyl halides can only do S_N2.

   f. A larger nucleophile/base favors E2 over S_N2.

7. For a strong base/good nucleophile and a 2° leaving group, what variable can push the reaction toward S_N2 vs. E2?

8. For a weak base/good nucleophile and a secondary alkyl halide, what variable can push the reaction to S_N2 vs. S_N1?

9. Give specific reagents/conditions that fit the following criteria: 2° alkyl halide and…
   a. strong base that would undergo S_N2

   b. weak base that would undergo S_N2

   c. good nucleophile that would undergo E2

   d. poor nucleophile that would undergo E2

   e. reagents that would undergo E1

   f. reagents that would undergo S_N1

Extend Your Understanding Questions (to do in or out of class)

10. In the laboratory, it is often impossible to avoid a mixture of substitution and elimination products. **Fill in the table with the phrases below without looking at the decision tree.**

Column 1	Column 2	Column 3	Column 4	Column 5	Column 6
Favored Mech.	**Base/Nuc**	**R—X**	**Rate dependent on [??]**	**Solvent**	**Temp**
S_N1					
S_N2					
E1					
E2					

Column 2: Strong base; Weak base and poor nucleophile; Weak base and OK nucleophile; Good nucleophile

Column 3: 2°, 3°, allylic, or benzylic only; 2°, 3°, allylic, or benzylic only AND must have H_β; methyl and 1° best but 2° OK; 1°, 2°, 3° allylic, and benzylic all OK but must have H_β

Column 4: R–X only (two boxes); R–X and Base; R–X and Nucleophile

Column 5: polar-protic required (two boxes); polar aprotic (two boxes)

Column 6: hot, cool

11. Explain why only S_N2 (no E2) is observed in Reaction A.

H_3C–OTs $\xrightarrow[\text{Reaction A}]{\text{KOH}}$

12. Explain why only E2 (no S_N2) is observed in Reaction B.

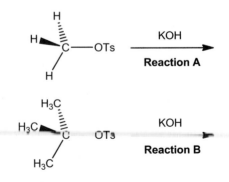

$(H_3C)_3C$–OTs $\xrightarrow[\text{Reaction B}]{\text{KOH}}$

Confirm Your Understanding Questions (to do at home)

13. For many pairs of reactants, at least two of the mechanisms S_N1, S_N2, E1, and E2 are in competition with one another. Each of the following pairs is an exception in that it goes almost exclusively by one of these four reaction mechanisms. For each reaction…

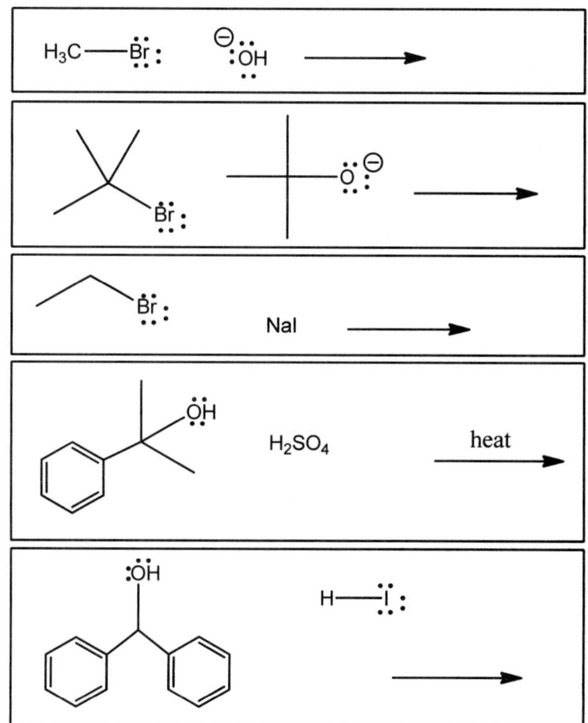

   a. Decide which reaction mechanism is most likely for each box, and write S_N1, S_N2, E1 or E2 over the reaction arrow.

   b. Use curved arrows to show the mechanism, and draw the product/s.

   c. Briefly explain why these starting materials are unlikely to go by each of the other three reaction mechanisms.

14. Agree or disagree with the following statements overheard during student discussions in class, and explain your reasoning.

   a. S_N1 and E1 cannot occur with a 1° leaving group

   b. S_N2 and E2 cannot occur with a 3° leaving group

15. For each of the following reactions, decide which, if any of the mechanisms: S_N2, S_N1, E1, or E2 is most likely, draw the resulting product (if any) and explain your reasoning.

   a. *t*-butyl chloride (2-chloro-2-methylpropane) mixed with aqueous HBr

   b. 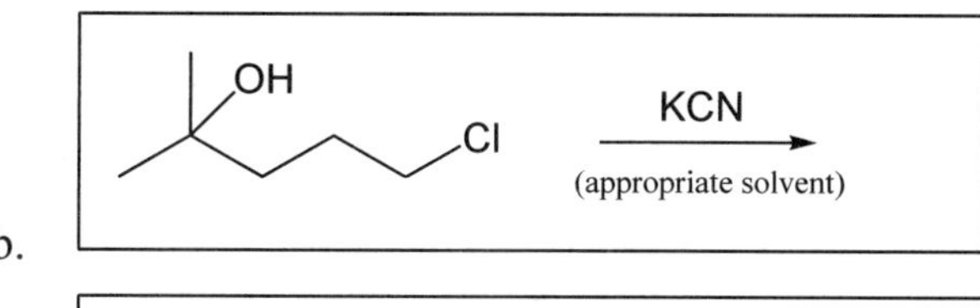

   c. 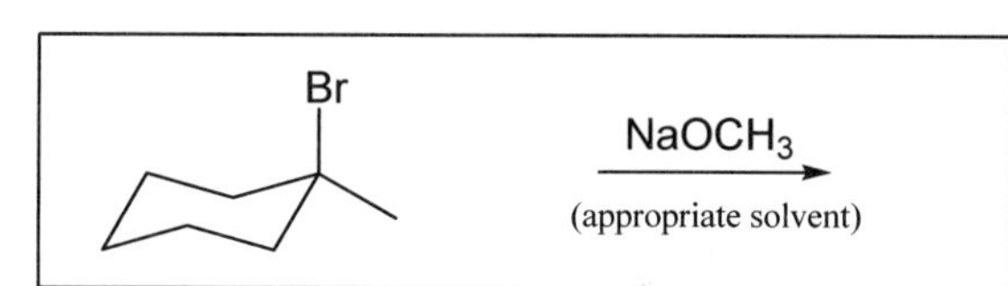

   d. *t*-butanol is heated in aqueous KBr

   e. Bromocyclohexane is heated in ethanol with NaOEt

16. The most common conditions for an E2 reaction are to heat an alkyl halide with an alkoxide base (e.g. NaOEt) using the corresponding alcohol as solvent (e.g. EtOH).
    a. The polar protic solvent (e.g. ethanol) will hydrogen bond to the base (e.g. NaOEt) and thereby reduce its effective basicity. Explain why heating the reaction can compensate for the decreased effective basicity.
    b. Polar protic solvents are NOT appropriate for S_N2 reactions. This is because a polar protic solvent will hydrogen bond to the nucleophile, dramatically reducing its nucleophilicity. What happens if you heat such a reaction to increase the activity of the nucleophile?
    c. The common conditions for E2 (alcohol/alcoxide) contradict the rule that you should use a polar protic solvent, such as ethanol, when you want to promote carbocation formation. Explain why S_N1 and E1 almost never compete with E2 in the presence of a strong base.

17. Take out a blank piece of paper, and **without consulting the decision tree** write down reagents/conditions that would yield… S_N2, S_N1, E1, or E2, respectively.

18. Take out a blank piece of paper, and try to generate from memory the decision tree for each methyl, primary, secondary, and tertiary leaving groups.

Read the assigned pages in the text, and do the assigned problems.

The Big Picture

Everyone finds this activity challenging. It takes a week or two of hard work, sifting through substitution and elimination reactions, to finally feel comfortable with the decision tree in Memorization Task 14.2. Do lots of problems, and constantly ask yourself or your study partner "why" for each branch point of the decision tree. (The only concept we are asking you to memorize without an explanation is that heat favors elimination—though this makes sense too, if you wish to look it up.) The others should all make sense after a careful study of the ChemActivities covering substitution and elimination.

Common Points of Confusion

- Do not blindly memorize the associations: tertiary leaving group → S_N1 and E1; and primary leaving group → S_N2 and E2. Only the first three-quarters of this rule are valid. Tertiary electrophiles cannot undergo S_N2 (too hindered), and (ordinary) primary and methyl electrophiles cannot undergo S_N1 or E1 (unfavorable carbocations). But the rule DOES NOT work for E2. Tertiary electrophiles commonly undergo **E2**.

Notes

ChemActivity 15: Retrosynthesis

(How can thinking backwards help me solve a synthesis problem?)

Model 1: Introduction to Multi-Step Organic Synthesis

The main work of most organic chemists is building new molecules for use as everything from food additives to materials for the space shuttle. A large portion of this work is focused on drug synthesis.

The end product of a synthesis is called a **target molecule**. Chemists rarely "hit" the target on the first try. In fact, a team of chemists may spend months or years trying to synthesize a particular molecule.

The total synthesis of Vitamin B12 led by Robert Woodward of Harvard University in the 1960s and 70s involved hundreds of graduate students and took over a decade. This amazing accomplishment ushered in the modern era of synthetic organic chemistry and convinced many that synthesis of any target was possible given enough time and resources.

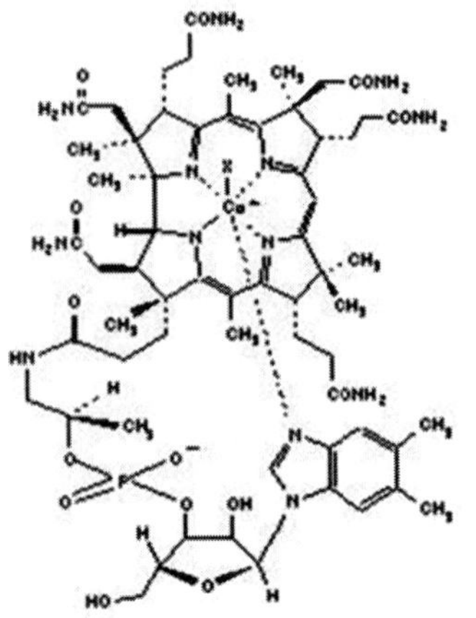

Figure 15.1: Vitamin B12

Our exploration of synthesis will focus on simple targets that are often **commercially available**. If we were really trying to obtain such a molecule we would order it from a chemical company with a facility for making it in large batches, allowing them to sell it for much less than it would cost us to make it in the laboratory.

Shown below is an example of a simple, linear, multi-step synthesis.

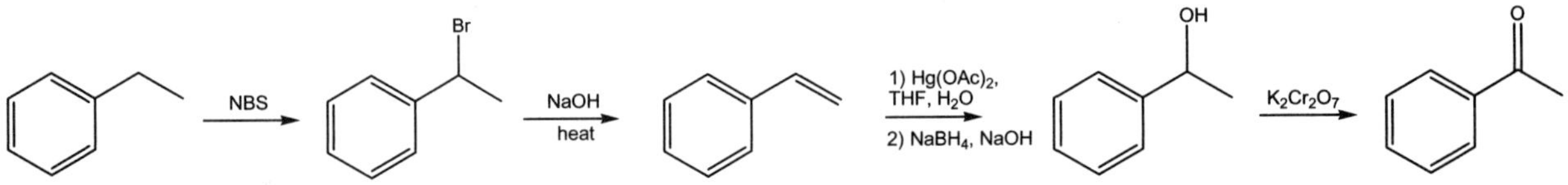

starting material — NBS → (Br) — NaOH, heat → — 1) $Hg(OAc)_2$, THF, H_2O 2) $NaBH_4$, NaOH → (OH) — $K_2Cr_2O_7$ → (O)

Construct Your Understanding Questions (to do in class)

1. Label the **target molecule** in the multi-step synthesis above.
2. Design a synthesis of the following target from the starting material.

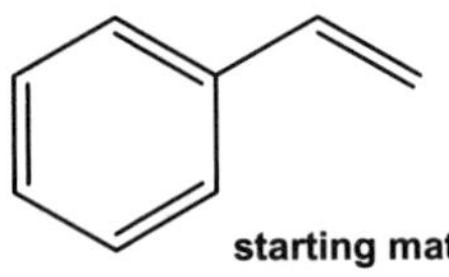

starting material

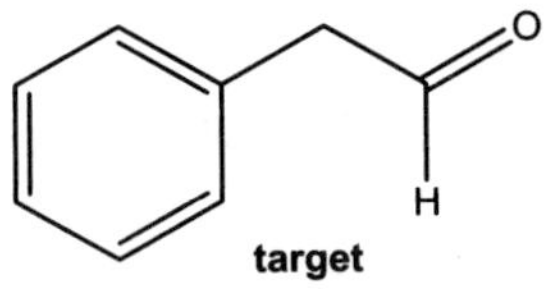

target (O, H)

Model 2: Retrosynthesis (Thinking Backwards)

Organic chemists use a process called retrosynthesis to "think backwards" from the target and figure out what **commercially available starting materials** are needed to make this target.

To solve the synthesis problem below using retrosynthesis you would start by asking yourself the question: "What molecules can be transformed into 2-bromo-3-methylbutane (4) in one synthetic step?" These molecules are called **precursors** of molecule **4**.

By convention, a **retrosynthesis arrow** is drawn from a molecule to its precursor (as shown below).

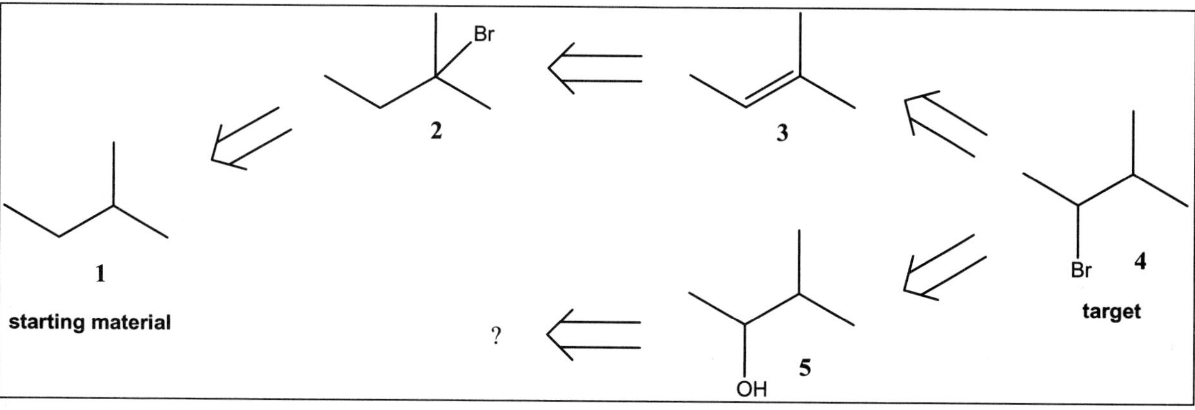

Figure 15.2: Retrosynthetic analysis of the target, 2-bromo-3-methylbutane (4)

Construct Your Understanding Questions (to do in class)

3. Identify the retrosynthesis arrows in Figure 15.2.
4. Label the two different precursors of molecule **4** shown in Figure 15.2.
5. Label the precursor of molecule **3** shown in Figure 15.2.
6. What reagents will transform **3** into **4**? (Draw a reaction arrow from **3** to **4**, and write these reagents above it.).
7. What reagents will transform **5** into **4**? (Draw a reaction arrow from **5** to **4**, and write these reagents above it.).
8. Are there reagents that will transform **1** into **5** in two or fewer synthetic steps?

9. Is your answer to the previous question consistent with the fact that the preferred route from **1** to **4** does not involve **5**?

10. You can deduce the structure of **2** by thinking of a precursor of **3**. Or, since the starting material is given, you can work forward one step from the beginning. This is especially easy when the starting material is an alkane because students typically learn only one type of reaction that applies to un-functionalized alkanes. Draw reaction arrows from **1** to **2**, and from **2** to **3**, and write reagents above them.

Memorization Task 15.1: Get Started with Radical Halogenation

When the starting material is an alkane, the first step in the synthesis is very often radical halogenation.

11. Use retrosynthesis to design a synthetic pathway from the starting material to the product.

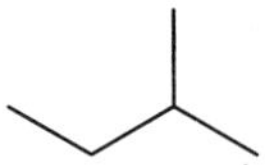

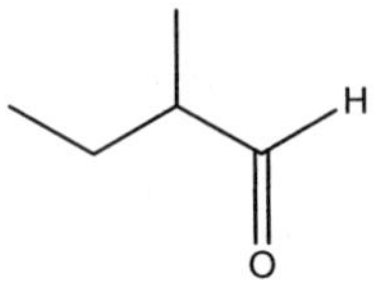

12. In the previous synthesis problem, did any carbon atoms need to be added to (or taken away from) the starting material to make the product?

13. In the next synthesis problem, do any carbon atoms need to be added to (or taken away from) the starting material to make the product?

14. Design a synthesis of the following target using the starting material and any reagents with one or fewer carbon atoms. (Work for five minutes on this question, then go onto the next section, which walks you part of the way through this problem.)

HC≡CH

starting material

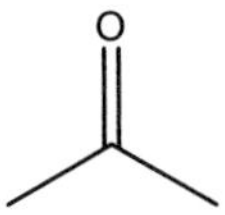

target

Model 3: Assembly of the Carbon Backbone

A key part of any synthesis is assembly of the **carbon backbone**. In the previous question the starting material has a two-carbon backbone, but the product has a three-carbon backbone.

This retrosynthetic analysis should likely start by deciding which two carbons in the product will come from the starting material. (In this case, because of symmetry, it does not matter if you choose the right two or left two.)

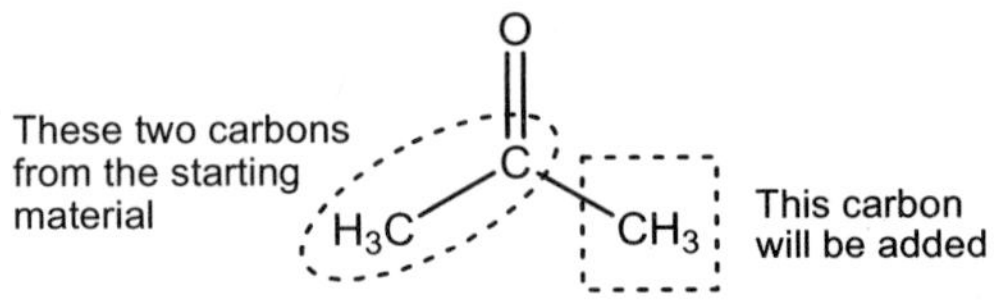

These two carbons from the starting material

H_3C C CH_3

This carbon will be added

During this synthesis we have to make a new carbon-carbon bond. For our retrosynthetic analysis (going backwards) we "cut" this new bond with a squiggly line to make two pieces. As you will see on the next page, **the challenge is to make these pieces and then bond them together**.

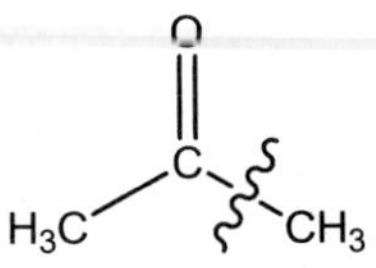

H_3C C CH_3

We know from our analysis on the previous page that at some point in this synthesis we must bond a two-carbon piece together with a one-carbon piece.

One way to bring these pieces together is to make one have some + charge and the other have some – charge. If we could magically create a perfect pair of pieces for the synthesis on the previous page, it might look like one of pairs in the boxes below. We call these idealized pieces **synthons**.

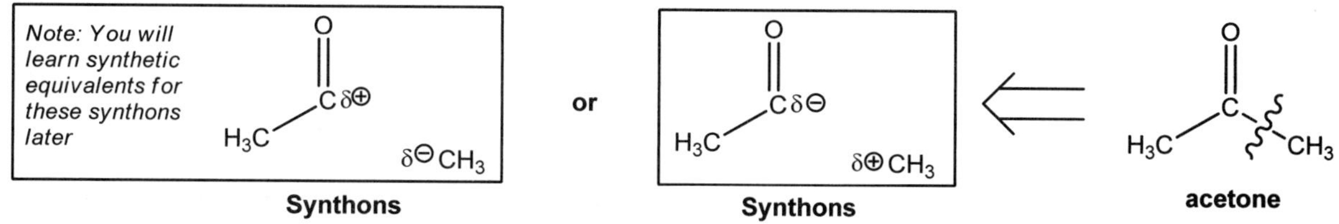

Figure 15.3: Two pairs of synthons for making the target acetone

Unfortunately, no real reagent looks exactly like any of these pieces, but we can often find real reagents that behave like a given synthon. A reagent that behaves like a synthon is called a **synthetic equivalent** of that synthon.

Construct Your Understanding Questions (to do in class)

15. In what way is the pair of synthons in the left in Fig. 15.3 the opposite of the pair on the right?

16. Which <u>one</u> of the following reagents is NOT a **synthetic equivalent** of the synthon $H_3C^{\delta+}$?

Br—CH_3 Cl—CH_3 I—CH_3 H—CH_3 TsO—CH_3

17. The previous question gives us a choice of several one-carbon electrophiles that can serve as a synthetic equivalent for $H_3C^{\delta+}$. Now we need a two-carbon nucleophile to serve as a synthetic equivalent of the synthon it is paired with in the box above. What reagent can transform the starting material (acetylene) into a nucleophile in one step? (*If you need a hint, look at Model 4.*)

18. <u>Draw the first two forward steps of this synthesis</u>. For each step, include a reaction arrow with the necessary reagents written above it.

HC≡CH

ethyne (or acetylene)

starting material

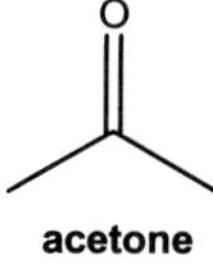

acetone

19. If you have studied alkynes already you likely learned that adding acid/water to propyne gives acetone. That is, HC≡C—CH_3 is a precursor of acetone. Add the third (and final) step to your synthesis above.

Model 4: Nucleophiles

Very Good Nucleophiles	Good Nucleophiles	Poor Nucleophiles
RS^-	Br^-	F^-
$N{\equiv}C^-$	R_2S	HCO_3^-
I^-	NR_3	R_2O
PR_3	Cl^-	*(water, alcohol, or ether)*
$*R_3C^-$	RCO_2^-	
$*R_2N^-$	N_3^-	*All strong acids are very poor nucleophiles*
$*RC{\equiv}C^-$		
$*RO^-$		

Extend Your Understanding Questions (to do in or out of class)

20. Look at Model 4 and find all three types of carbon nucleophiles listed there.

   a. The nucleophile R_3C^- is a topic for later in the course. For now, we will focus on the other two carbon nucleophiles from the table. What do they have in common?

   b. In the next question, how many carbons need to be added in going from starting material to target?

21. Use retrosynthesis (work backwards) to solve this synthesis problem in two different ways.

   a. First, solve the problem using the one-carbon nucleophile in Model 4. (Note that hydrolysis of $R\text{-}C{\equiv}N$ with acid/water gives a carboxylic acid, $R\text{-}CO_2H$).

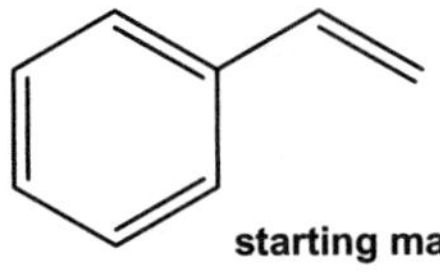

starting material

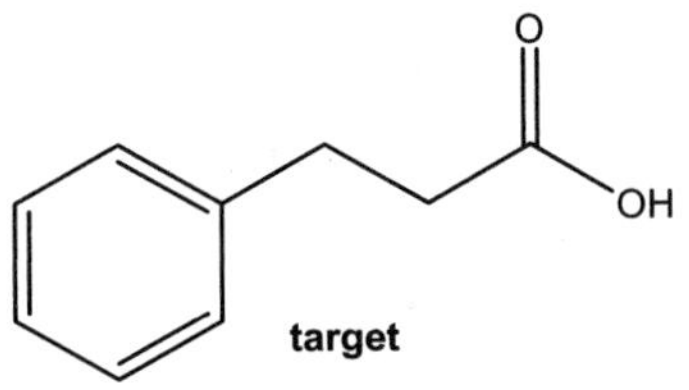

O

OH

target

   b. Now solve it using an acetylide ion ($RC{\equiv}C^-$). [*Hints: First adding carbons, then remove all but one. What reactions do you know for breaking carbon-carbon bonds?*]

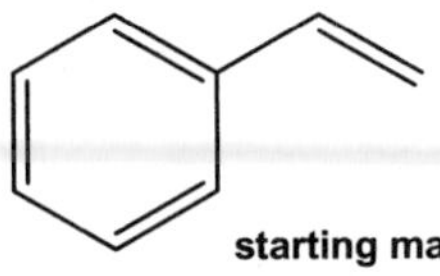

starting material

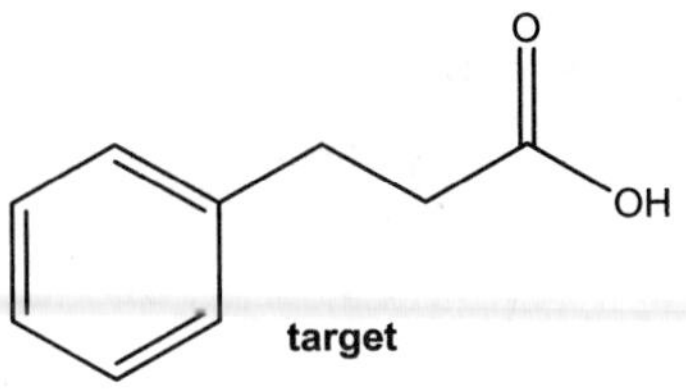

O

OH

target

Model 5: Functional Group Approach

One way to organize a textbook or course is by **functional group**. For example, you can find a chapter in most other textbooks called "Alcohols" that has sections on naming alcohols, making alcohols, and reactions of alcohols.

The chapters in this workbook are instead organized around the main reaction types: electrophilic addition, nucleophilic substitution, elimination, (and in Volume II) nucleophilic addition, addition-elimination, and electrophilic aromatic substitution.

However, now when you are thinking about organic synthesis it is critical to know the names and characteristics of the main organic functional groups. More and more we will be asking questions about functional group transformations. (e.g., What reagent will transform an alcohol to an alkyl bromide?) Mechanisms for many reactions will still be important, but the emphasis is shifting toward synthesis.

Extend Your Understanding Questions (to do in or out of class)

22. Below are examples of molecules containing various functional groups.
    a. Match the name of the functional group with the appropriate molecule. Choose from: Amine, Alcohol, Aldehyde, Alkane, Alkene, Alkyl Halide, Alkyne, Carboxylic Acid, Epoxide, Ether, Ketone, Nitrile (cyano), Nitro, Phenyl Group (benzene ring) and Thiol.
    b. Next to the carboxylic acid, draw its conjugate base (a **carboxylate ion**).
    c. Next to the amine, draw its conjugate acid (an **ammonium ion**).

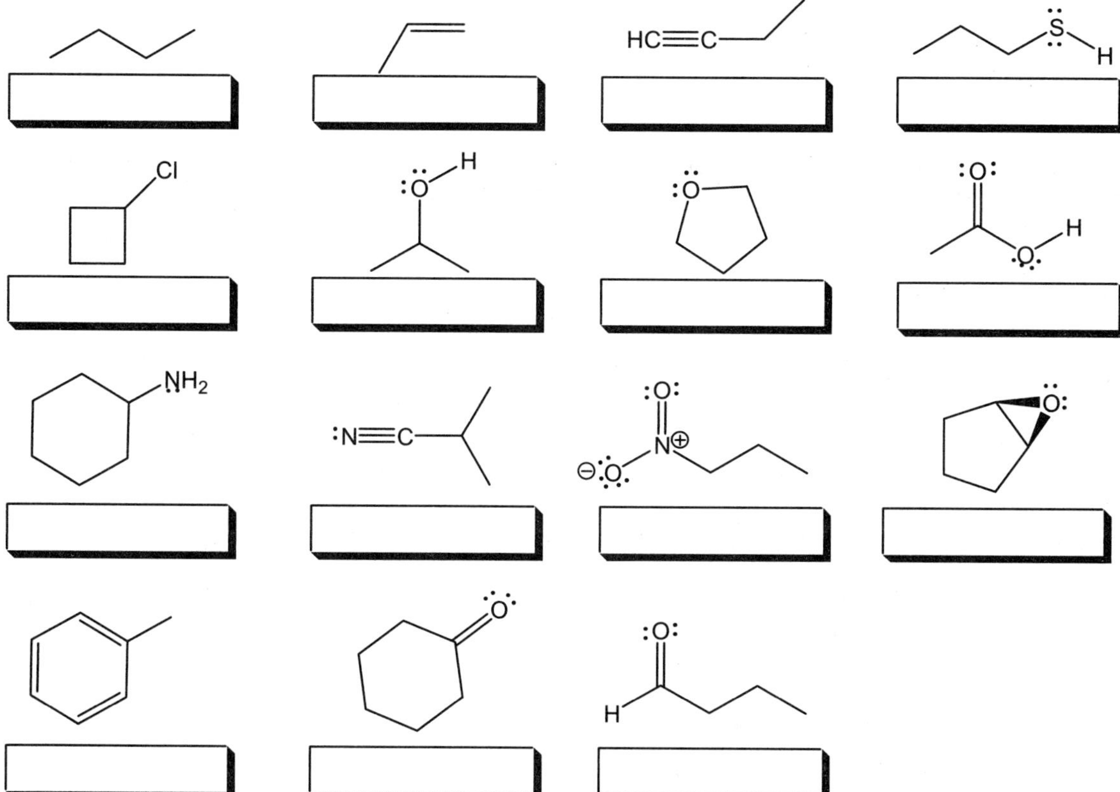

Confirm Your Understanding Questions (to do at home)

23. Draw at least two precursors of ….
    a. an alkene
    b. an alkyl bromide
    c. an alcohol
    d. a diol

24. Consider the acid-base reactions at right.
    a. Use curved arrows to show each reaction.
    b. Draw the products of each reaction.
    c. For each reaction, estimate ΔH_{rxn} using pKa values

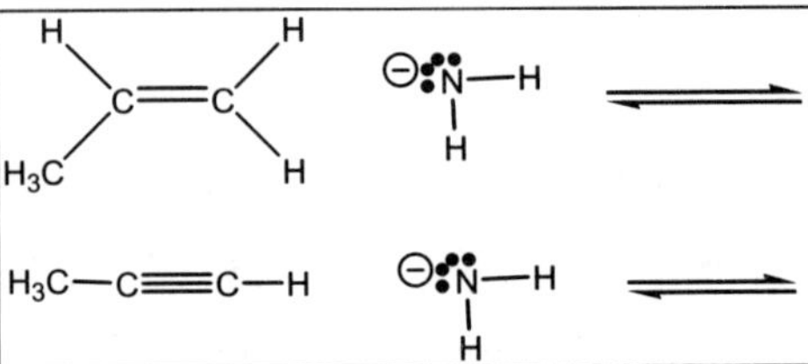

    d. Which reaction is downhill (favorable), and which reaction is uphill (unfavorable)?
    e. Explain why the methyl H's are not removed in the case of the terminal alkyne.

25. Heating a **geminal dihalide** in base gives an alkyne. Show the mechanism and products of the following reaction.

NaOH heat

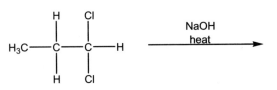

26. Use retrosynthesis to design a synthesis of each target using the starting material and any reagents containing one carbon or less.

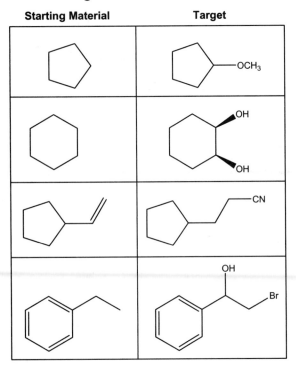

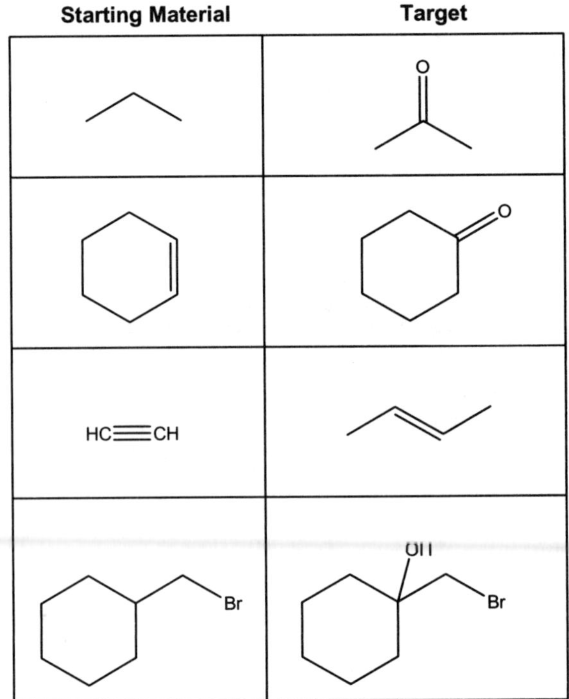

27. Use retrosynthesis to design a synthesis of each target using the starting material and any reagents containing three carbons or less.

Starting Material	Target
	CO_2H
	O

Starting Material	Target
OH	O, H
	O
	O, O
	O, OH

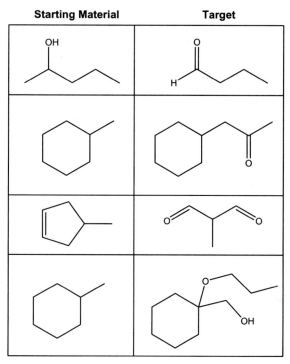

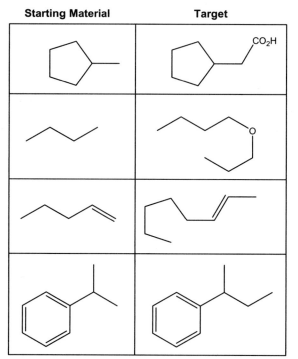

28. Explain why an electron pair in an *sp* hybrid orbital is lower in potential energy than an electron pair in an sp^2 hybrid orbital.

29. Explain why ethyne is more acidic than ethene or ethane.

30. Construct an explanation for why the C—H bond in ethyne is shorter than the C—H bond in ethene or ethane (even though it is easier to break with a base).

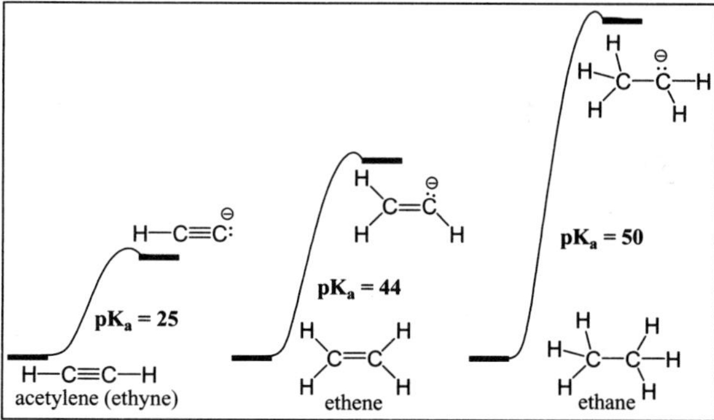

31. True or False: It is easier to reduce the first π bond of a triple bond than the second π bond. Cite evidence from one or more Synthetic Transformations to support your choice.

Read the assigned pages in the text, and do the assigned problems.

The Big Picture

Most organic chemistry students find synthesis to be the most challenging aspect of the course, but it is potentially the most fun and creative. Synthesis problems are this course's best approximation of the open-ended problems that dominate research and medicine. Often the hardest part of a synthesis problem is getting started and developing a plan. The Common Points of Confusion section outlines some study strategies that will help you do this, making many synthesis problems trivial and the others an exciting challenge.

Going forward, synthesis is going to appear on every exam and in every chapter of this book and is the work of the majority of organic chemists, those called synthetic organic chemists.

Common Points of Confusion

As with every topic in this course, the situation is not nearly as bad as you think. Do not let the reputation of this course or your anxiety get in the way of enjoying one of the most creative and potentially fun challenges you will encounter in all of science: organic synthesis.

- **Memorize all the Synthetic Transformations.** The number one reasons students find synthesis difficult is that they have not spent sufficient energy learning the synthetic transformations!
- Many students find it useful to make note cards, several for each synthetic transformation. For example, on the front of the first card put the starting material and product, and on the back put the reagents. On the front of the next card put the starting materials and reagents, and on the back the product. You must know these forward and backward. Put some time and creativity into making the cards and the act of making them will take you halfway to your goal of memorizing them.
- It is far more beneficial to make your own notecards, but once you have made and memorized them, swap your cards with a group mate. Working through someone else's cards is a good way to test your understanding and fill in any gaps.
- Make a stack of the notecards that you do not know yet, then put them in your back pocket. Review them whenever you have two spare minutes, in the bathroom or even waiting to cross the street.
- Do synthesis problems working both from the starting material forward and, using retrosynthesis, from the target backwards.
- On a test, don't waste time staring at a blank page. Keep your pencil moving and your brain will follow. List the possible first steps forward from the starting material AND the possible last steps, thinking backwards (retrosynthetically) from the target.
- Many two-to-four-step synthesis problems can be solved in the forward direction without using retrosynthesis, but retrosynthesis is an invaluable tool for developing a plan for solving more complicated syntheses. If you do not understand the point or methods of retrosynthesis, discuss it with a person who seems to like synthesis.
- Chess masters study games played by other masters to learn common sequences of moves. Study the solutions to synthesis problems to build your repertoire of common sequences of transformations.
- Make up your own synthesis problems, and share them with your study partner.

Notes

ChemActivity 16: Carbon (^{13}C) NMR

BUILD MODELS: 1-bromo-2,3,3-trimethylbutane

(What can a ^{13}C NMR spectrum tell you about the structure of a molecule?)

This activity often requires more than one class period, and can be completed at home or in lab.

Model 1: Spectroscopy Using Radio Waves

In the presence of an external magnetic field, a ^{13}C atom will absorb light in the radio frequency range. The precise frequency of the light absorbed tells us about the atoms near that carbon atom.

For example: the five different carbon atoms labeled **a** to **e** below absorb light of different frequencies. Frequency is measured along the x axis on the chart in a ratio of units called **ppm**. Each cluster of peaks (labeled with a letter) indicates a unique carbon that absorbs light at the frequency indicated on the chart.

When the light is absorbed, it causes what is called a **nuclear spin flip**, so this type of spectroscopy is called **nuclear magnetic resonance** (or **NMR**) spectroscopy.

NMR is the most widely used tool for determining the structure of molecules.

It turns out that the physics of NMR spectroscopy is quite complex. Since its discovery in the 1940s it has grown to become its own subfield of chemistry. The simplified explanation above will be enough for us to learn to interpret NMR spectra.

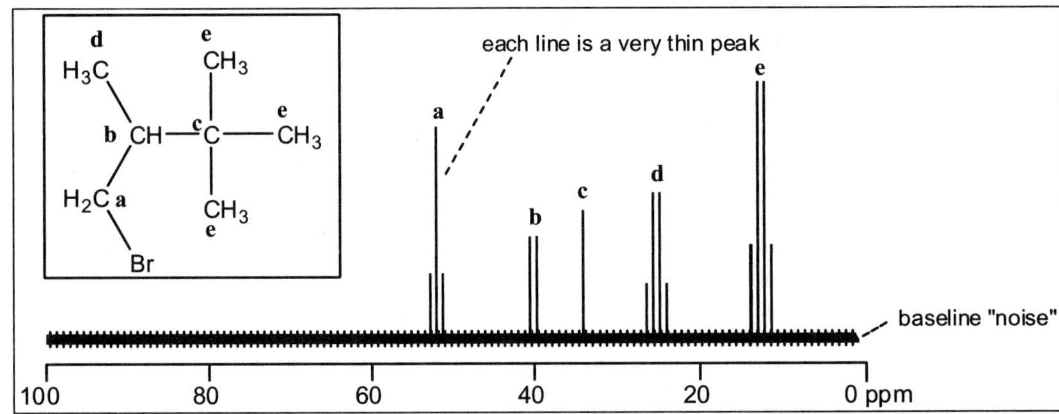

Figure 16.1: Cartoon of (coupled) ^{13}C NMR Spectrum of 1-bromo-2,3,3-trimethylbutane

Construct Your Understanding Questions (to do in class)

1. Next to each carbon in the structure, indicate the number of hydrogens attached to that carbon. Also write this number next to the corresponding **peak cluster** on the spectrum.

2. Construct an explanation for why the three carbon atoms labeled "**e**" are called chemically equivalent (or NMR equivalent) carbons. A model may help.

3. Is there a relationship between the number of hydrogens attached to a given carbon and the number of peaks in the cluster associated with that carbon? If so, what is it?

Memorization Task 16.1: Peak Multiplicity = [Number of H's] + 1

Chemists say that "the two H atoms attached to C_a (we will call them H_a) **'split'** the signal of C_a into three peaks." The **multiplicity** (number of peaks in the cluster) of the C_a and C_b peaks is illustrated in the diagram below. The dotted line illustrates what the peaks would look like without any splitting.

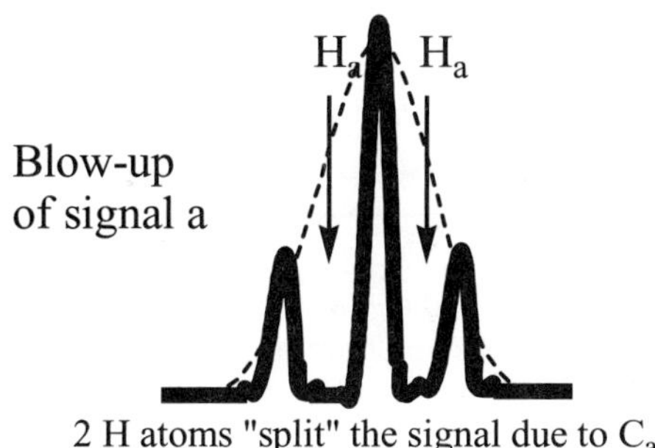

2 H atoms "split" the signal due to C_a

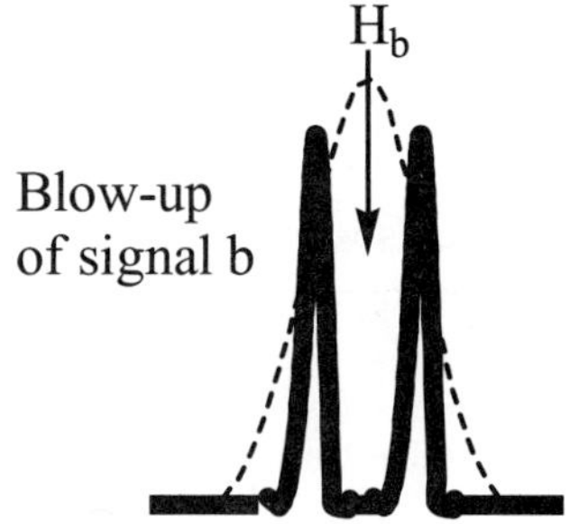

1 H atom "splits" the signal due to C_b

A peak cluster with…

- **one** peak is called a **singlet (s)**
- **two** peaks is called a **doublet (d)**
- **three** peaks is called a **triplet (t)**
- **four** peaks is called a **quartet (q)**
- **five or more** peaks is called a **multiplet (m)**

Construct Your Understanding Questions (to do in class)

4. Label each peak cluster in Figure 16.1 with s, d, t, q, or m.
5. Which of the following best explains the placement of peak clusters along the x axis?
   a. The number of hydrogens attached to that carbon atom.
   b. The total number of bonds to that carbon atom.
   c. Distance from the Br atom (Number of bonds away from Br atom).

Memorization Task 16.2: Memorize the following ^{13}C NMR ppm ranges

The property measured by the x axis is called **chemical shift** (think of it as frequency). It is measured in special units called **ppm**.

The chemical shift of an atom is related to the density of the electron cloud around that atom.

Chemical shift is hard to predict precisely. Memorize these **NMR ppm ranges (^{13}C)**.

Chemical shift factors can be synergistic. *e.g. an alkene* ***C*** *attached to a halogen (C=****C****–Br) could be found above 150 ppm.*

Type of Carbon	Examples (R = H or alkyl)	ppm Range
C with all single bonds to C's or H's	CH_4 R_3**C**H	5-60
C with a single bond to O, N, or a halogen	R_3**C**—OH R_3**C**—Br	20-90
C with double/triple bond to C (e.g. C=C)	R_2**C=C**R_2	110-150
C with double bond to O (**C**=O)	R_2**C**=O R**C**O_2H	150-220

6. Consider the following ^{13}C NMR spectrum of 2-bromobutane. Label each peak cluster with a number (1, 2, 3, or 4) indicating its assignment to a specific carbon and a letter (s, d, t, or q).

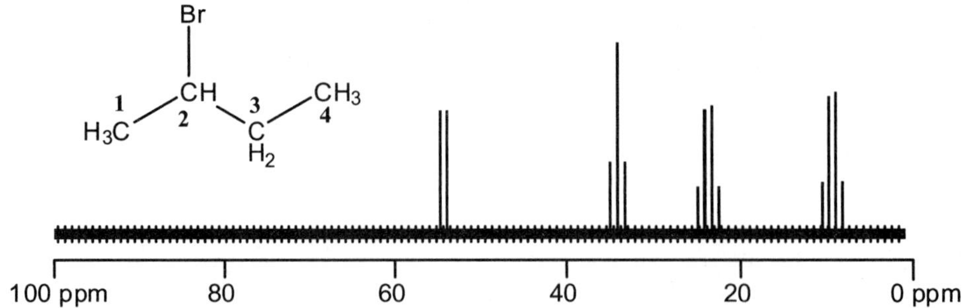

Figure 16.2a: Cartoon of (coupled) ^{13}C NMR Spectrum of 2-bromobutane

Model 2: Coupled versus Decoupled NMR Spectra

If you have a complex molecule with several similar types of carbon atoms, the peak clusters often overlap. This makes interpretation very difficult. To solve this problem, chemists "**decouple**" the H's from the C's. A **decoupled** ^{13}C NMR spectrum of 2-bromobutane is shown below.

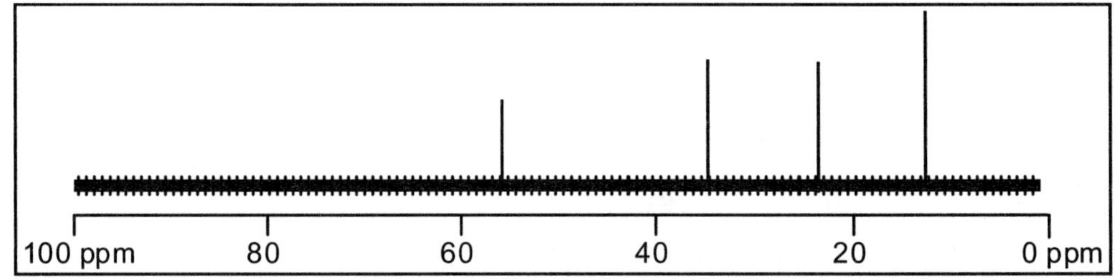

Figure 16.2b: Cartoon of *Decoupled* ^{13}C NMR Spectrum of 2-bromobutane

Think of each C–H bond as a phone link. In the coupled spectrum, each H is "talking to" the nearest carbon, "splitting" the carbon signal through this "phone link." To make a decoupled spectrum, the instrument produces lots of static on the "phone line" so each carbon cannot "hear" the H's attached to it. Consequently, each carbon shows up as a singlet (one peak). You may never see a coupled ^{13}C NMR spectrum outside this activity, though they are presented here as a simple example of splitting in NMR.

Construct Your Understanding Questions (to do in class)

7. What structural information is lost when we decouple the C's from the H's?

8. What basic structural information does a decoupled ^{13}C NMR spectrum still convey?

9. **T** or **F**: In ^{13}C NMR the height of a peak is exactly proportional to the number of equivalent carbons represented by that peak.

10. Sketch a reasonable (coupled) ^{13}C NMR spectrum for 3-methyl-1-pentene. Then, below it, sketch the decoupled spectrum. Don't worry about exact placement or relative order of the peak clusters. Just make sure each peak cluster is within the correct range based on Mem. Task 16.2.

11. The NMR spectra in the previous questions are cartoons. Shown below is a real decoupled ^{13}C NMR spectrum of 2-bromobutane. Label some differences between the computer-generated decoupled spectrum (Figure 16.2b) and the real version (Figure 16.3, below).

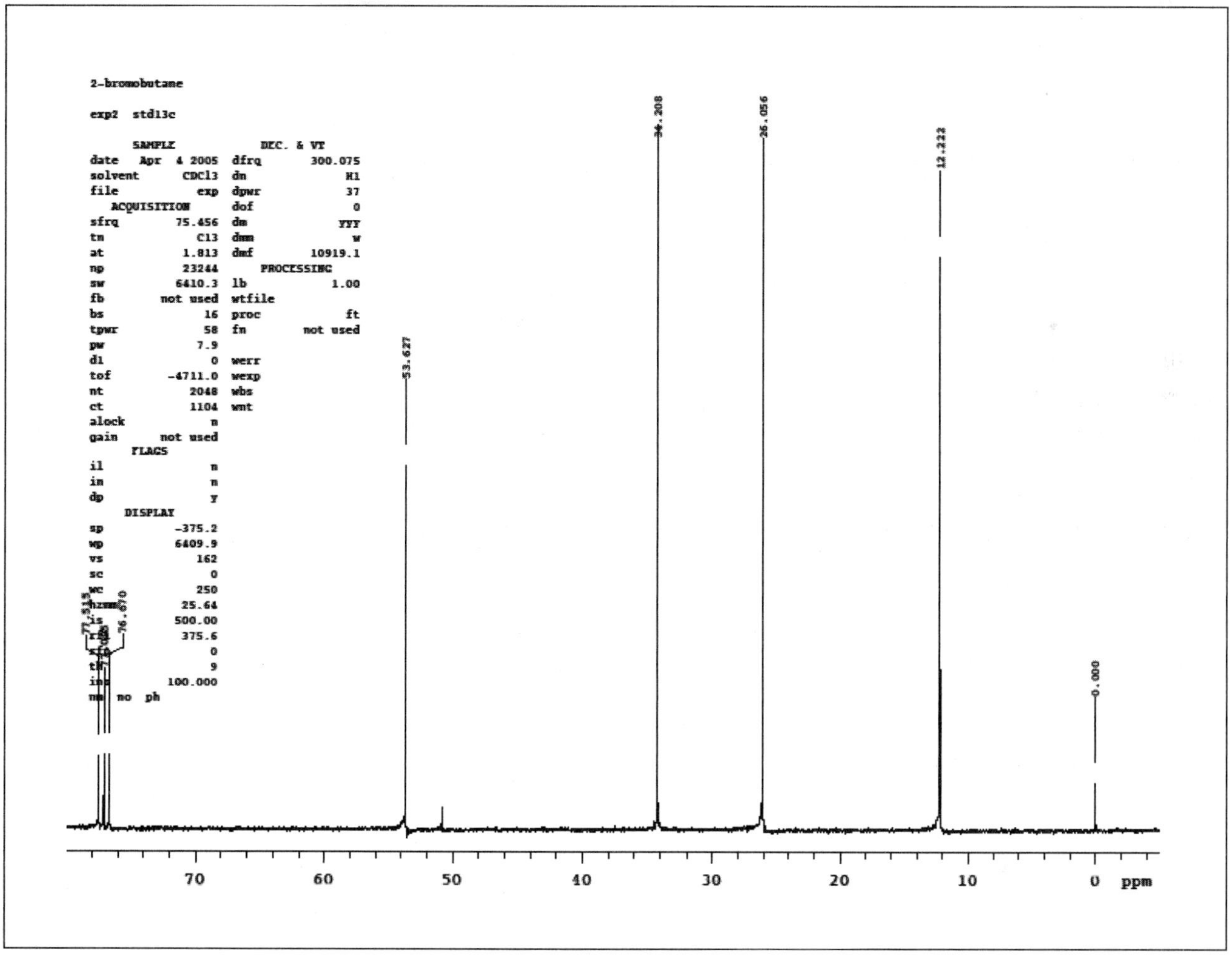

Figure 16.3: Decoupled ^{13}C NMR Spectrum of 2-bromobutane

Memorization Task 16.3: NMR Reference and Solvent Peaks

On a ^{13}C NMR spectrum (also called a CMR spectrum) you will often see a peak at 0 ppm and three peaks centered at 77.0 ppm. **These peaks are associated with additives, not the analyte molecule.** The most common additives are **TMS** (a reference compound) and **$CDCl_3$** (a solvent).

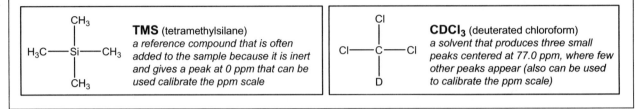

$(CH_3)_4Si$	**TMS** (tetramethylsilane) *a reference compound that is often added to the sample because it is inert and gives a peak at 0 ppm that can be used calibrate the ppm scale*
$CDCl_3$	**$CDCl_3$** (deuterated chloroform) *a solvent that produces three small peaks centered at 77.0 ppm, where few other peaks appear (also can be used to calibrate the ppm scale)*

12. Label the peaks in Figure 16.3 due to TMS and $CDCl_3$ with these terms. Also label the impurity peak near 50 ppm that is not associated with the solvent, reference, or the analyte (2-bromobutane).

Model 3: DEPT NMR

Figure 16.4 is a **DEPT** NMR spectrum of 2-bromobutane. A DEPT spectrum consist of **four spectra:** the top spectrum shows only carbons with three H's; the second shows only carbons with two H's; the third shows only carbons with one H; and the bottom shows all carbons with one or more H.

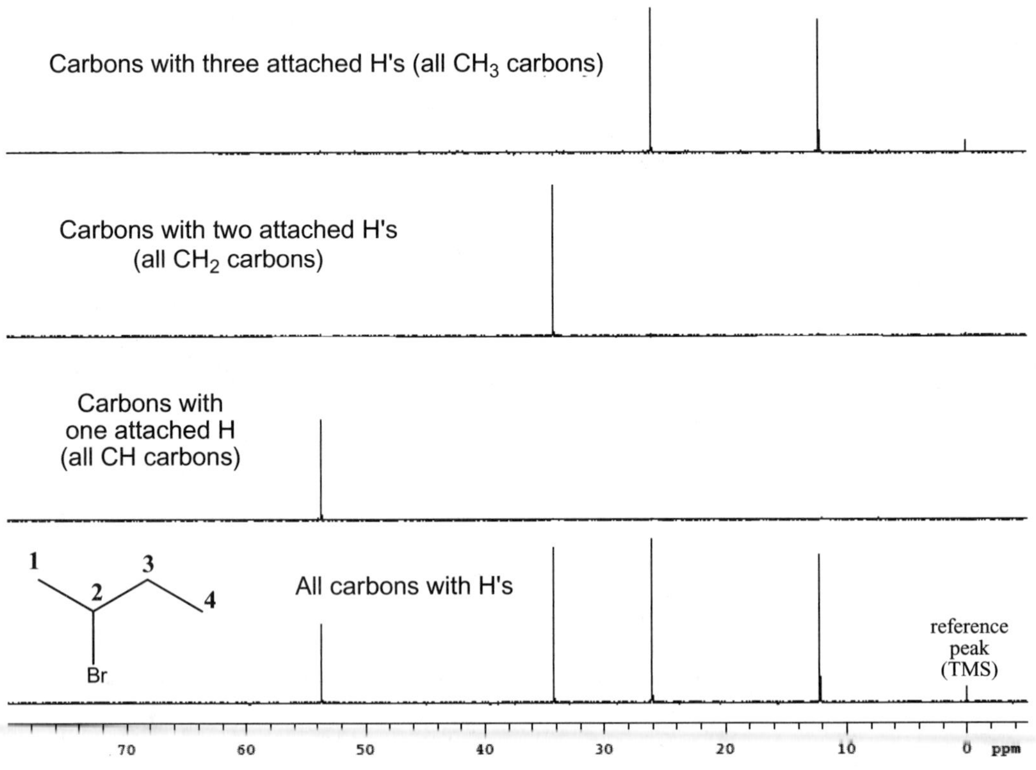

Figure 16.4: DEPT NMR Spectrum of 2-bromobutane

Extend Your Understanding Questions (to do in or out of class)

13. Even though each peak cluster on a DEPT spectrum appears as single peak, you can tell by the level on which it appears if it is a quartet, triplet, doublet, etc. Label each peak with the letter q, t, d, or s, and assign the peaks to carbons 1-4 on the structure of 2-bromobutane on the spectrum.

14. Look at the structure of TMS in Memorization Task 16.3 and explain why the TMS peak appears on the bottom and top levels of the DEPT spectrum.

Memorization Task 16.4: Singlets do not appear on a DEPT Spectrum

By convention, only carbons with H's appear on a DEPT spectrum. To see all the peaks including the singlet peaks you must look at an ordinary ^{13}C NMR spectrum.

15. Label each carbon on the structure of 2-butanone with a letter (s, d, t, or q) indicating the type of peak you expect for that carbon

    a. Label each peak on the DEPT spectrum of 2-butanone (below) with a number (1, 2, 3, or 4) assigning it to a specific carbon on the structure.

    b. Which carbon of 2-butanone does not show up on the DEPT spectrum?

    c. In what ppm range on what type of spectrum do you expect to find the peak for carbon 2? (Check your work: See Figure 16.5, *next page*.)

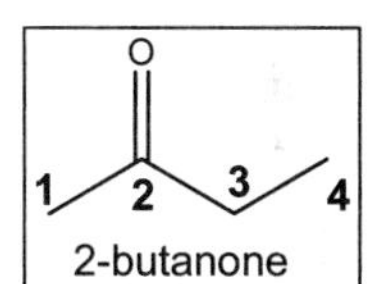

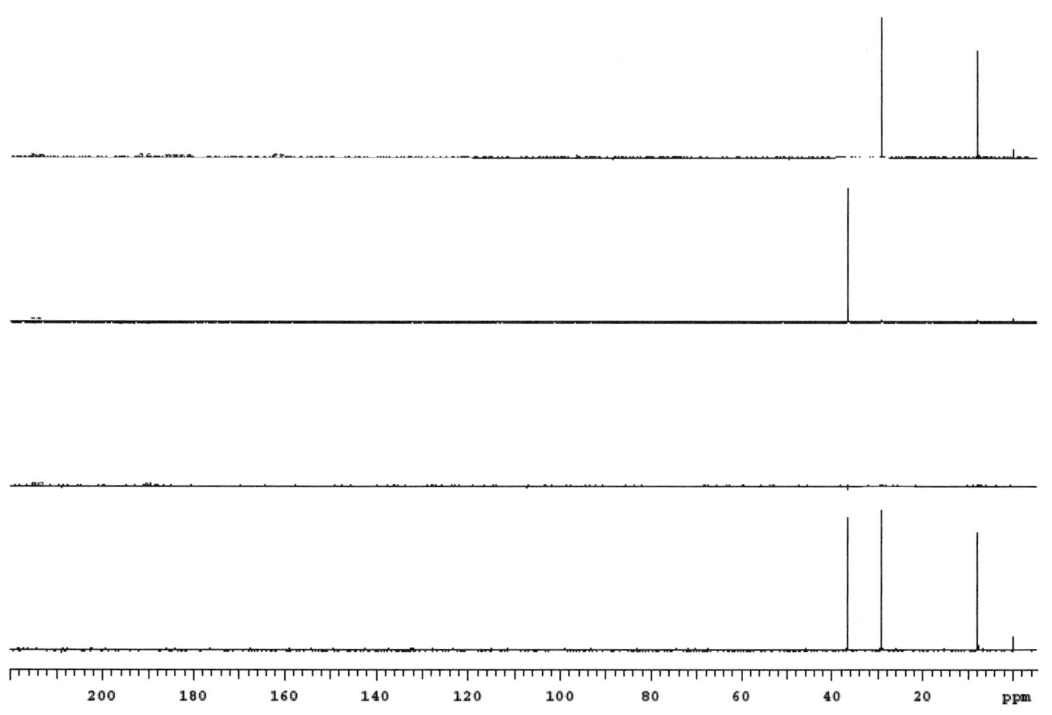

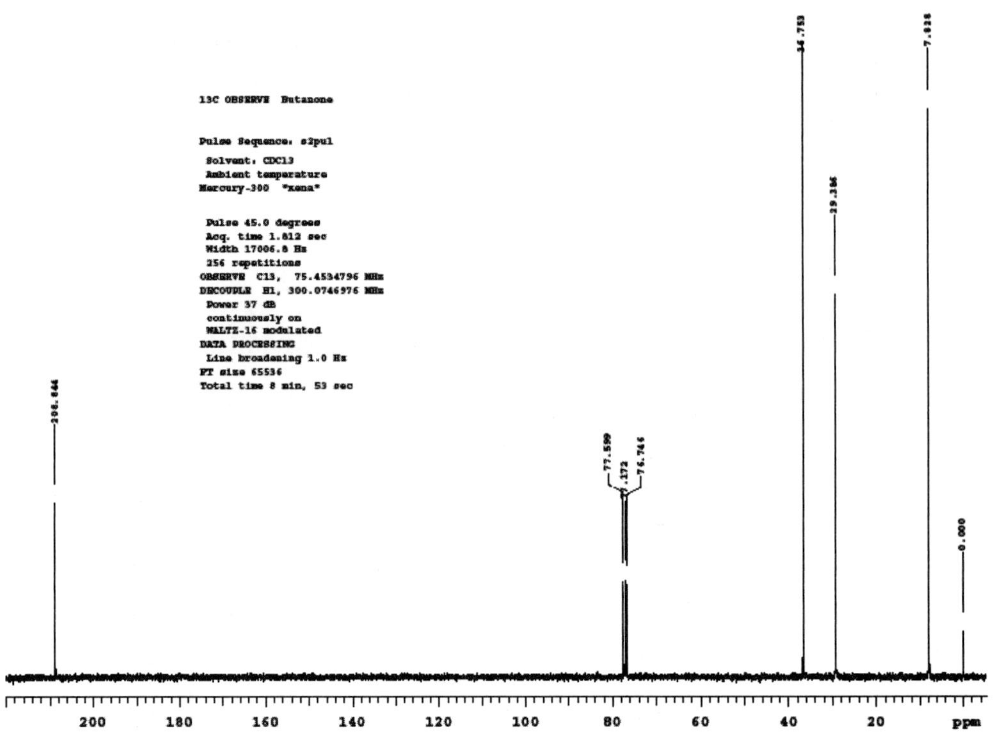

Figure 16.5: Ordinary ^{13}C NMR Spectrum of 2-butanone

Model 4: Mirror Planes

Like all two-dimensional objects, Objects 1-3 have a plane of symmetry (also called a **mirror plane**) in the plane of the paper.

For Objects 1-3, some planes of symmetry are shown with dotted lines. Some are not.

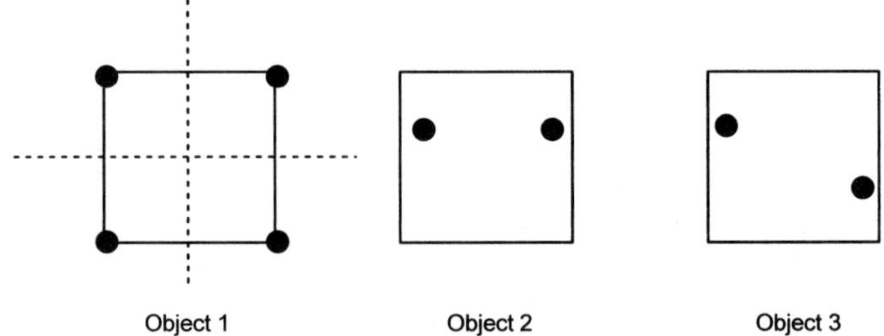

Figure 16.6: Flat Objects Marked with Dots

Extend Your Understanding Questions (to do in or out of class)

16. Mark the planes of symmetry that are NOT marked or noted in Model 4.
    a. Object 1 has two additional planes of symmetry. Mark them.
    b. Object 2 has one additional plane of symmetry? Mark it.
    c. Does Object 3 have any additional planes of symmetry? If so, mark them.

17. Mark each plane of symmetry in the following molecules. For each, indicate if the plane of the paper is a plane of symmetry by writing "POP".

Because of single bond rotation, OH sweeps out a symmetrical region of space

Note: Consider the average of both resonance forms and the average of single bond rotations when considering symmetry.

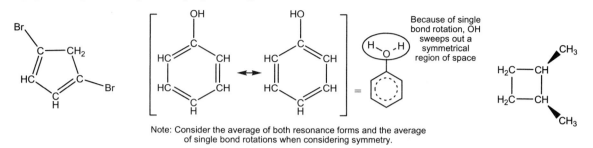

Figure 16.7: Molecules with at Least One Symmetry Plane

Model 5: Identifying NMR Equivalent Carbons Using Symmetry

The general test to determine if two atoms (C_A and $C_{A'}$) are NMR equivalent is as follows: draw the structure, but replace C_A with an X. Now do the same replacing $C_{A'}$ with an X. If the two structures are the same *or enantiomers* then C_A and $C_{A'}$ are NMR equivalent.

Example: The three circled carbons are equivalent because replacement of any one with X gives the same molecule. (That is, if X = Cl, all three replacements would give 2-chloro-2-methylpentane)

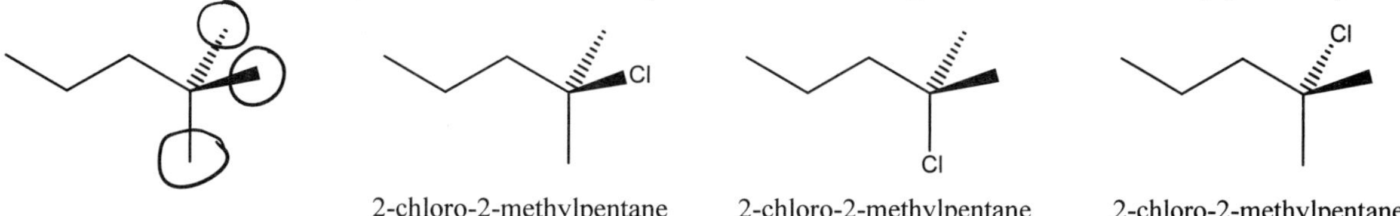

For many molecules, it is easiest to identify equivalent carbons using symmetry: Two carbons are considered NMR equivalent if they can be equated using a mirror plane.

Extend Your Understanding Questions (to do in or out of class)

18. Make a model of 2,2-dimethylpentane (from Model 5), and put it in a conformation so there is a mirror plane equating two of the three circled carbons. Now change the conformation so that a different pair of circled carbons is equated.

19. Number the carbons in each molecule in Figure 16.7 to show which ones are equivalent or distinct. That is, give equivalent carbons the same number. (When you number a molecule to assign its carbons to an NMR spectrum you do not have to follow IUPAC numbering rules.)

20. For the following molecules, confirm that the plane of the paper (POP) is NOT a symmetry plane.

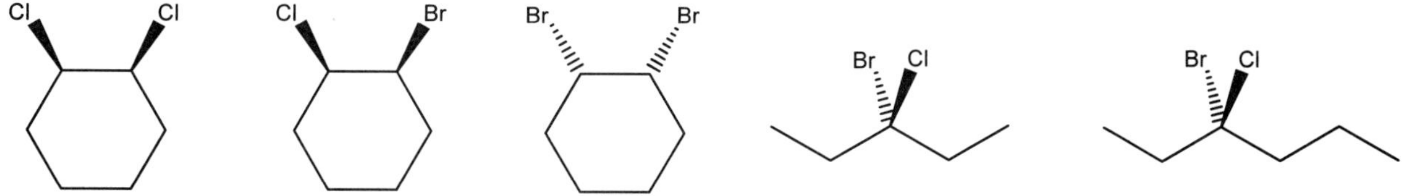

a. Circle the two molecules that DO NOT contain any symmetry plane. For the others, mark all planes of symmetry using a dotted line.

b. Number the carbons to indicate which carbons are identical and which are distinct.

Check Your Work: Tips for finding symmetry planes in a molecule

Use the following tips to check your answers to the previous two questions.

- If there are two of something (e.g., Br groups or methyl groups in Figure 16.7), look for a plane of symmetry halfway between them.
- If there is only one of something (e.g., OH group—see Figure 16.7), look for a plane of symmetry that contains that group.

Confirm Your Understanding Questions (to do at home)

21. For each structure:
    a. Find each mirror plane.
    b. Number the carbons, giving the same number to equivalent carbons.
    c. Label each carbon with a letter to indicate the multiplicity of its ^{13}C NMR peak (s, d, t, or q), and where it would appear on a DEPT spectrum.

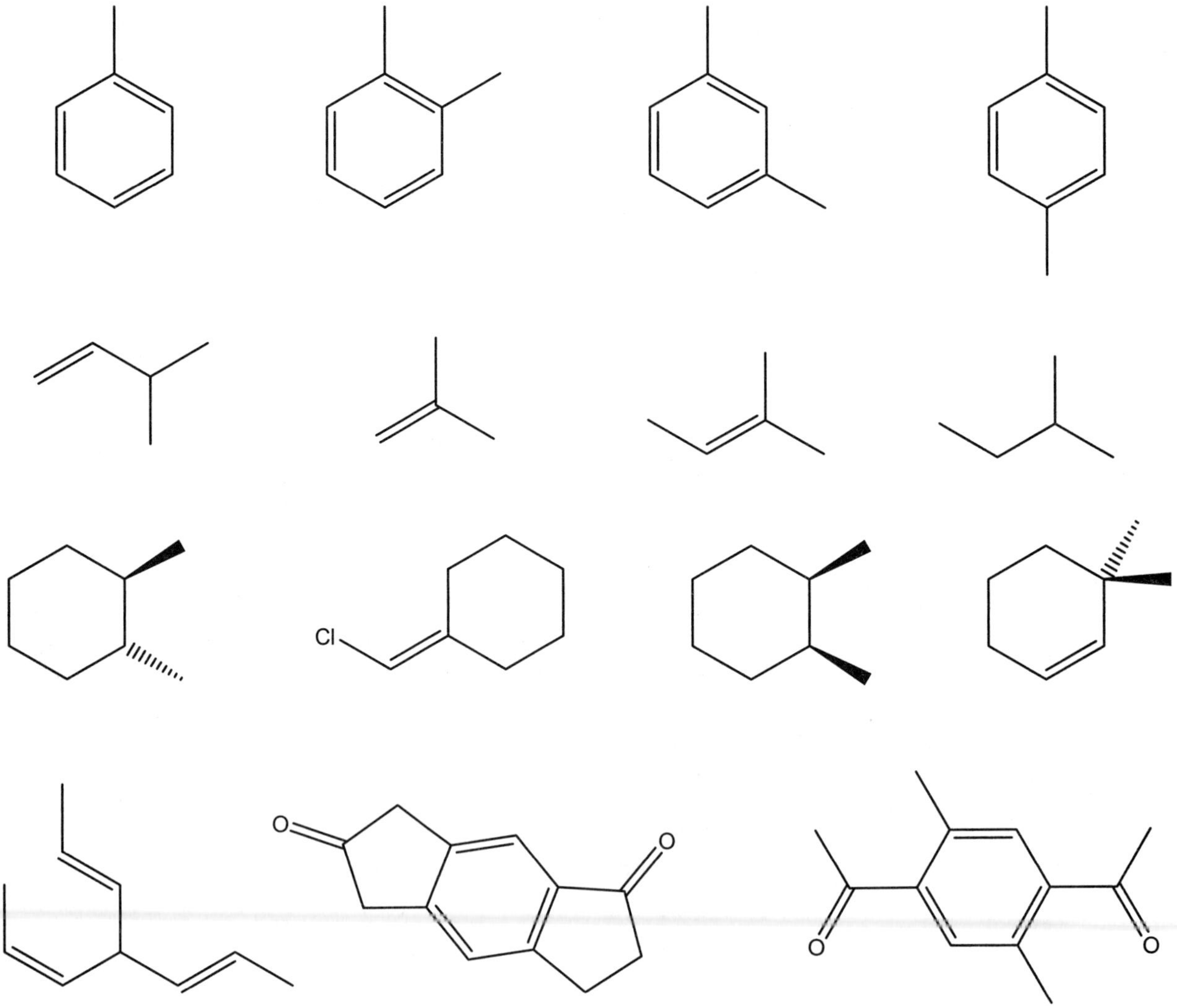

Cl

O O

O O

22. Complete the sentence: In coupled ^{13}C NMR, the number of peaks in a peak cluster ("multiplicity of the peak cluster") tells you…

23. Complete the sentence: In ^{13}C NMR, the location of the peak cluster along the x axis (ppm value) tells you…

24. Complete the sentence: In decoupled ^{13}C NMR, the number of peaks tells you…

25. Complete the sentence: In decoupled ^{13}C NMR, the height of a peak tells you…

26. Chemists rarely use proton-coupled ^{13}C NMR spectra. Explain why.

27. **T** or **F**: In decoupled ^{13}C NMR each peak cluster is reduced to a singlet (a single peak).

28. The <u>decoupled</u> ^{13}C NMR spectrum of a molecule with the molecular formula C_6H_{12} is shown at right. On the basis of this spectrum, propose a structure for this molecule.

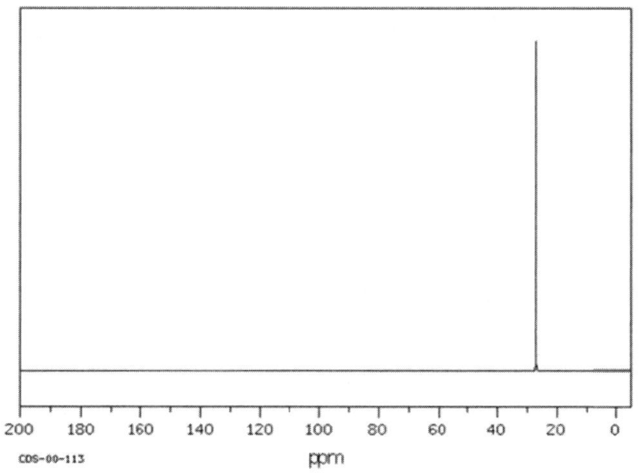

29. For each structure:

- Number the carbons, giving the same number to equivalent carbons.
- Label each carbon with a letter to indicate the multiplicity of its ^{13}C NMR peak (s, d, t, or q), as it would appear on a coupled CMR spectrum.

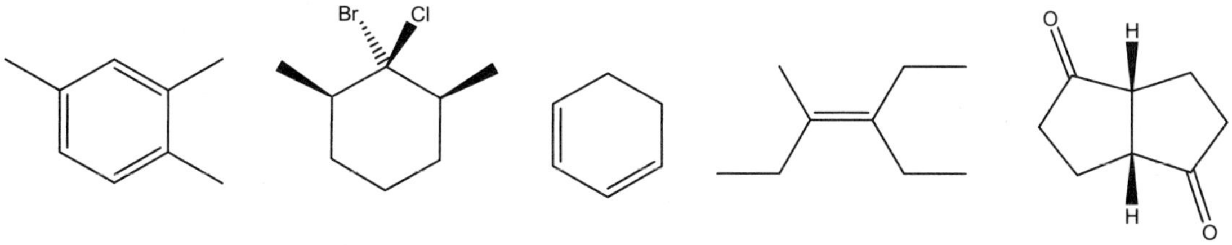

Br Cl

O H H O

30. Draw the structure of the molecule with molecular formula $C_6H_6O_2$ that is expected to have only the following two carbon NMR peaks: 140.6 ppm, s; and 117.8 ppm, d.

31. Below is the CMR spectrum of a molecule with molecular formula $C_5H_{10}O$. Draw a likely structure of the molecule. The DEPT shows the peaks (from left to right) to be: s, t, q.

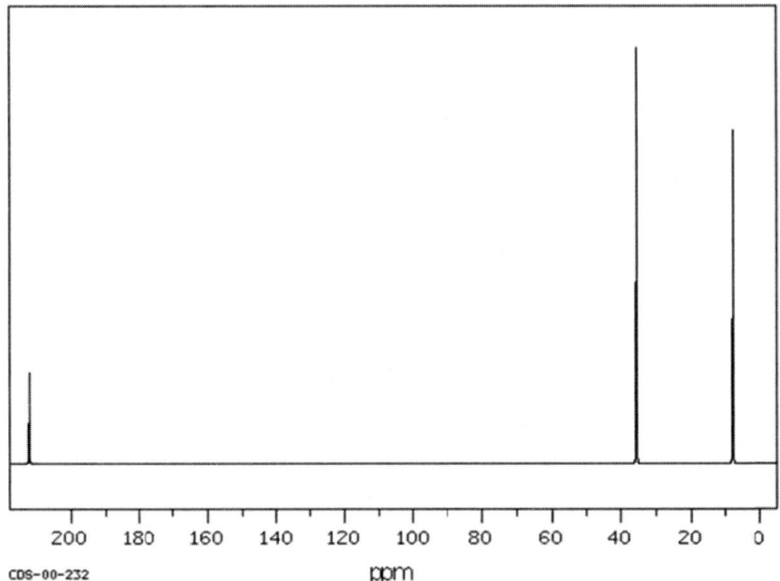

32. The following is a CMR of an isomer of the compound above. Draw a possible structure.

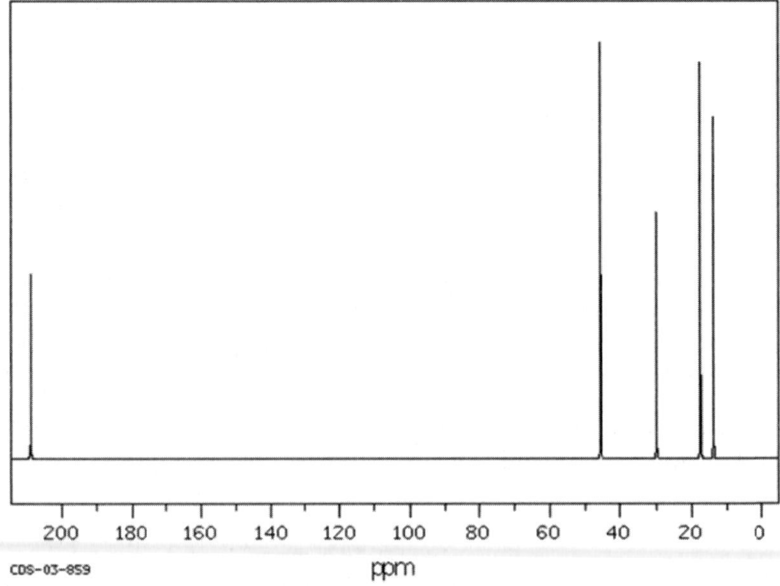

Read the assigned pages in the text, and do the assigned problems.

The Big Picture

NMR (nuclear magnetic resonance) spectroscopy is the most powerful tool that scientists have for looking at the structure of organic molecules. Medicine makes extensive use of this technique–though they drop the word "nuclear" and call it MRI (magnetic resonance imaging). The physics behind NMR is very complex, although it is essentially similar to other spectroscopy such as IR—except that radio frequency light is used instead of infrared light. The very low energy radio waves excite a property called **nuclear spin**. Though many organic chemists do not have a deep understanding of the theory and physics behind NMR, **it is critical that an organic chemist be an expert at interpreting NMR spectra**. Therefore, interpretation of NMR spectra is the focus of this activity and the next.

Interpretation of a (decoupled) carbon NMR spectrum involves two variables: number of peaks, and peak location. The challenge is figuring out how many peaks are expected for a candidate structure; in other words, figuring out which carbons are unique and which carbons are equivalent. The two key tools for this are identifying mirror planes visualizing the molecule in motion.

This activity covers carbon NMR. The next activity covers hydrogen NMR (also called ^{1}H NMR, proton NMR or PMR). In proton NMR, each peak is due to a unique **hydrogen** in the molecule. Organic chemists use the powerful combination of carbon NMR, proton NMR, and MS to deduce the structure of most unknown molecules.

Common Points of Confusion

- **ppm** or **chemical shift**: If a C is close to an electronegative element or involved in a multiple bond, or both, you will find the corresponding peak at higher ppm (farther left on the spectrum). Each chemically distinct C should have a unique chemical shift, though in practice different peak clusters sometimes overlap just by coincidence. This can make interpretation of coupled spectra very difficult. This is why decoupled spectra are usually taken.
- You will not be asked to give the *exact* ppm value associated with a carbon on a given structure. You need only know the ppm ranges. If you are asked to draw a possible structure for a molecule (as in Question 10) just be sure that each peak is within the specified range. You may not even be able to predict the correct order of the peaks within a given range.
- **Peak height** is not a reliable measure of the number of carbons of a given type in C-13 NMR. It is a function of many things, one being whether the C is 1°, 2°, 3°, or 4°.
- **Multiplicity** in a proton-coupled ^{13}C NMR spectrum tells you the number of H's attached to a given carbon. For example, a C that produces a doublet (two peaks) must have one H attached to it. Very often chemists record "proton-decoupled" ^{13}C NMR spectra. Such spectra have a singlet for each chemically unique carbon. This is useful for complex molecules for which the peak clusters would overlap. A decoupled ^{13}C NMR spectrum tells you the number of different carbon atoms present in the sample, but not the multiplicity. This is where a DEPT spectrum can be very useful as it tells you the multiplicity of each peak appearing on the ^{13}C NMR spectrum.
- **DEPT** spectra do not include singlets. This means it is usually necessary to use a DEPT in conjuction with an ordinary ^{13}C NMR spectum. Using the DEPT, you can assign each peak on the ordinary ^{13}C NMR spectum as a siglet, doublet, triplet, or quartet.
- Keep in mind that an NMR spectrum is a "snapshot" of the molecule over a number of seconds. Molecular motion such as rotation of single bonds gets averaged. The second example in Question 17 highlights this. This means the OH group does not break the symmetry of the molecule.

Notes

ChemActivity 17: Proton (^{1}H) NMR

(What can a ^{1}H NMR spectrum tell you about the structure of a molecule?)

This activity often requires more than one class period, and can be completed at home or in lab.

Model 1: Proton NMR (or Hydrogen NMR or ^{1}H NMR)

^{1}H NMR signals are generated by **hydrogen nuclei** and most appear between 0–12 ppm.

[Recall that ^{13}C NMR peaks are generated by carbon-13 nuclei and most appear between 0–220 ppm.]

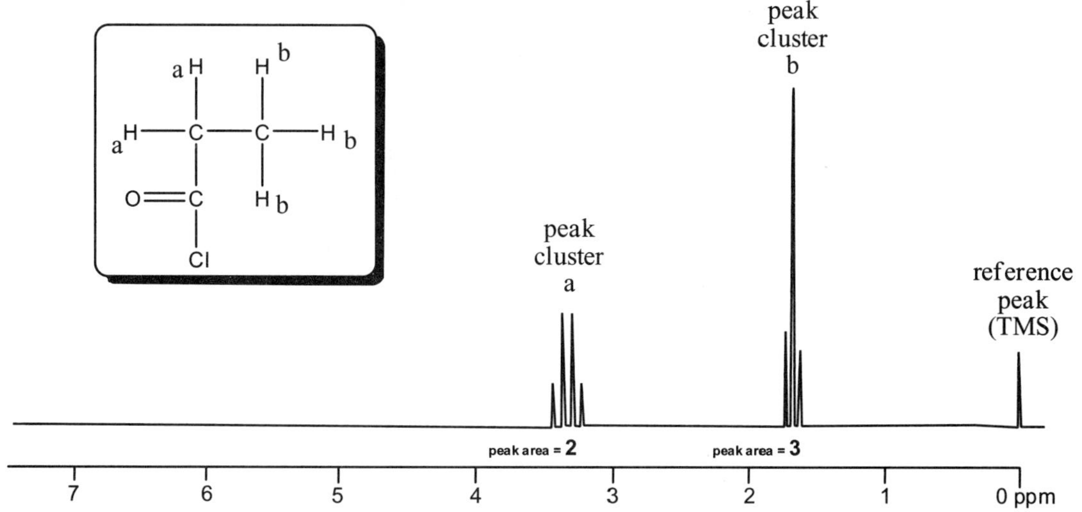

Construct Your Understanding Questions (to do in class)

1. How many peak clusters (excluding the reference peak) are there on the spectrum in Model 1?
2. How many different types of chemically distinct H's are there on the structure in Model 1?
3. Complete the following sentence: In a proton NMR spectrum there is one peak cluster for each chemically distinct type of **H** or **C** [circle one].
4. The number listed below a peak cluster gives you the **area (size) of the peak cluster**.
   a. Which peak cluster **a** or **b** [circle one] is bigger (as measured by area)?
   b. What is the ratio of the size of peak cluster a : size of peak cluster b?
   c. How many H's are there of **type a**? … **type b**? Does this match the ratio above?

   d. Hypothesize: what structural information is conveyed by peak cluster areas in ^{1}H NMR?

Memorization Task 17.1: Integration

Peak cluster area, also called the **integral**, **integration**, or **integrated area**, tells you the relative number of hydrogens associated with a given peak, and is often represented by a number written above or below a peak cluster. *Careful! Peak area (not height) tells you the number of H's associated with a peak cluster.*

5. **For now, don't worry about the *number* of peaks in a peak cluster.** We will deal with that on the next page. Based only on the integration of each peak cluster (shown as a bold number below the cluster) assign a letter (**a** or **b**) to each peak cluster, matching it to a type of H on the structure.

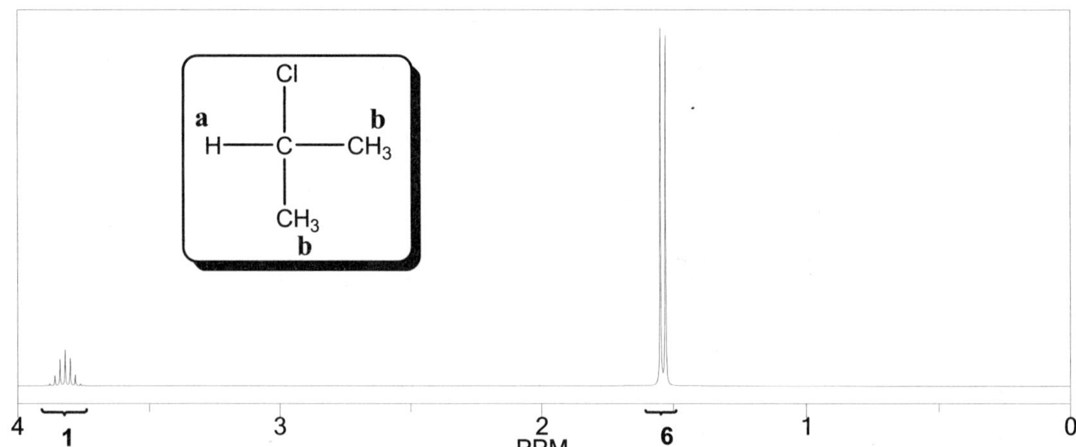

6. Hydrogen atoms attached to the *same* carbon are nearly always equivalent.

   a. Confirm that in each case on this page of two or three hydrogens attached to the same carbon, the same letter assignment is made.

   b. Not all CH_3 (or CH_2) groups are equivalent! Construct an explanation for why the two CH_3 groups on the structure above are equivalent (making all 6 H's labeled "**b**" equivalent), but the two CH_3 groups on the structure at right are NOT equivalent.

a H_3C — **b** H_2C — **c** CH(Cl) — **d** CH_3

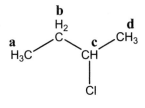

Memorization Task 17.2: General Method of Testing if Two Atoms are NMR Equivalent

As with ^{13}C NMR, symmetry is an excellent way to identify NMR equivalent H's. The general test to determine if two atoms (H_A and $H_{A'}$) are NMR equivalent is also the same as for carbon NMR:

Draw a structure, but replace H_A with an X, then do the same for $H_{A'}$. If the two structures are the same *or enantiomers* then H_A and $H_{A'}$ are NMR equivalent.

Example: The three circled hydrogens are equivalent because replacement of any one with X gives the same molecule. (That is, if X = Cl, any one replacement gives 1-chloro-3-methylpentane)

2-methylpentane

1-chloro-3-methylpentane

1-chloro-3-methylpentane

1-chloro-3-methylpentane

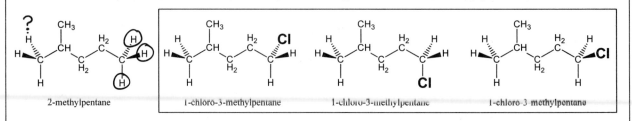

7. Draw and name the structure that results if the H with a "?" above it is replaced with a Cl. Based on this, decide if the "? H" is equivalent to a circled H, and explain your reasoning.

8. (Check your work) The "? H" is not identical to the circled H's, but it is identical to five other H's on 2-methylpentane. Identify these five H's that are equivalent to the "? H."

9. **Review Memorization Task 16.1 in the previous ChemActivity**, and answer the following questions about the spectrum from Model 1, reproduced below.

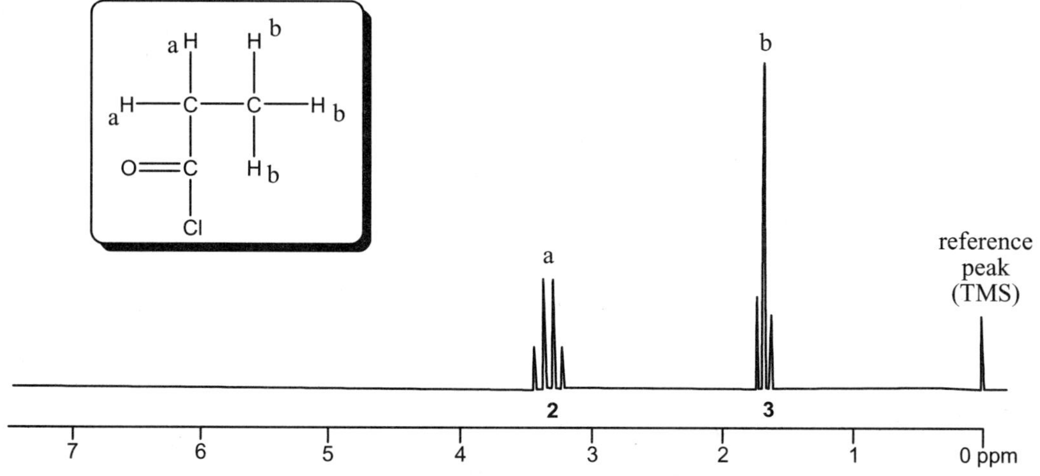

a. Label peak clusters **a** and **b**, above, with s, d, t, q, or m, as appropriate.

b. According to the language in **Mem. Task 16.1 in the previous ChemActivity**, how many times was **peak cluster b** "split/cut" to produce the **three** peaks shown?

10. Just as in ^{13}C NMR, splitting in ^{1}H NMR is caused by nearby H's. Shown at right is a cartoon of peak cluster b from the spectrum above. The dotted line shows the peak as it would be without splitting. Identify the **two** H's on this molecule that are likely splitting this peak into a triplet.

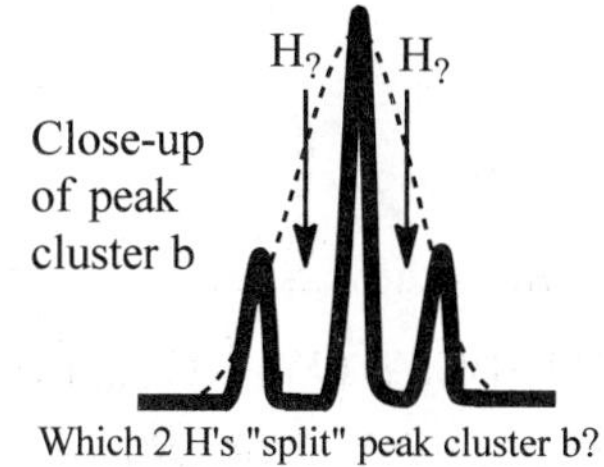

Which 2 H's "split" peak cluster b?

11. It turns out that in ^{1}H NMR counting peaks in a cluster tells you the number of neighbor hydrogens within three bonds. (Note equivalent H's do not split each other.)

a. (Check your work) Replace each question mark on the figure above, right with a letter "a" and check that this consistent with your answer to the previous question.

b. Identify the **three** neighbor H's on the structure above which split **peak cluster a** into four peaks.

c. Complete the table at right describing the relationship between peak type and number of neighbor hydrogens within three bonds.

Number of peaks in a peak cluster (peak type)	Number of neighbor H's within 3 bonds
1 (s)	
2 (d)	
3 (t)	
4 (q)	
5 (m)	4 or more

Memorization Task 17.3: Multiplicity in Proton NMR

In ^{1}H NMR, H's split the signals of foreign (= non-equivalent) hydrogens within three bonds.

This means chemically equivalent H's DO NOT split each other.

12. For each individual drawing, indicate whether the circled H will split the signal of the boxed H. (When counting bonds for splitting, ignore whether a bond is a single, double, or triple bond. For example, the circled and boxed H's in the first column are 3, 4, and 3 bonds apart, respectively.)

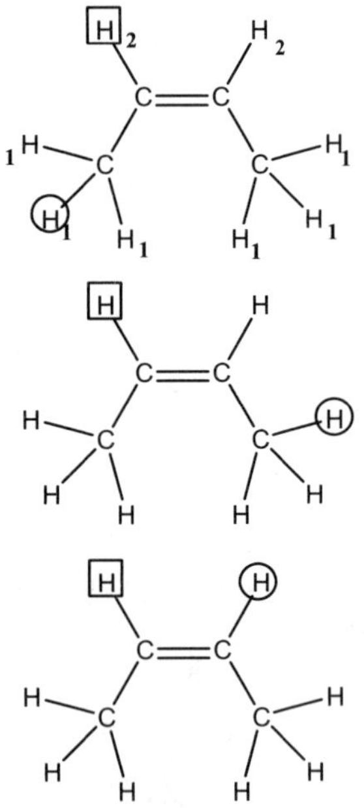

Molecule I 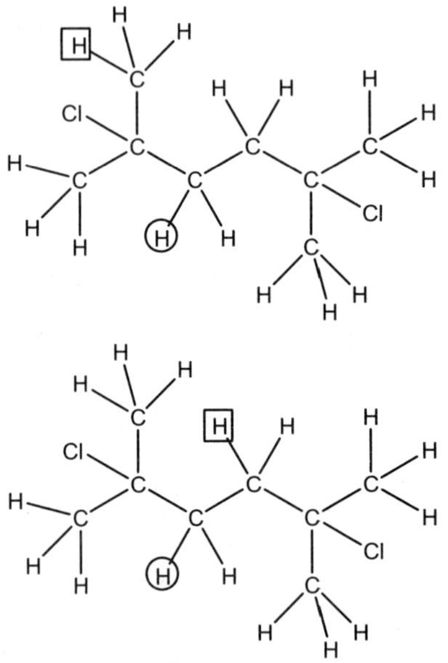 **Molecule II** 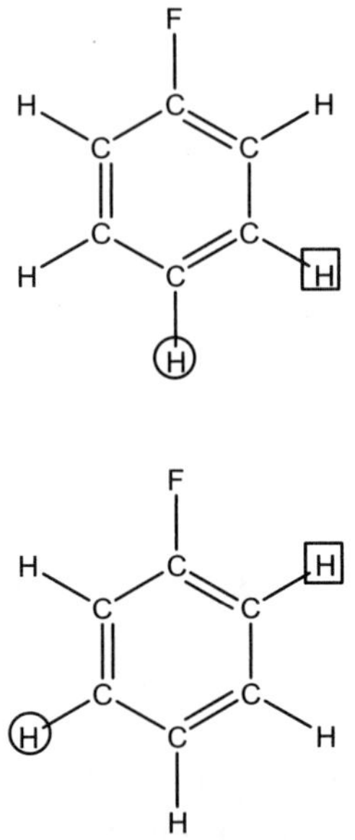 **Molecule III**

13. On the first row of structures above…

a. Label each hydrogen on the structure with a number, giving identical hydrogens the same number. (The first one is done for you.)

b. For molecules II and III, complete a table, like the one for Molecule I, below, by reporting the **integration** and **multiplicity** of the peak due to each type of hydrogen.

Molecule I	Integration	Multiplicity
H_1	6*	d
H_2	2*	q

*Ratio of H_1 to H_2 could also be reported as 3:1

14. (Check your work) Explain why, for Molecule II, the H's on carbons 3 and 4 do not split one another even though they are only three bonds apart.

15. (Check your work) There are no CH_2 (methylene) groups on Molecule III, yet the signal due to the H on carbon 4 is a triplet. Explain how a triplet can occur with no CH_2 group next door.

16. On a proton NMR spectrum, the chemical shift (ppm/placement along the x axis) conveys similar information as in ^{13}C NMR. Based on this, construct an explanation for why **peak cluster a** is farther to the left (higher ppm) than **peak cluster b** on the spectrum in Model 1.

Memorization Task 17.4: Memorize Key ^1H NMR Chemical Shifts (ppm ranges)

Type of Hydrogen	Examples (Marked H's are within the given NMR ppm ranges)	ppm Range
H three or more bonds from a functional group	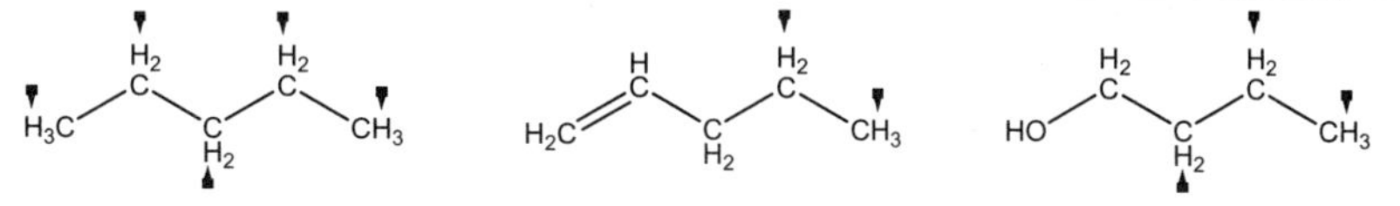	1 - 2
H two bonds from π bond (C=C or C=O)	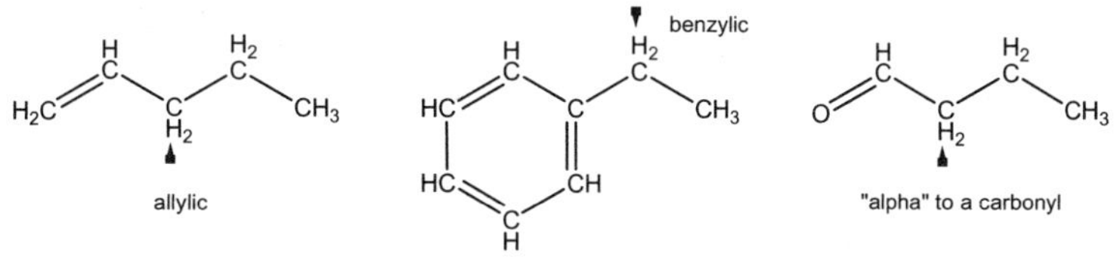 allylic; benzylic; "alpha" to a carbonyl	2 – 3 (1.8-2.6)
H two bonds from O, N, or halogen	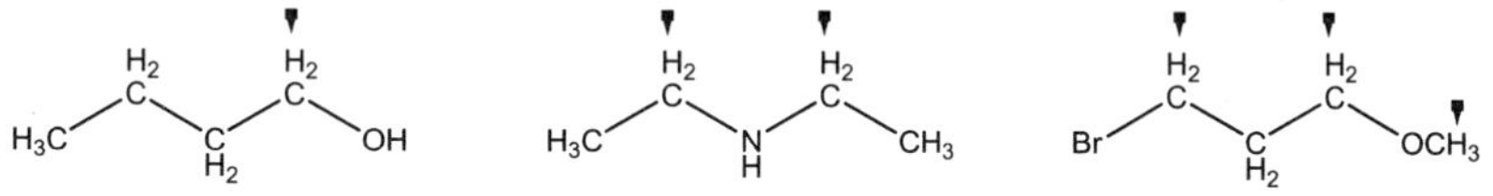	3 – 4 (2.5-4.5)
H attached to C=C (not benzene)	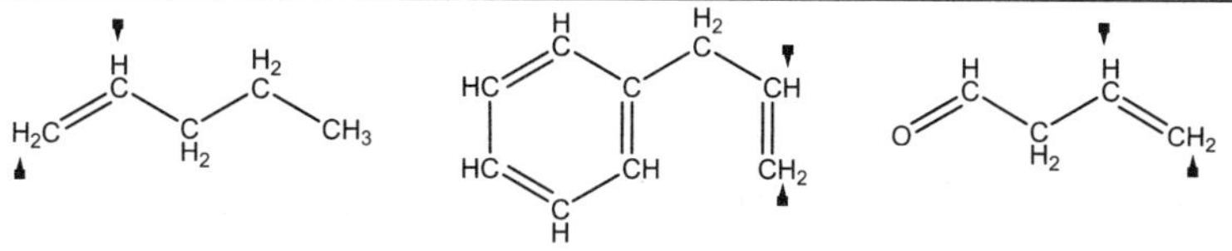	4 – 7 (4.5-7.0)
H attached to benzene ring	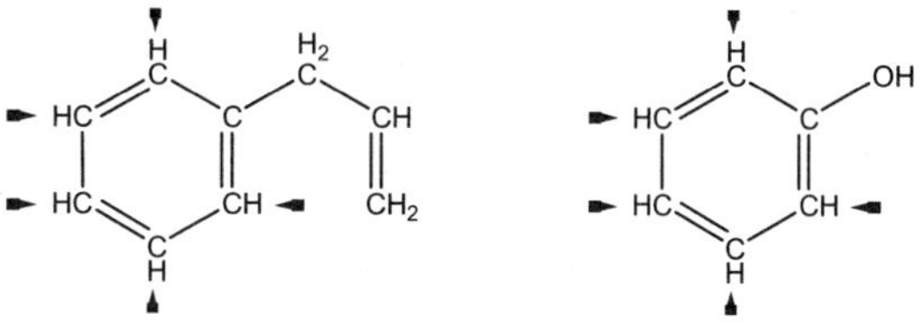	7 – 9 (6.5-8.5)
H attached to carbon of C=O (carbonyl)	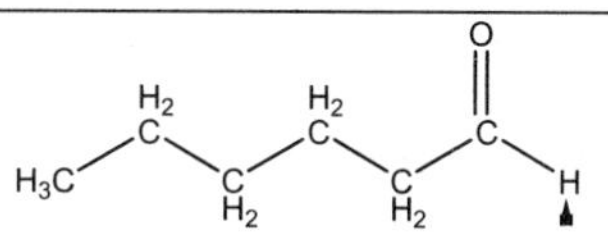	10 (9.7-10)
H attached to O or N	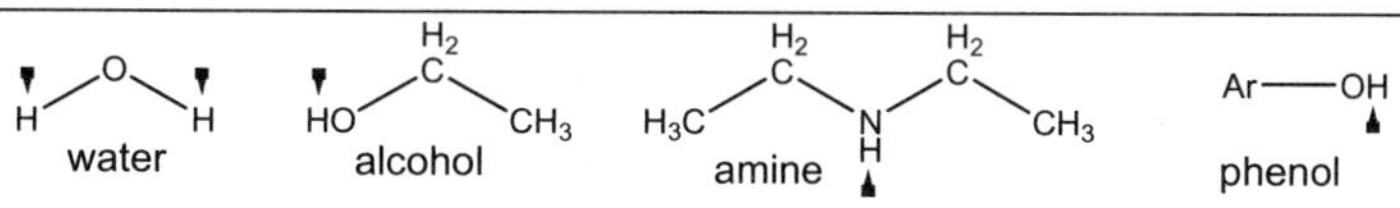 water; alcohol; amine; phenol	1 - 8
	carboxylic acid 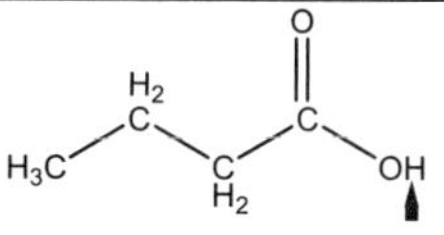	10 - 13

Extend Your Understanding Questions (to do in or out of class)

17. Alcohols do not always follow the rules for predicting multiplicities. Based on integration and chemical shift alone, assign each of the three (non-reference) peaks in the proton NMR spectrum of ethanol at right.

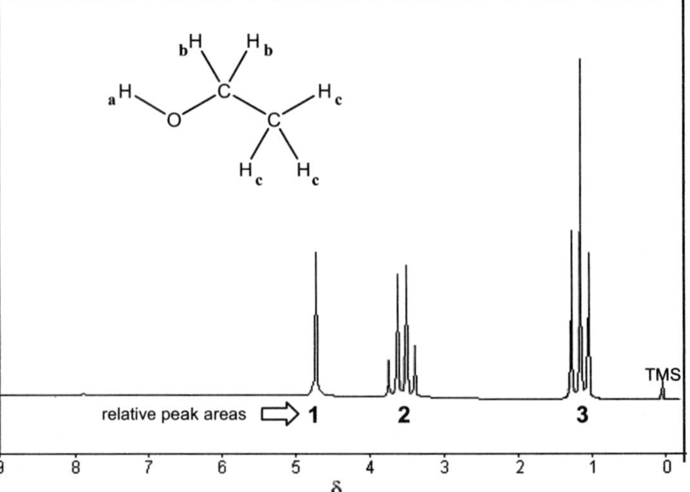

18. Write the expected multiplicity above each peak cluster associated with ethanol. Which clusters have a multiplicity that does not fit the rules we have learned so far?

Memorization Task 17.5: Special Considerations for H Attached to N or O

- Under normal conditions, the H of an alcohol or amine does not participate in splitting.
- When D_2O is added to a NMR sample, a peak due to an H attached to O or N will disappear.

19. Does Memorization Task 17.5 explain the inconsistencies you noted above? Explain.

20. In a typical proton NMR sample, the solvent is present in much higher concentrations than the molecule being analyzed. Construct an explanation for why this means that any solvent with a hydrogen (such as $CHCl_3$, C_6H_6, CH_2O, etc.) is *not* useful as a proton NMR solvent.

21. $CDCl_3$ (deuterated chloroform, or chloroform-d) is the most common NMR solvent because it is cheap relative to other deuterated solvents, dissolves many organic molecules, and has no H's. Most $CDCl_3$ is contaminated with a very small amount of $CHCl_3$. (The H of $CHCl_3$ appears as a singlet at 7.24 ppm.) Explain how this peak removes the need to add TMS to the sample.

22. Cross out the two molecules below that cannot serve as a proton NMR solvent.

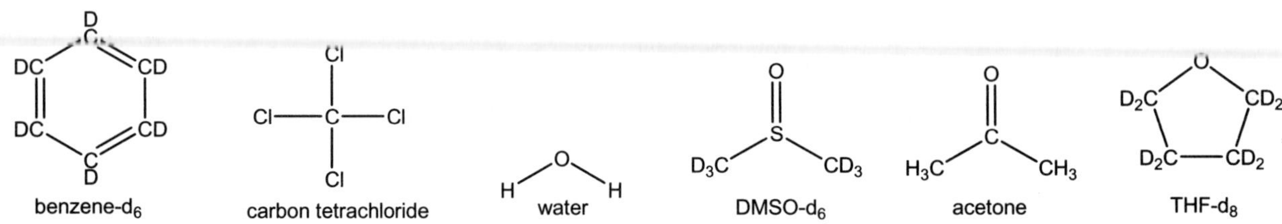

Confirm Your Understanding Questions (to do at home)

23. Identify/assign each peak (including any solvent or reference peaks) on the following spectrum.

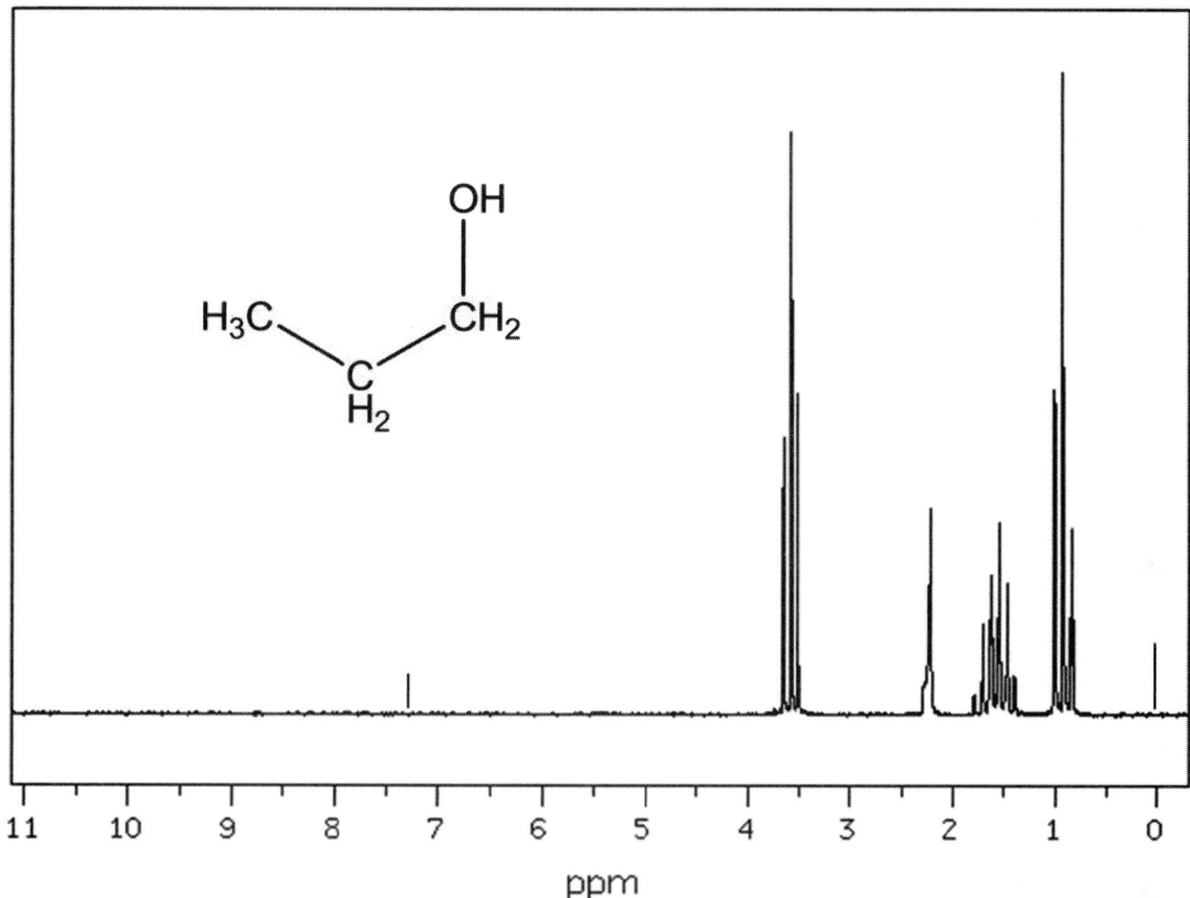

24. For the spectrum above, does the alcohol H participate in splitting (as described in Memorization Task 17.5)?

25. (Check your work) Are your assignments above consistent with the fact that only the peak at 2.23 ppm disappears when the sample is treated with D_2O? Explain.

26. Assign each peak cluster on the following spectrum, and predict the integration value for each.

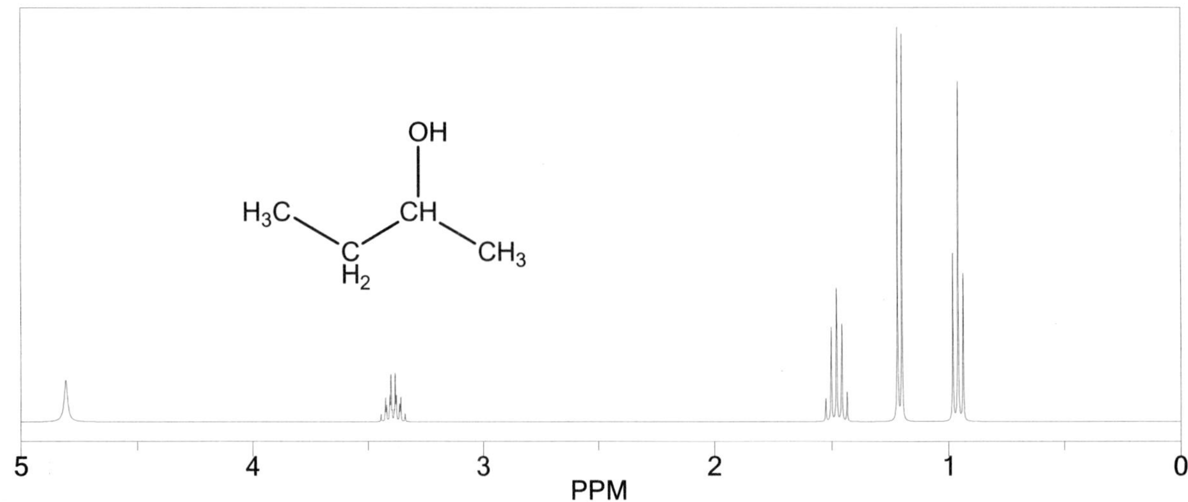

27. The proton NMR data for 1-bromopropane are as follows: H_a: triplet (2H) 3.32ppm; H_b: multiplet (2H) 1.81ppm; H_c: triplet (3H) 0.93ppm. (Relative integrations shown in parentheses.)
    a. Through how many bonds can a hydrogen split another hydrogen?
    b. According to this splitting rule, does H_a split H_c?
    c. Is your answer in part a) consistent with the multiplicity listed for peak clusters a and c?
    d. How many hydrogens split H_b?
    e. Upon <u>very close</u> inspection of the proton NMR spectrum of 1-bromopropane, you would find that peak cluster b has at least six peaks. Is this consistent with your answer in part d)?
    f. Speculate as to why any peak cluster with more than four peaks is listed simply as a "multiplet."

28. For each structure below, use letters or numbers to indicate chemically equivalent and distinct hydrogens, and make a table showing the predicted integration and multiplicity of each peak cluster.

```
      H   H   H            H   Cl  H            H   H   H
      |   |   |            |   |   |            |   |   |
Cl————C———C———C————H   H———C———C———C———H   Cl———C———C———C———Br
      |   |   |            |   |   |            |   |   |
      H   H   Cl           H   Br  H            H   H   H
```

29. For each structure below, use numbers to indicate chemically equivalent and distinct hydrogens, and make a table showing the predicted integration and multiplicity of each peak cluster.

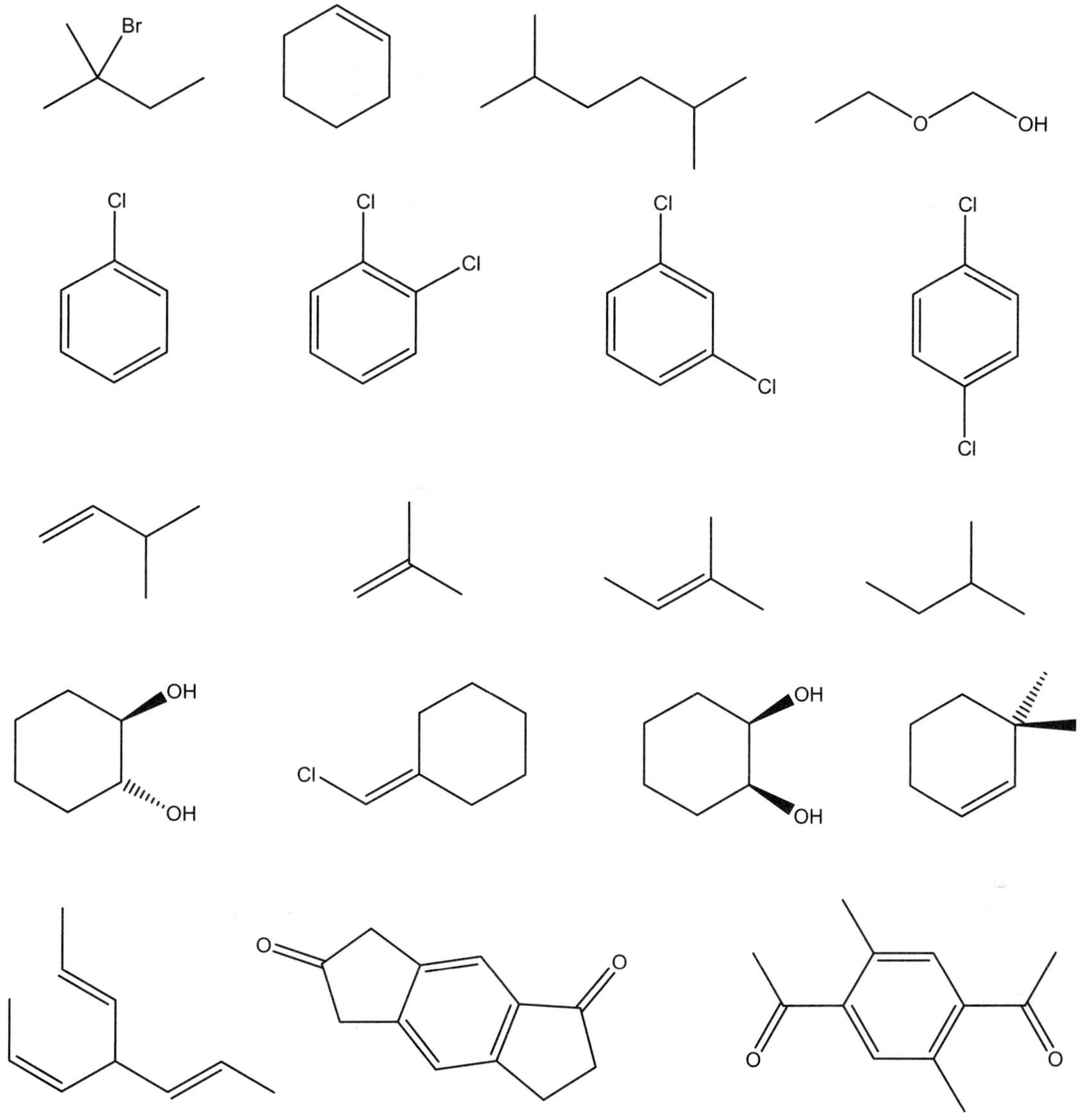

Br

O OH

Cl Cl Cl Cl Cl Cl Cl Cl

OH OH Cl OH OH

O O O O

30. Imagine you have two bottles: one with (R) and the other with (S)-2-bromobutane. Unfortunately, your lab partner messed up and labeled both bottles simply "2-bromobutane." Can you use NMR to sort out this problem? Explain why or why not.

31. A researcher wants to take a proton NMR of a molecule that dissolves only in water. What solvent should she use to make the NMR sample?

32. Based on the ppm range in which the peaks appear on the spectrum below, what can you infer about the molecule associated with the spectrum?

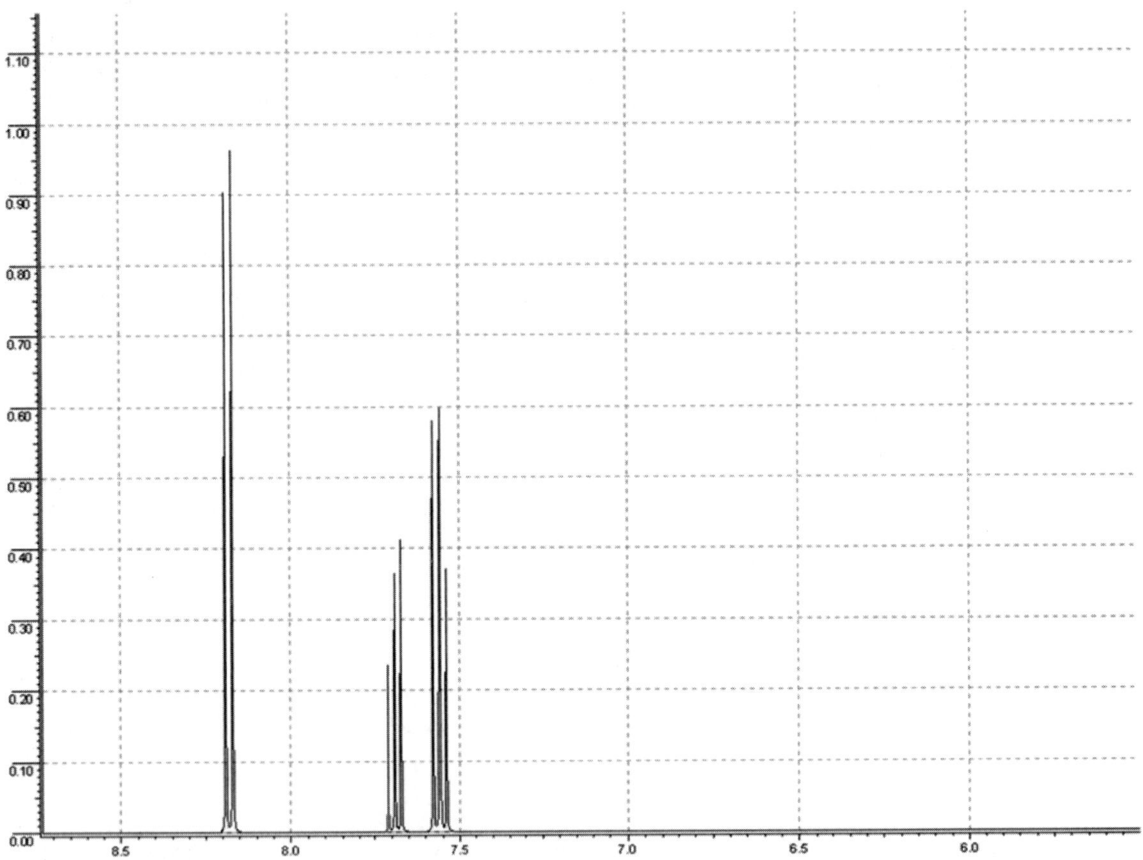

33. Propose a structure to go with the spectrum above assuming the unknown has molecular formula $C_6H_5NO_2$, and the integrations of the peaks are 2:1:2, from left to right.

34. Propose a structure to go with the spectrum below assuming the unknown has molecular formula C_4H_8O, and the integrations are as shown. (*Hint*: First determine the degrees of unsaturation.)

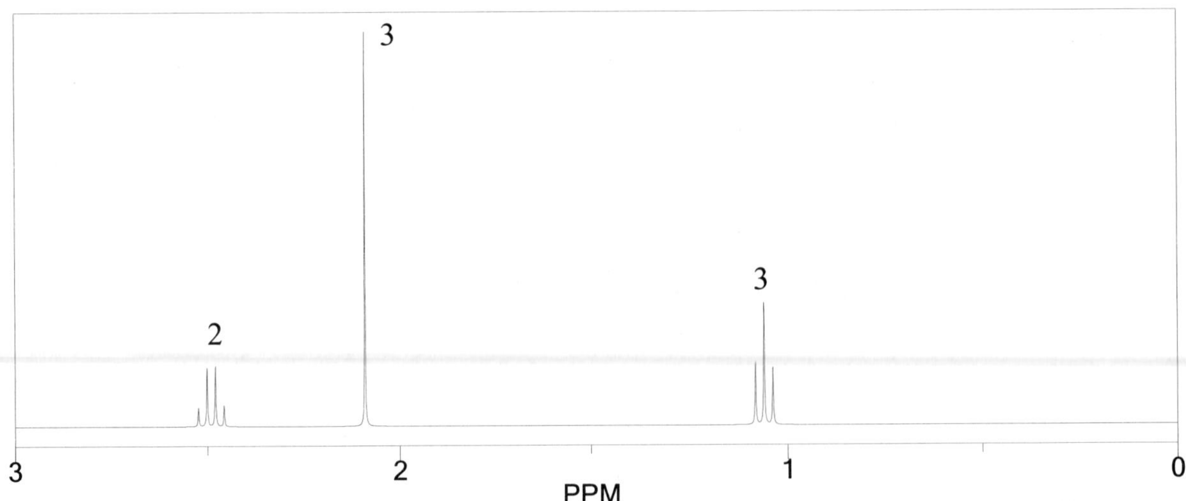

35. (Check your work) The peaks at 1.1 and 2.5 ppm on the previous page represent the signature pattern of an ethyl group. This pattern (a quartet worth 2H's and a triplet worth 3 H's) is so common it is worth memorizing.

    a. Find this pattern on the proton NMR spectrum of ethanol in CTQ 20.

    b. Sketch the proton NMR spectrum expected for diethyl ether ($CH_3CH_2OCH_2CH_3$).

36. Propose a structure to go with the proton NMR spectrum and the following mass spectral data: $[M]^+ = 136$ (100), $[M+1]^+ = 137$ (3.3), $[M+2]^+ = 138$ (97.3)

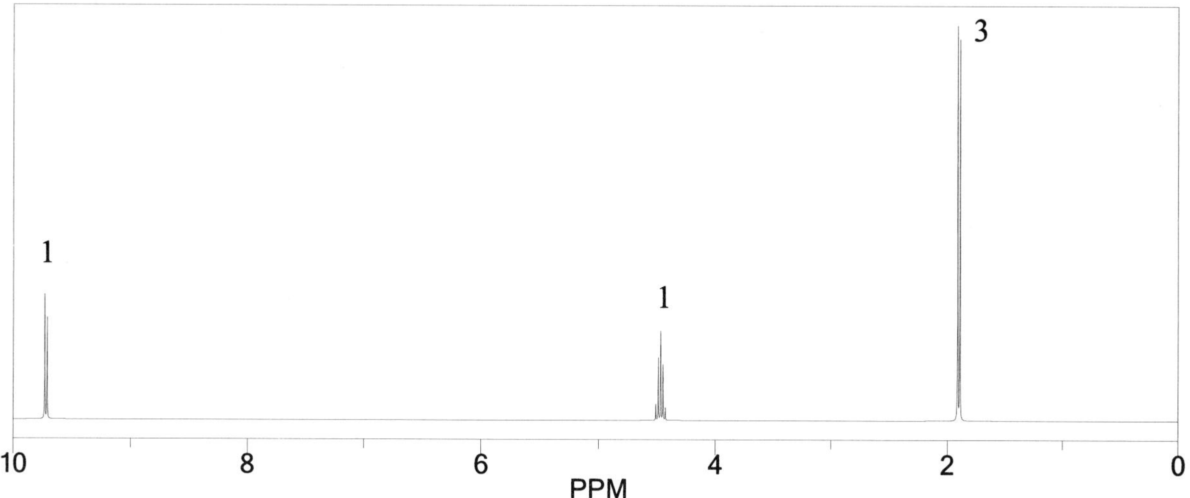

37. Construct an explanation for why the peak in the spectrum above found at 4.5 ppm is outside the typical range for a hydrogen alpha to a carbonyl (2-3) *and* the typical range for a hydrogen bound to the same carbon as a halogen (3-4).

38. Draw the structure of an unknown to go with this proton NMR and the following mass spectral data: $[M]^+ = 166$ (50), $[M+1]^+ = 167$ (5.5), $[M+2]^+ = 168$ (16.1)

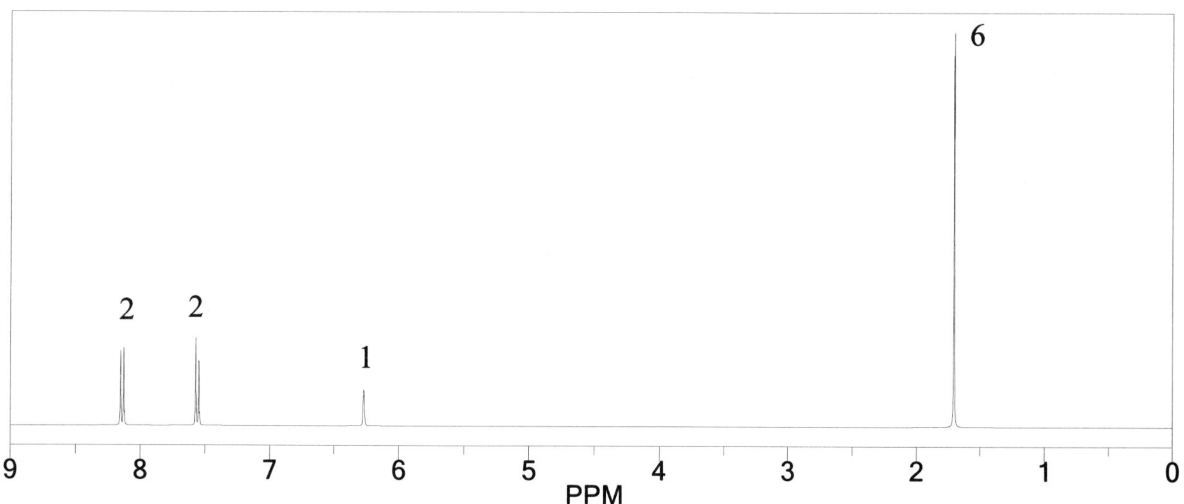

39. Peak clusters in the 7-9 ppm region are almost always indicative of H's attached to a benzene ring. (As we will learn later, the region from 7-9 ppm is called the aromatic region, and molecules containing a benzene ring are by far the most common type of aromatic molecule).

    Peak clusters in the aromatic region often overlap one another. For example, the mass of peaks between 7.3 and 7.4 ppm on the spectrum below is actually two overlapping peak clusters: a doublet of doublets worth 2H's, overlapping with a triplet worth 1H.

    Draw one structure that goes with both spectra on this page.

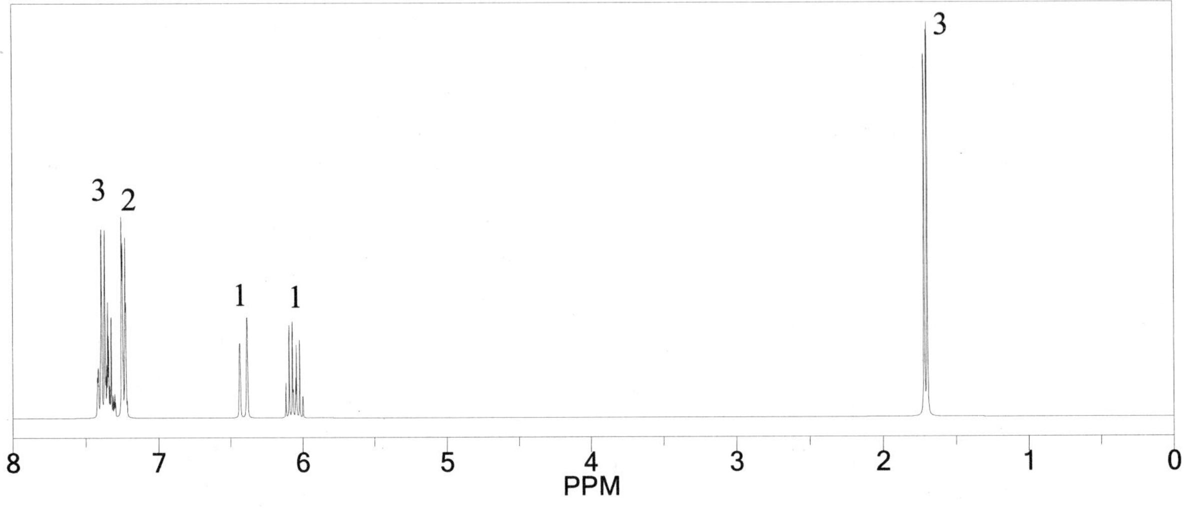

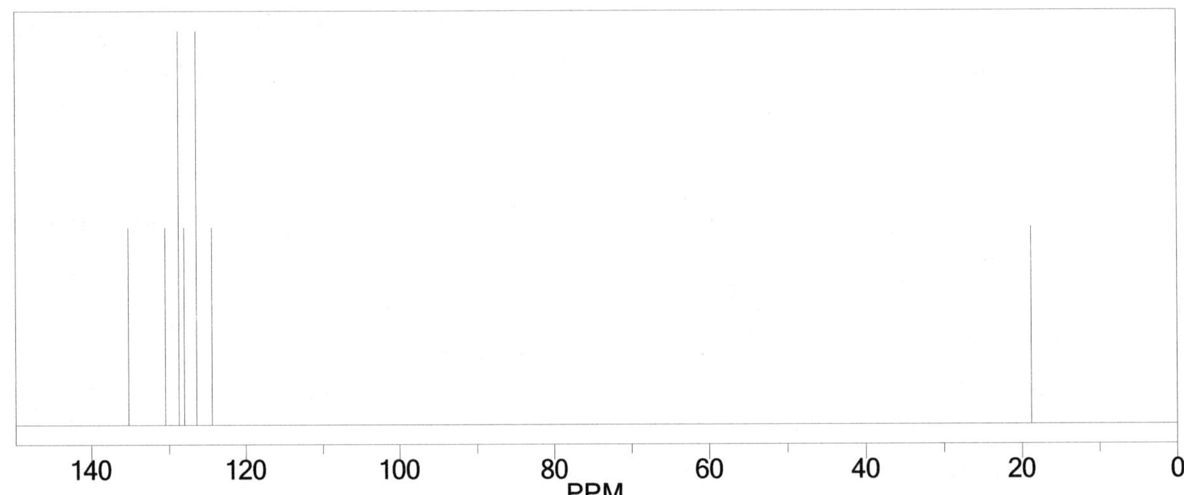

40. (Check your work) How can you tell the top spectrum on this page is a proton NMR spectrum and the bottom spectrum is a carbon NMR spectrum?

41. (Check your work) According to the CMR spectrum, how many unique C's are there?

42. (Check your work) Explain why, on the proton NMR spectrum, the peak at 6.1 is expected to be a doublet of quartets.

43. The spectrum below goes with one of the diasteriomers shown. The other diasteriomer is the answer to the previous question.

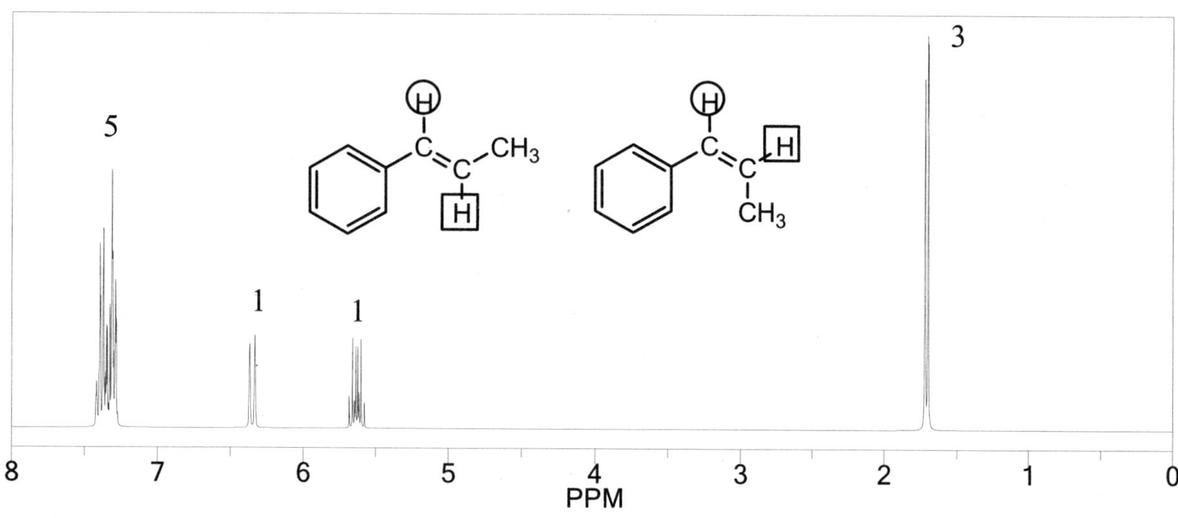

Memorization Task 17.6: Splitting between *trans* H's is larger than between *cis* H's

For reasons we will not discuss here, splitting between two H's that are *trans* to each other is larger than splitting between two H's that are *cis* to each other. This is often used to identify whether a molecule is *E* or *Z*.

a. Splitting between the circled and boxed H's below can be measured by looking at either the peak cluster of the circled H or the boxed H; however, the cluster associated with the boxed H is much harder to interpret since it is split both by the circled H and the methyl group. Identify the peak cluster that goes with the circled H, and confirm that ONLY the boxed H is near enough to split this peak.

b. On which proton NMR spectrum (the one above or the one on the previous page) is the splitting between the circled and boxed H's largest?

c. Cross out the structure above that does NOT go with the spectrum, and explain your reasoning.

d. Number the H's on the remaining structure and use these to assign each unique type of H to a cluster on the spectrum. (Note that there are some overlapping clusters near 7.4 ppm.)

e. (Check your work) The structure that you crossed out above goes with the NMR spectra on the previous page. Draw this structure on the previous page, and assign each H to a peak cluster on the proton NMR spectrum and each C to a peak on the carbon NMR spectrum.

Read the assigned pages in the text and do the assigned problems.

The Big Picture

Proton NMR works in much the same way as carbon NMR, but again interpretation of the spectra is a much more important skill for an organic chemist than understanding the complex physics behind the instrument. The key elements of a proton NMR spectrum are reviewed below.

Common Points of Confusion

- Equivalent H's do NOT split each other. The rule itself is not hard to follow, but students sometimes forget to check for symmetry and so do not realize that two neighboring H's are equivalent. (*See* Molecule II in Question 12)
- Peak area, not peak height, tells you the relative number of H's associated with a peak cluster. This leaves open the possibility that a tall skinny peak could be smaller (in terms of area) than a short fat peak.
- Splitting in carbon NMR tells you the number of H's attached to a given carbon. For this reason, students incorrectly assume that splitting in proton NMR tells you the number of H's associated with a peak cluster. In fact, it is a bit more complicated (see last bullet).

The following is a summary of the key elements of proton NMR:

- **ppm** or **chemical shift** (*given by location along the x axis*) is a function of the amount of electron density around an H. The closer the H is to an electronegative element, the more "**deshielded**" it is and therefore the higher the ppm number of its peak cluster (farther left on the spectrum). Multiple bonds also cause the signal of nearby H's to be shifted to the left. Memorization Task 17.4 gives chemical shifts for common functional groups. Each chemically distinct H is expected to have a unique chemical shift, though in practice different peak clusters sometimes overlap just by coincidence. This is a bigger problem in proton NMR than in carbon NMR (especially in the so-called "aromatic region" from 7-9 ppm), since most proton NMR peaks are squeezed into just an 8 ppm range (1-9 ppm).
- **Integration** or **peak cluster area** *(given by a number above or below a peak cluster, or by a line stepping up from left to right called the integration line)* tells you the relative area of each peak and therefore the relative number of equivalent H's represented by each peak. Note that integration (especially the integration line) gives you only a ratio of peak areas. This means the same integration may be reported, for example, on the spectrum of a molecule with 1H to 3H ratio and a 2H to 6H ratio.
- **Multiplicity** is the number of peaks in a peak cluster (also called **splitting** or **proton-proton coupling**). It tells you the number of nonequivalent neighbor H's within three bonds. For example, a doublet (two peaks) tells you there is exactly one non-equivalent H within three bonds of the H responsible for this signal.

Notes

Nomenclature Worksheet 1

NAMING ALKANES AND CYCLOALKANES

Model 1: Alkane Nomenclature

alkane = molecule consisting entirely of carbon and hydrogen atoms connected by single bonds

Questions (to do in or out of class)

1. Cross out each molecule below that is NOT an alkane.

CH_4 (H–C–H with H above and below) H_3C—CH_3 H_3C–CH_2–CH_3 H_3C–CH(CH_3)–CH_2–CH_3 H_3C–CH_2–CH_2–CH_2–CH_3

H_3C–C(CH_3)(Br)–CH_3 H_3C–CH(OH)–CH_3 CH_3Cl H_3C–C(CH_3)=CH–CH_2–CH_3 H_3C–C(=O)–CH_3

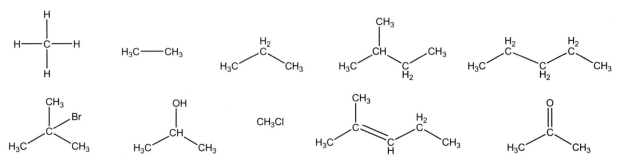

Memorization Task NW1.1: Memorize the names of the straight-chain alkanes below

#C's	Condensed Structure	Name
1	CH_4	**meth**ane
2	CH_3CH_3	**eth**ane
3	$CH_3CH_2CH_3$	**prop**ane
4	$CH_3CH_2CH_2CH_3$	**but**ane
5	$CH_3CH_2CH_2CH_2CH_3$	**pent**ane
6	$CH_3CH_2CH_2CH_2CH_2CH_3$	**hex**ane
7	$CH_3CH_2CH_2CH_2CH_2CH_2CH_3$	**hept**ane
8	$CH_3CH_2CH_2CH_2CH_2CH_2CH_2CH_3$	**oct**ane
9	$CH_3CH_2CH_2CH_2CH_2CH_2CH_2CH_2CH_3$	**non**ane
10	$CH_3CH_2CH_2CH_2CH_2CH_2CH_2CH_2CH_2CH_3$	**dec**ane

2. (E)Write a correct name below each of the unbranched alkanes in Question 1.
3. (E)What suffix do all the names in Model 1 have in common with each other?
4. (E)What name is used to designate a chain of eight carbons?

5. **parent chain** = longest continuous chain found in a branched molecule. Circle and name the parent chain in each molecule below. The first one is done as an example.

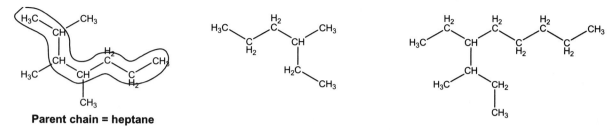

Parent chain = heptane

6. A molecule can have more than one parent chain of equal length. Mark the alternate seven-carbon parent chain in the heptane example molecule above. *(In this case, the two parent chains are equivalent, but as we will see later on, it sometimes matters which parent chain you choose.)*

Model 2: Naming Branched Alkanes (Alkyl Groups)

The parent chain name serves as the main part of the name of a branched alkane. For example:

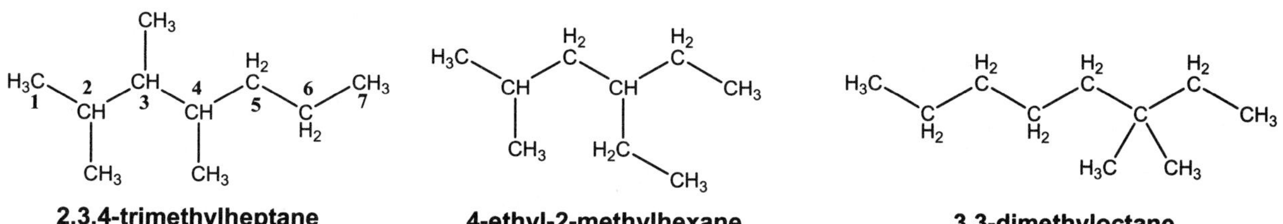

2,3,4-trimethylheptane **4-ethyl-2-methylhexane** **3,3-dimethyloctane**

Questions (to do in or out of class)

7. Number the carbons in each parent chain in the structures in Model 2. By convention, numbering of a parent chain starts from the end nearest a branch. (The first one is done for you.)

8. Underline the parent name in each chemical name in Model 2, and confirm that the parent chain matches this name. (The first one is done for you.)

9. According to the examples in Model 2, what word in a name signifies a…

   a. one-carbon branch?

   b. two-carbon branch?

10. Use Model 1 to propose names for three-, four-, five-, and six-carbon branches that follow the same pattern as "methyl" and "ethyl" for one- and two-carbon branches, respectively. *(Note: The names of seven-, eight-, etc. carbon branches follow the same pattern, but branches of such length are rare since more than five carbons in a row usually constitutes the parent chain.)*

11. What information do the numbers in the names in Model 2 convey?

12. What do the words "di" and "tri" in the names in Model 2 convey?

13. (Check your work.) Do your answers to Q10 fit the data in Memorization Task NW1.2 below?

Memorization Task NW1.2: Names of commonly found branches ("alkyl groups")

# C's	Structure	IUPAC Name
1	$-CH_3$	**meth**yl
2	$-CH_2CH_3$	**eth**yl
3	$-CH_2CH_2CH_3$	**prop**yl or *n*-propyl
3	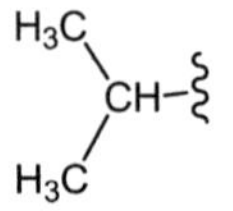 H_3C / CH– / H_3C	iso**prop**yl
4	$-CH_2CH_2CH_2CH_3$	**but**yl or *n*-butyl
4	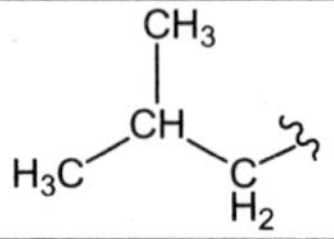 CH_3 / CH / H_3C / C / H_2	iso**but**yl
4	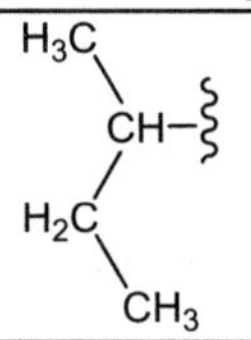 H_3C / CH– / H_2C / CH_3	*sec*-**but**yl
4	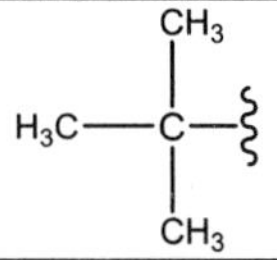 CH_3 / H_3C—C— / CH_3	*tert*-**but**yl or *t*-butyl *(looks like a T!)*
5	$-CH_2CH_2CH_2CH_2CH_3$	**pent**yl

Multiple different branches are listed in a name **alphabetically** (not including prefixes di, tri, *sec*, or *tert*).

14. Name the following alkanes.

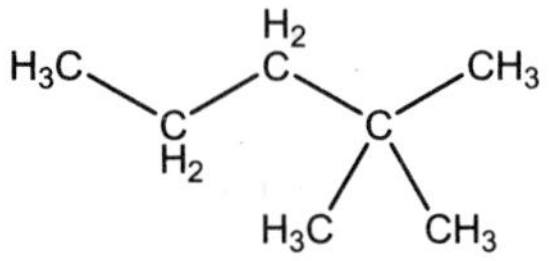

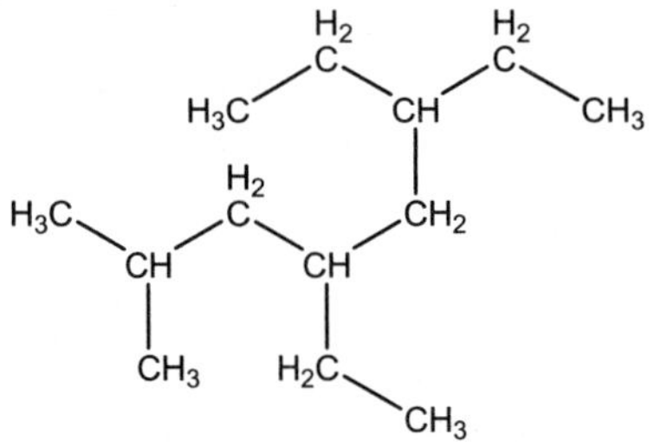

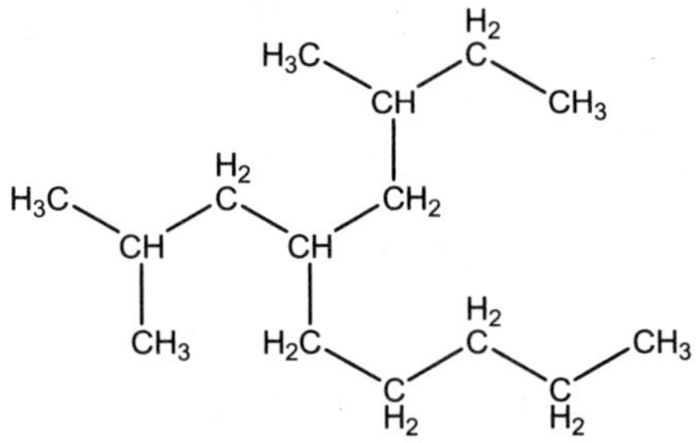

15. (Check your work.) Explain what is wrong with each of the following names for the first molecule above:

a. 4,4-dimethylpentane

b. 1,1,1-trimethylbutane

c. 2,2-methylpentane

d. 2-dimethylpentane

16. (Check your work.) Explain what is wrong with each of the following names for the third molecule in Questions 14.

    a. 6-ethyl-2-methyl-4-pentyloctane

    b. 3-methyl-5-isobutyldecane (*Hint*: Read the rule at the bottom of Memorization Task NW1.2)

17. Draw the following alkanes

    a. 2,3,3-trimethylpentane

    b. 3-ethyl-2,5-dimethylhexane

Model 3: Cyclic Alkanes

- If a ring is present in a molecule, this ring almost always is considered the parent chain.
- Number ring carbons starting at a branch point (alkyl group), and number around the ring in a direction toward the closer group.
- If there are exactly two alkyl groups, decide which gets "1" based on alphabetical order. *(Alphabetical order if often used as a tiebreaker when all other factors are equal.)*

Examples

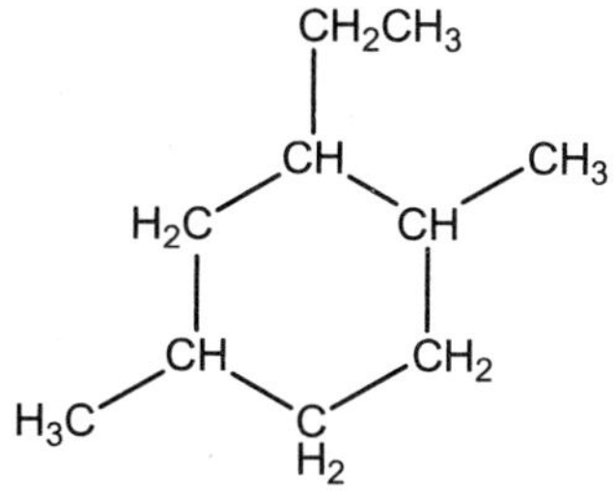

2-ethyl-1,4-dimethyl**cyclohexane** 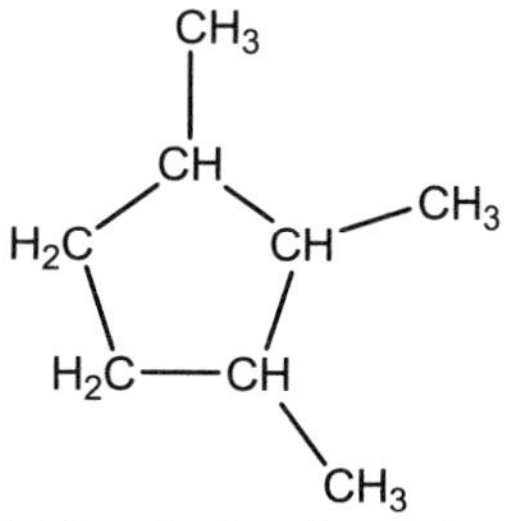 1,2,3-trimethyl**cyclopentane** 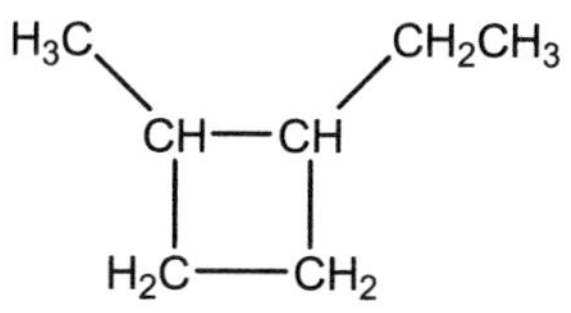 1-ethyl-2-methyl**cyclobutane**

Questions (to do in or out of class)

18. What prefix is used to indicate that a parent chain (e.g., hexane) is cyclic?

19. Draw structures that correspond to the following names:

    a. 3-methylhexane b. 1,1-diethylcyclobutane c. 1-*t*-butyl-4-methylcyclohexane

20. For mono-substituted cycloalkanes the "1" is not included. That is, the name 1-methylcyclohexane is not used. The name is simply "**methylcyclohexane**".

a. Draw methylcyclohexane.

b. Explain why adding a "1" to methylcyclohexane does not add any new information.

21. Write a correct name for each of the following structures:

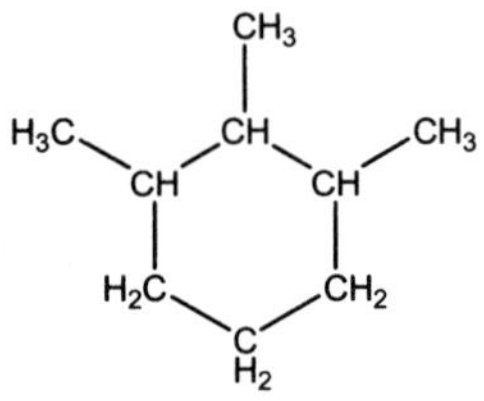

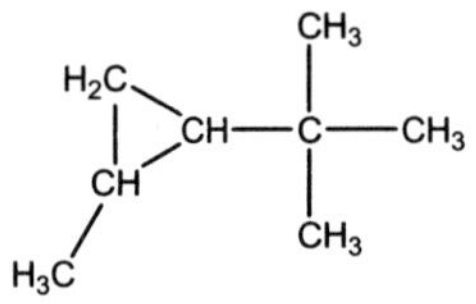

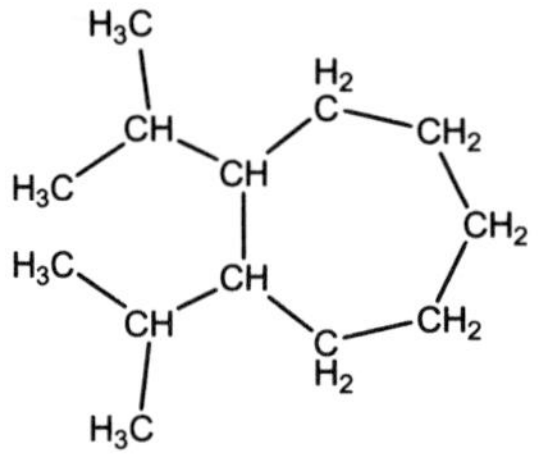

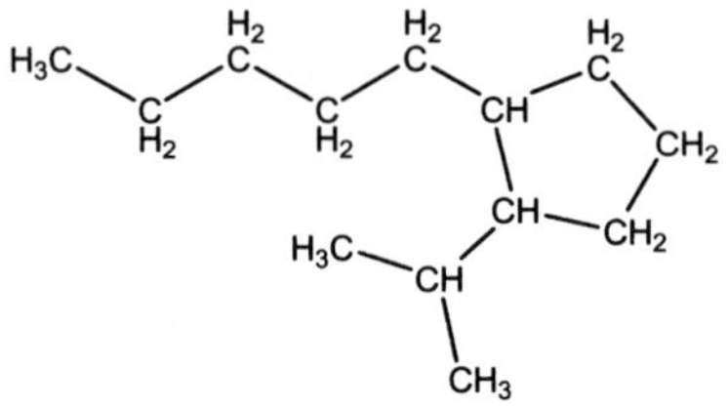

Confirm Your Understanding Questions (to do at home)

22. Construct an explanation for why the following molecule can be called methylpropane instead of 2-methylpropane.

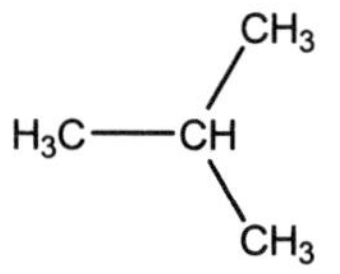

methylpropane

23. Draw the following molecules:

   a. 2,2,3,3-tetramethyl-6-propyldodecane
   b. 3-ethyl-4-isobutyl-2-methyldecane
   c. *tert*-butylcyclopentane
   d. 4-*t*-butyl-2,6-dimethylheptane
   e. 4-isopropylheptane
   f. 2,3-dimethyl-4-propyloctane
   g. 1,3,5-trimethylcyclohexane

24. Name each of the following structures.

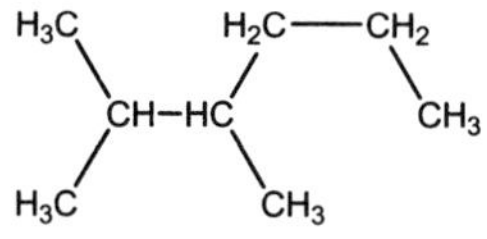

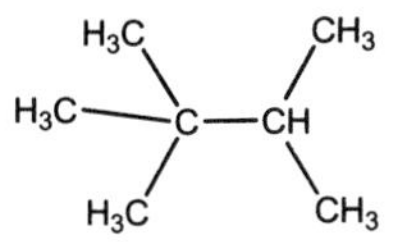

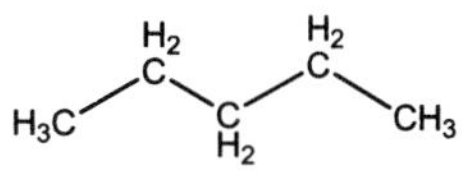

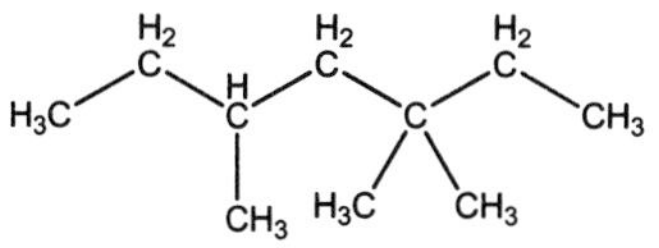

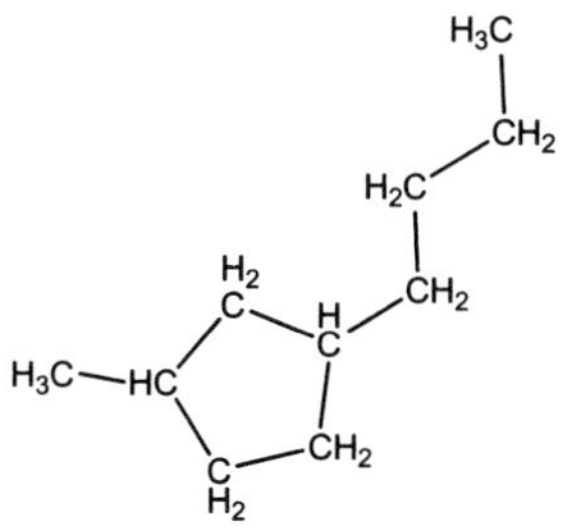

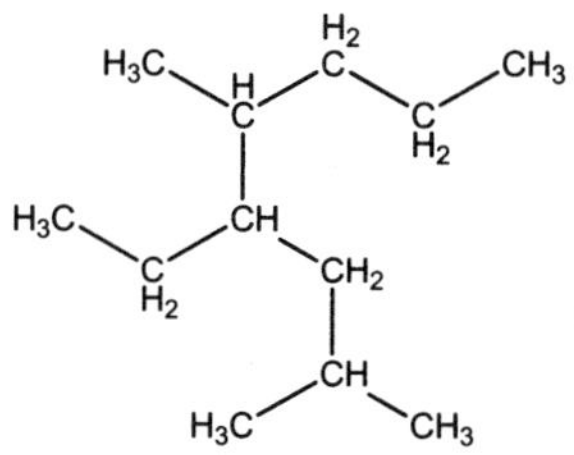

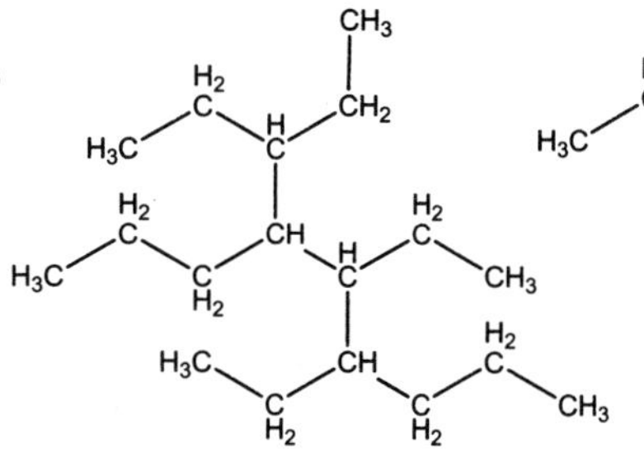

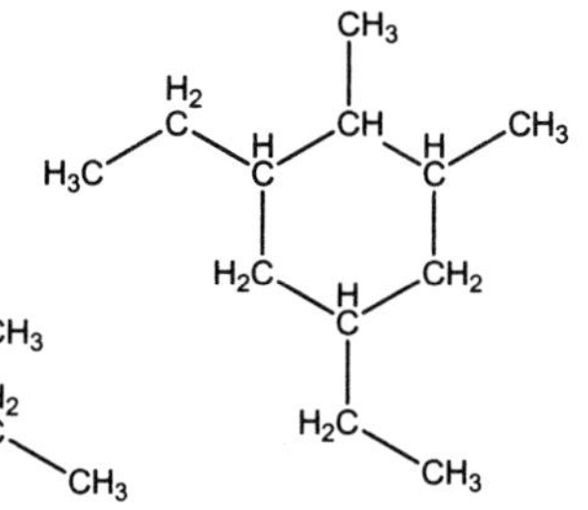

Notes

Nomenclature Worksheet 2

INTRODUCTION TO NAMING FUNCTIONAL GROUPS

Model 1: Alkene Nomenclature

alkene = molecule containing a carbon-carbon double bond (C=C)

Naming an alkene is the same as naming an alkane except the suffix "ane" is replaced with "**ene**" and...

- The parent chain must contain the double bond
- The parent chain is numbered to give the carbons in the double bond the lowest possible numbers (the lower of these two numbers is placed before the suffix, for example, but-1-ene)
- For cyclic alkenes, give the double-bond carbon with the most branches the lowest possible number; usually a "1" (see second example below).
- The endings **diene, triene** and **tetraene** are used to signify **two, three,** or **four** double bonds, respectively. (e.g., butane becomes buta**diene**; and hexane becomes hexa**triene**, etc.)

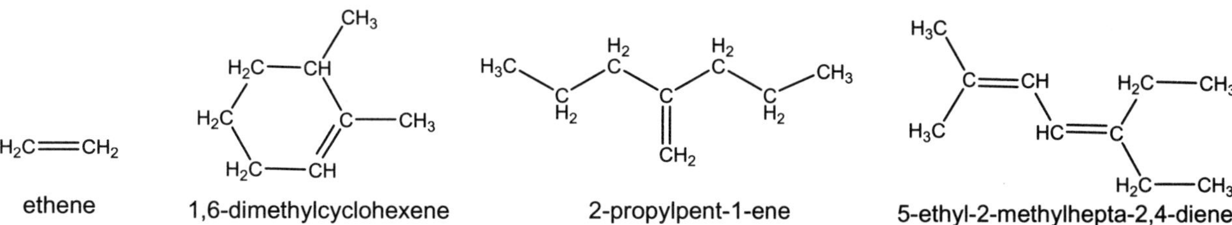

ethene | 1,6-dimethylcyclohexene | 2-propylpent-1-ene | 5-ethyl-2-methylhepta-2,4-diene

Questions (to do in or out of class)

1. A student names the second structure above 2,3-dimethylcyclohex-1-ene. What rule does this violate?

2. In each name in Model 1, circle the number(s) that designate the location of a double bond.

3. Name each of the following alkenes:

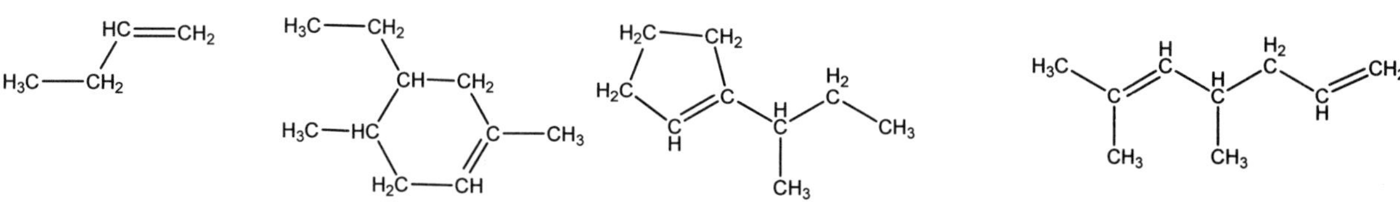

4. Draw the following alkenes:

   a. 2,3-dimethylbut-2-ene
   b. cyclopenta-1,3-diene
   c. penta-2,3-diene

Model 2: Alkyne Nomenclature

alkyne = molecule containing a carbon-carbon triple bond (C≡C)

ethyne 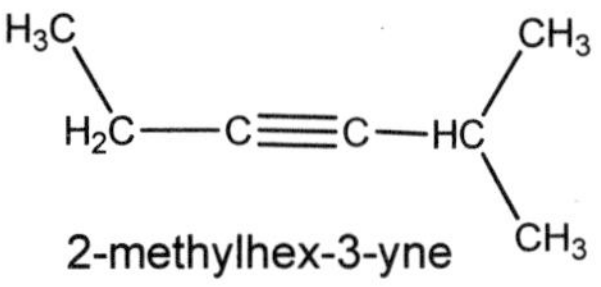 | 2-methylhex-3-yne 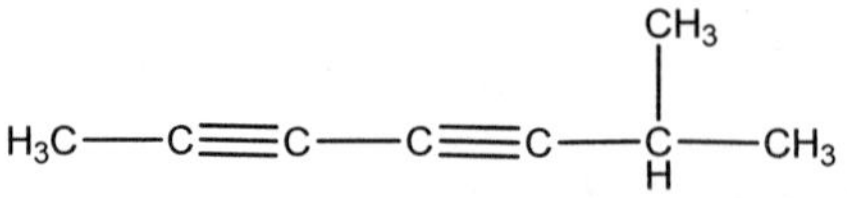 | 6-methylhepta-2,4-diyne 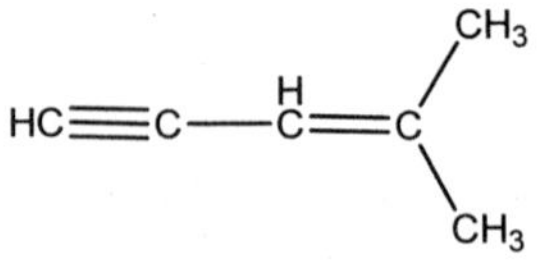 | 4-methylpent-3-en-1-yne

Note: "a" is inserted in names where consonants appear in a row. e.g., 6-methylhept<u>a</u>-2,4-diyne, not 6-methylhept-2,4-diyne

Memorization Task NW2.1: Know the common name of ethyne (HC≡CH) → "acetylene."

Historical Note: IUPAC Name vs. Common Name

Until the last century, chemists had little idea of the structure of chemicals. Consequently, newly discovered chemicals were named at the whim of their discoverers. For example, ethyne can be transformed into vinegar so it was first called **acetylene** after the Latin word for vinegar *acetum*.

Many common names have been incorporated into the official International Union of Pure and Applied Chemistry (IUPAC) naming rules you are learning in this activity. For example, "**but**" (pronounced *beut*), is the IUPAC name for a four-carbon chain. It is derived from the Latin word for butter (*butyrum*) because a four-carbon molecule (butyric acid) is responsible for the smell in rancid butter.

Another example is "**acet**" (from acetylene, HCCH), which is an alternate name for a **two-carbon group**.

Questions (to do in or out of class)

5. Naming an alkyne is the same as naming an alkene except the suffix "ene" is replaced with what suffix?

6. Draw the following alkynes:

   a. pent-2-yne

   b. 2,5-dimethylhex-3-yne

   c. 5-methylhex-1,3-diyne

7. Write the IUPAC name for each of the following structures:

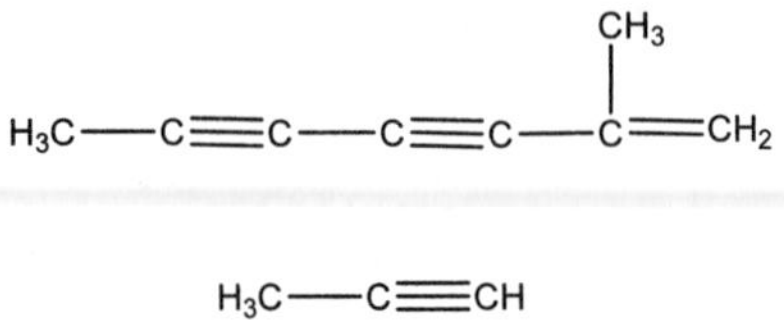

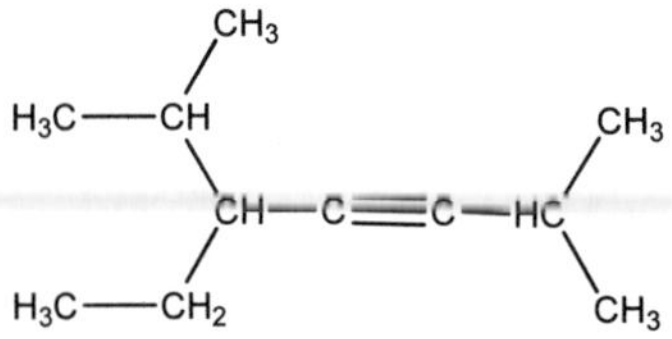

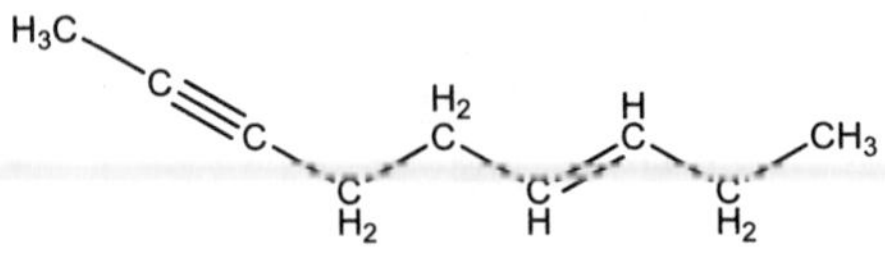

Functional Group = the general term used for a site on a molecule where a reaction is likely to occur. Often a functional group contains a heteroatom (N, O, halogen, etc.); however, a double or triple bond is also considered a functional group because such bonds are sites of reactivity within a molecule. The following Models contain instructions for naming molecules with various functional groups. These functional groups are also sometimes called **substituents**.

Model 3: Naming Haloalkanes (alkyl halide nomenclature)

haloalkane = molecule containing a halogen (normally F, Cl, Br, or I)

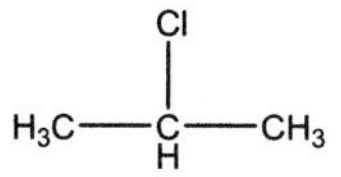

2-chloropropane 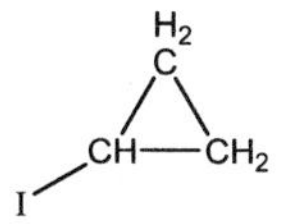 — iodocyclopropane — dichloromethane (CH_2Cl_2) — 1,5-dibromohex-2-yne 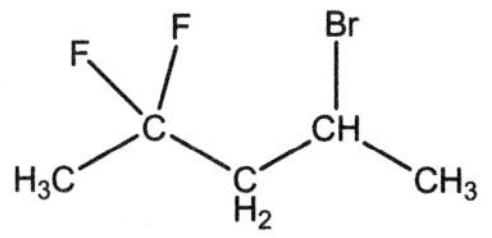 — 4-bromo-2,2-difluoropentane

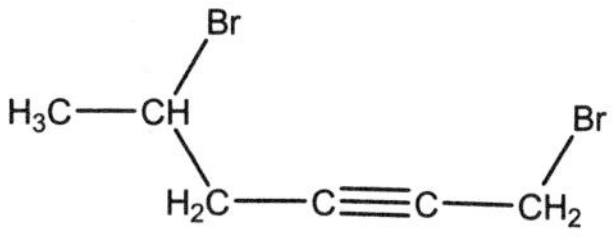

Questions (to do in or out of class)

8. Based on the examples in Model 3, write the IUPAC name for each structure.

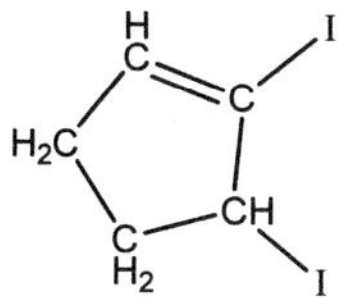

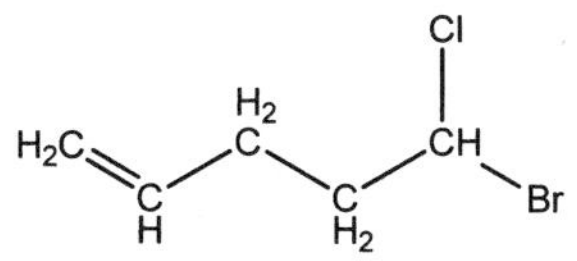

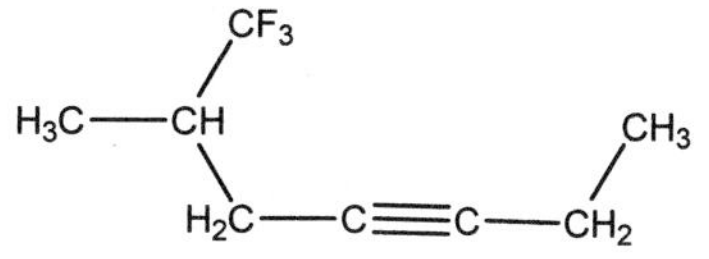

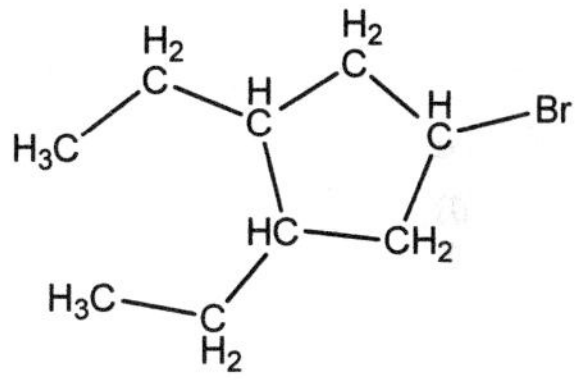

Model 4: Alcohol Nomenclature

alcohol = molecule containing an OH group

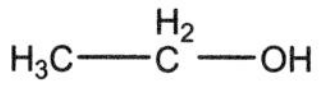

ethanol 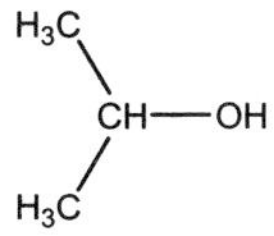 — propan-2-ol 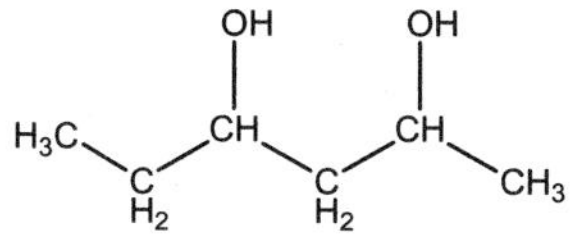 — hexane-2,4-diol 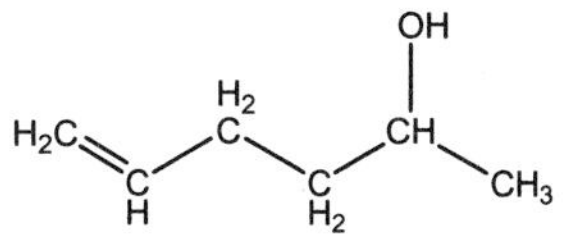 — hex-5-en-2-ol 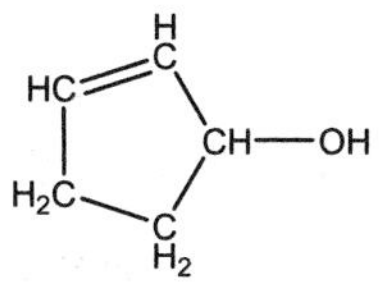 — cyclopent-2-enol

Questions (to do in or out of class)

9. Which two structures in Model 4 reflect that fact that OH groups get numbering priority over other functional groups? *(Note that an OH on a ring always gets a "1" so **cyclopent-2-en-1-ol** is redundant.)*

10. Based on the examples in Model 4, write the IUPAC name for each structure.

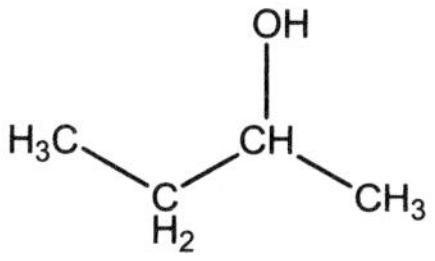

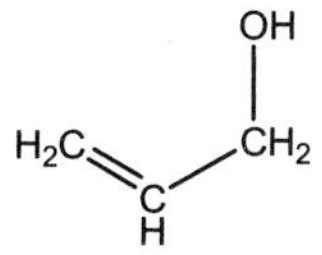

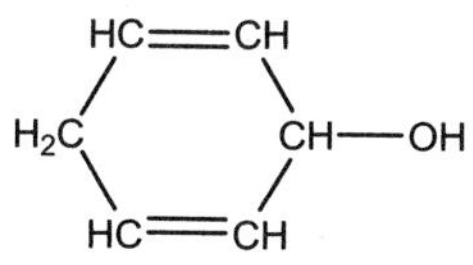

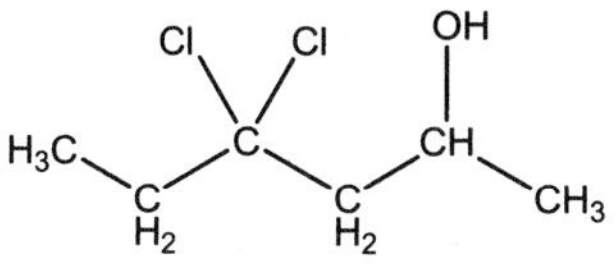

CF_3OH

11. Instead of using the ending "**ol**" an OH can be named as if the OH group were a halogen using the term "**hydroxy**." For example, hexane-2,4-diol can be called 2,4-dihydroxyhexane. Write an equivalent name for 1,2,3-cyclohexanetriol using the term "hydroxy."

Model 5: Ether Nomenclature

ether = molecule containing R—O—R (where R ≠ H)

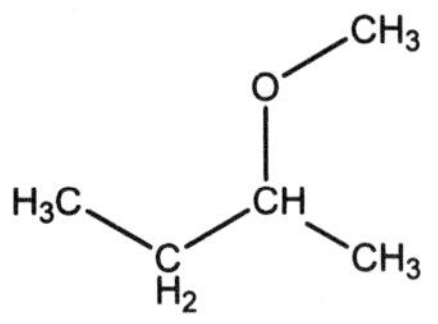

2-methoxybutane 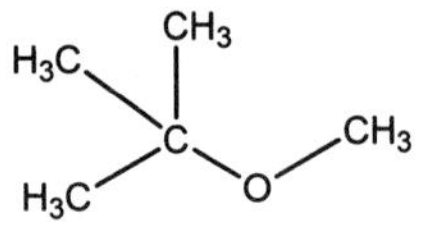 | 2-methoxy-2-methylpropane 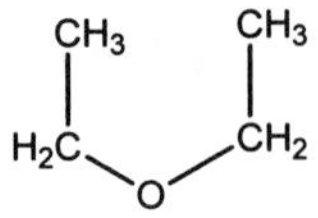 | ethoxyethane 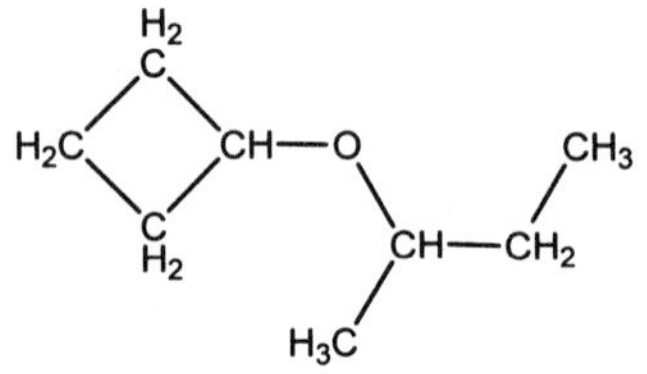 | *sec*-butoxycyclobutane

Ethers are named by designating one R group as the parent chain and treating the remaining OR group as a substituent. To indicate the oxygen, the suffix "**oxy**" replaces "**yl**" in the substituent name. For example: meth**yl** becomes meth**oxy**; eth**yl** becomes eth**oxy**, and isopro**yl** becomes isoprop**oxy**, etc.

Questions (to do in or out of class)

12. [R]Circle the "R" group that is considered the parent chain in each ether in Model 5.

13. Based on these examples, which takes precedence as the parent chain, a ring or non-ring chain?

14. Based on the examples in Model 5, write the IUPAC name for each structure.

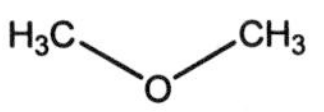

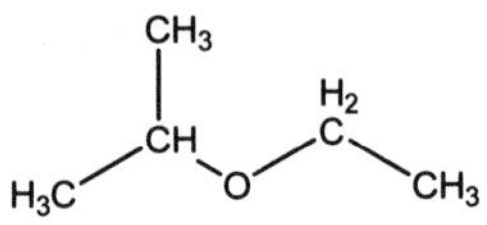

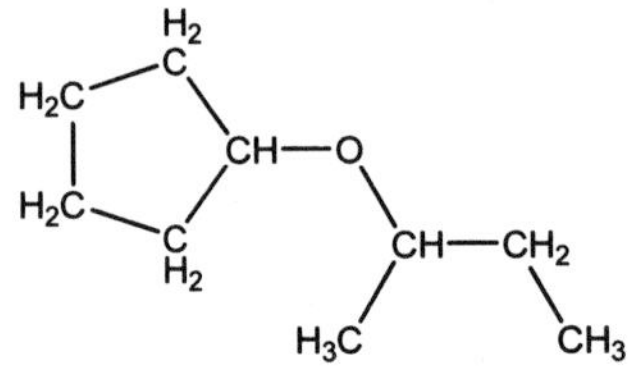

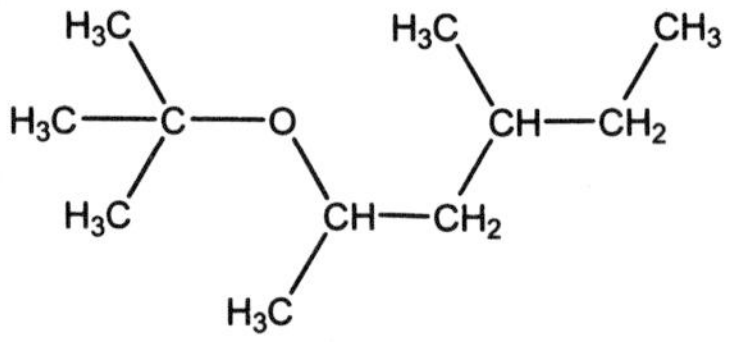

Memorization Task NW2.2: Ether Solvents and an Alternate Way of Naming Simple Ethers

Ethers are polar but do not react with many molecules, so they often are used as solvents.
(A solvent is an inert liquid in which two molecules can be dissolved so they can collide and react.)

Simple ethers can be named by listing the two alkyl groups followed by the word ether, as with the first two examples below. **Know both the common name (in parenthesis) and the IUPAC name of these three common ether solvents**.

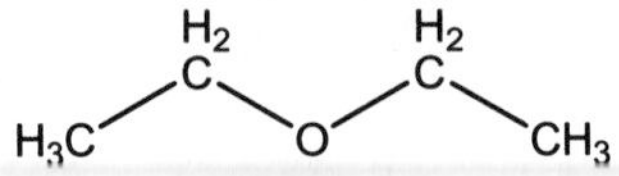

diethyl ether (ether) | methyl *tert*-butyl ether (MTBE) | tetrahydrofuran (THF)

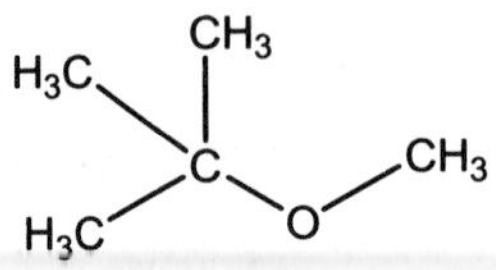

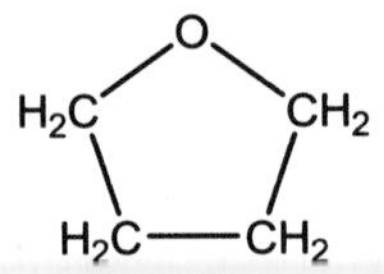

Model 6: Amine Nomenclature

amine = molecule containing NR_3 (where R = H or alkyl)

One way to name an amine (there are many acceptable ways) is to generate a parent name by replacing the last **e** in the corresponding alkane parent name with the suffix "**amine**" (e.g., propan<u>e</u> becomes propan**<u>amine</u>**), then inserting a number to indicate the point of attachment to N (e.g., propan-1-amine).

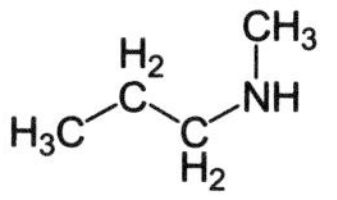

N-methylpropan-1-amine

Other alkyl groups attached to N are listed at the front using an *N* (e.g., *N*-methylpropan-1-amine, *above*).

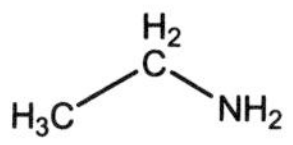

ethanamine 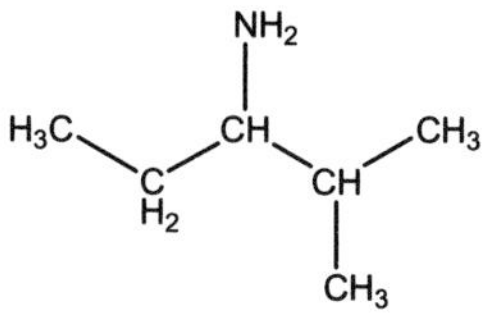 2-methylpentan-3-amine 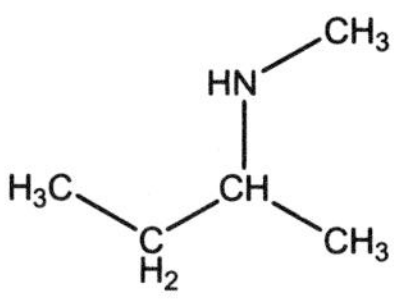 *N*-methylbutan-2-amine 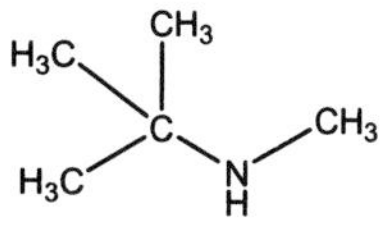 *N*,2-dimethylpropan-2-amine 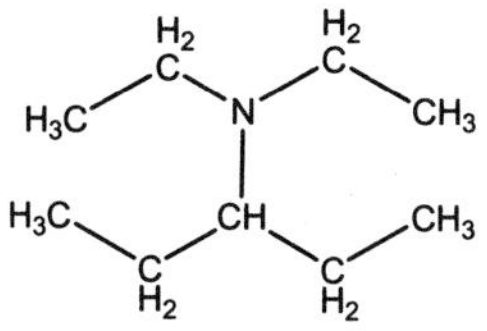 *N*,*N*-diethylpentan-3-amine

Questions (to do in or out of class)

15. Simple amines can be named by turning one alkyl group into the parent name and listing other alkyl groups attached to N at the front, as in ethylamine, triethylamine, or methyl ethylamine. Draw each of these simple amines.

16. [(E)]Circle the parent chain in each molecule in Model 6.

17. What information about the structure does the "3" in 2-methylpentan-<u>**3**</u>-amine convey?

18. [(E)]Put a box around each alkyl group (other than the parent chain) that is attached to each N.

19. What information does the *italicized* capital *N* convey in the name of an amine?

20. Based on the examples in Model 6, write an IUPAC name for each structure.

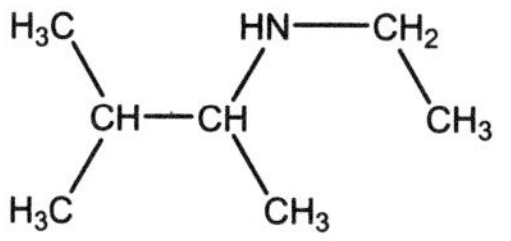

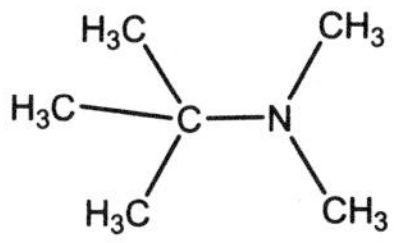

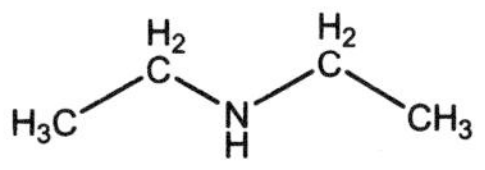

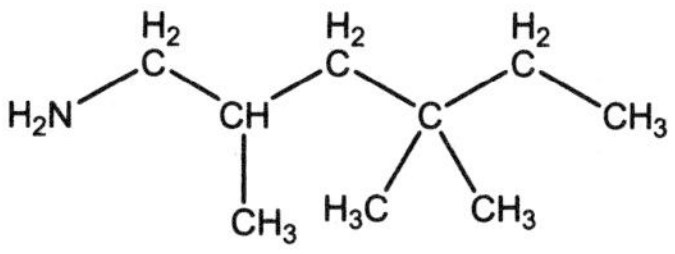

21. Amines containing NH_2 (especially diamines, triamines, etc.) can be named by calling each NH_2 an **amino** group. For example the first two molecules in Model 6 can be named **aminoethane**, and **3-amino-2-methylpentane**, respectively. Name the following molecules using the word **amino**.

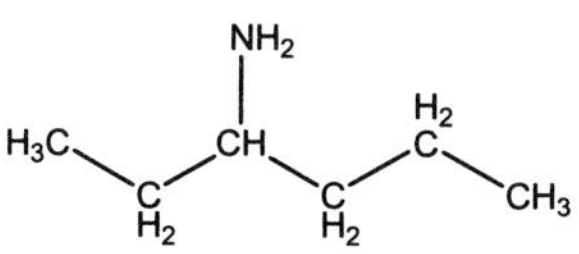

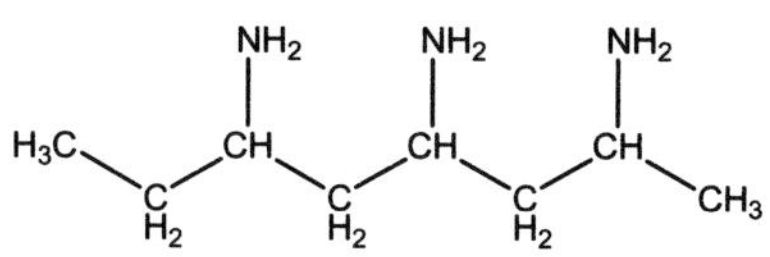

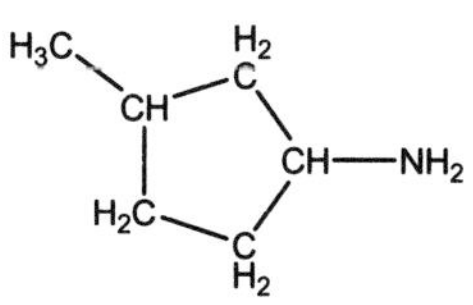

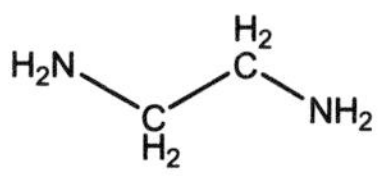

Confirm Your Understanding Questions (to do at home)

22. Draw the following molecules:

a. 2-methylhept-2-ene
b. 5-butyl-6-isopropyldec-5-ene
c. 1-butylcyclopent-1-ene
d. but-3-enylcyclopentane
e. hexa-1,2-dien-5-yne
f. nona-1,8-diene
g. non-4-yne
h. octa-1,3,5,7-tetrayne
i. 2,2-dibromo-6-methylheptane
j. 2-bromo-4-ethyl-2-methyloctane
k. cycloprop-2-enol
l. 2-methoxy-2-methylbutane
m. 3-propylhexan-2-ol
n. 3-isopropylpentane-1,4-diol
o. 1,1,1-trifluoropentan-2-ol
p. cyclohexa-2,4-dienol
q. 2-propoxypentane
r. diisopropyl ether
s. 1-*sec*-butoxyhexane
t. 2-methyl-1-propoxyprop-1-ene
u. octan-1-amine
v. *N*-methylbutan-1-amine
w. *N*,3,4-trimethylhexan-2-amine
x. 1,1-diamino-2-methylpropane
y. 6-amino-5-chlorocyclohex-2-enol
z. 3-amino-2,2-dimethylpentane

23. Write an IUPAC name for each of the following structures:

(a) (b) 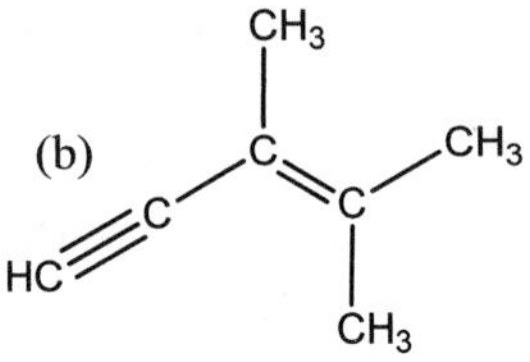 (c)

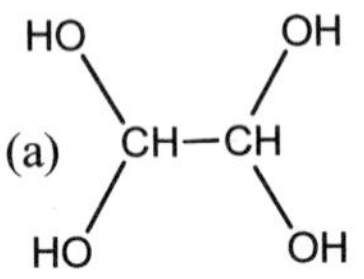

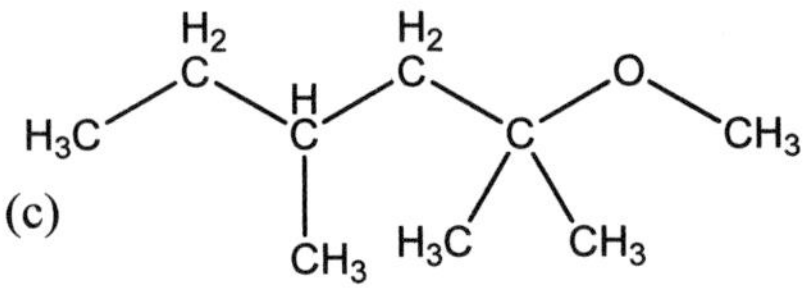

(d) 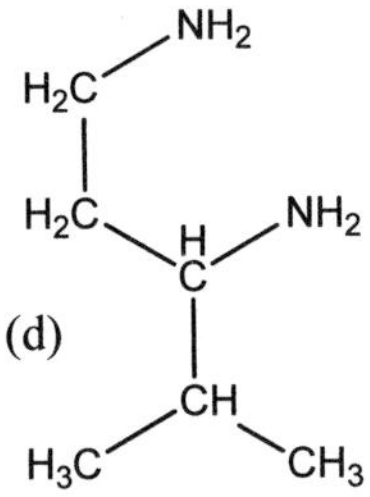 (e) 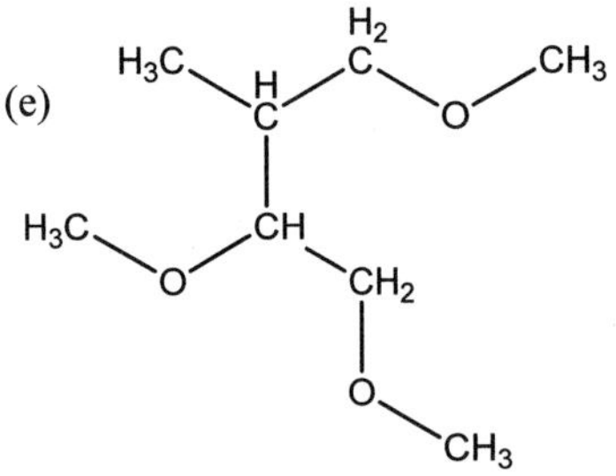 (f)

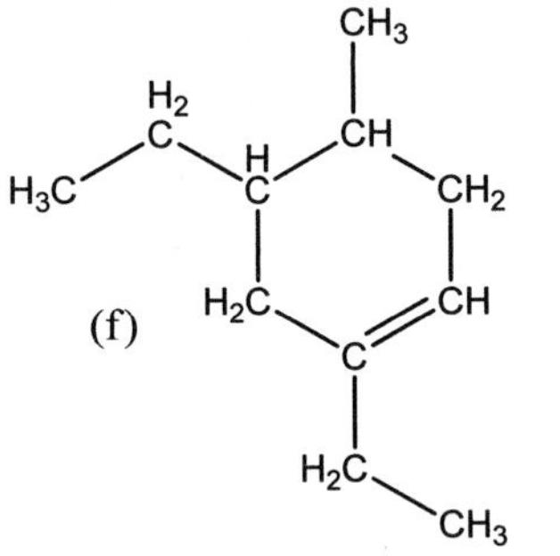

Notes

Notes

Notes

Notes

Notes

Notes